Darwin for Today

Darwin for Today

THE ESSENCE OF HIS WORKS

Edited, and with an Introduction, by

Stanley Edgar Hyman

The Viking Press · New York

CONTENTS

Darwin for Today

Editor's Introduction

Charles Darwin tells the story of his life in one of the most charming of autobiographies (pp. 325–404), written in his old age. Until a few years ago it was known only in the version, heavily censored by the family, printed by his son Francis in *The Life and Letters of Charles Darwin*. Francis Darwin used some of the suppressed material elsewhere in his book, and the manuscript has been available to other biographers. In 1958 Darwin's granddaughter Nora Barlow published a scholarly edition of *The Autobiography of Charles Darwin* with all the omissions, almost six thousand words, restored and identified, and with an extensive appendix and notes. For this, as for so much other Darwin research, we are greatly in her debt.

The suppressed material no longer seems very shocking. It consists in part of Darwin's franker statements about the Christian religion: that the creed is "unintelligible"; that everlasting damnation is "a damnable doctrine"; that a belief in God may be compared to a monkey's "instinctive fear and hatred of a snake." The rest of it consists mainly of Darwin's personal comments on some of his contemporaries. These were suppressed where unfavorable (the statement about Samuel Butler's "insane virulence"), neutral (the anecdote about Charles Babbage's plan for fire prevention), or even favorable (the characterizations of Sir Joseph Hooker and Thomas Henry Huxley, Darwin's best friends).

The *Autobiography* needs to be supplemented with some information that Darwin omitted for a variety of reasons. He did not know it, he forgot it at the time of writing, or he thought it unnecessary or out of place in the work. Some of this is quite trivial, such as the fact that he had a slight stammer, a difficulty with "w," all his life. At least one story that Darwin tells only in part because he knew it only in part is utterly fascinating: his relationship with Robert Fitz-Roy, the captain of H.M.S. *Beagle*. Darwin remarks in the *Autobiography:* "The voyage of the *Beagle* has been by far the most important event in my life and has determined my whole career; yet it depended on . . .

1

such a trifle as the shape of my nose." The shape of Darwin's nose, which had almost convinced Fitz-Roy (an amateur physiognomist) that Darwin's character was too weak for the long voyage, determined Fitz-Roy's whole life too, in a tragic fashion. Darwin remarks: "He was afterwards very indignant with me for having published so unorthodox a book (for he became very religious) as *The Origin of Species*." Darwin also mentions that Fitz-Roy's "end was a melancholy one, namely suicide." Darwin surely did not know how closely those two facts were related. Fitz-Roy was not only indignant at Darwin, he was obsessed with his own responsibility for Darwin's views, since he had taken him on the *Beagle* voyage that began everything. At the famous Oxford meeting of the British Association in 1860, the first public debate on the evolution of species by natural selection, Fitz-Roy appeared, waving a Bible and shouting that he had warned Darwin against holding views contrary to the word of God. In 1865, still brooding over his responsibility for evolution, Fitz-Roy went into the bathroom one morning and cut his throat.

The *Autobiography* briefly mentions Darwin's lifelong ill health but does not explain it. Darwin never knew its cause, nor is there yet any generally accepted explanation, beyond a considerable agreement that it was functional or neurotic rather than organic. A book and a number of articles on the subject are synopsized and discussed in a note in Nora Barlow's edition of the *Autobiography* and in my *The Tangled Bank* (p. 12).* Darwin observes in the *Autobiography* that although his illness frequently incapacitated him for scientific work, it also aided that work. "Even ill health, though it has annihilated several years of my life," he writes, "has saved me from the distractions of society and amusement." It also provided Darwin with a laboratory subject when he turned to ethology, the study of behavior. He studied his own face in the mirror while he was vomiting and retching, and managed to discover that his orbicular muscles were strongly contracted and that tears were produced.

Darwin says enough about his wife and children in the *Autobiography* to make it clear that he had an exceptionally happy

* *The Tangled Bank: Darwin, Marx, Frazer and Freud as Imaginative Writers*, by Stanley Edgar Hyman, New York, 1962.

marriage and very loving relations with his children. He does not mention an extraordinary memorandum (reprinted in another note in the Barlow edition) that he wrote to himself a year or two before his marriage. It lists the arguments for and against marriage, proposes that despite many freedoms and advantages, the bachelor's life is that of "a neuter bee," and concludes: "Marry—Marry—Marry." Since the *Autobiography* was addressed to his children, it does not mention one of the happiest features of Darwin's life, the involvement of the children in his scientific work. In the case of one child at least, Darwin kept a daily record of his behavior, published as "Biographical sketch of an infant." He carefully studied the behavior of all his children at any opportunity. When they grew up, they helped him in every fashion they could. Francis, who became a plant physiologist, was his father's assistant at Down and in Darwin's later years did a considerable part of the botanical experimentation. George, who became a mathematician and astronomer, made calculations and did botanical drawings for his father, who drew very badly. The other sons, William (a banker), Leonard (a eugenist), and Horace (an inventor), collected specimens and made observations for their father, who scarcely left home in his later years. Darwin's last book, *The Formation of Vegetable Mould through the Action of Worms* (a chapter is included, pp. 407 through 428) is touching in the number of acknowledgments to the sons for help that it records.

Darwin writes in the *Autobiography* with his characteristic frankness: "I became convinced from various small circumstances that my father would leave me property enough to subsist on with some comfort, though I never imagined that I should be so rich a man as I am." This seems to suggest that Darwin's fortune was all inherited, and until recently that was the belief of Darwin scholars. With the publication of Sir Arthur Keith's *Darwin Revalued* in 1955, it became known that most of Darwin's fortune was self-made through investment, that he was in fact what Sir Arthur calls "a great financier." Darwin's father left him about £45,000 when he died in 1849 and had probably given him a total of about £20,000 before that. Darwin often spoke with admiration, his son reports, of a relative who had doubled his fortune. When Darwin died in 1882, *he* had more than quadrupled his inherited fortune, mostly by shrewd investments in railroads, and he left an estate of approximately £282,000. He studied his investments at night, after his scientific work was finished.

The most engaging of Darwin stories is the story of his involvement with Anthony Rich. Rich was an aptly named total stranger who wrote Darwin in 1878 that he and his sister, at seventy-four and seventy-three, were the last of their family, and that he had made a will leaving his property to Darwin, in recognition of what Darwin had done for the benefit of mankind. In 1881, Darwin told Rich that he really did not need Rich's fortune and suggested that Rich alter his will and leave the property elsewhere. Rich refused to alter his will, protesting that Darwin had five sons to provide for. When Darwin died the next year, his children became Rich's heirs. *They* then wrote him that his generous intentions were enough honor for their father without the money, and again suggested that he alter his will. Rich again refused, pleased to think that his property would descend "to the worthy children of so noble a man." When Rich died nine years later, the Darwin children got his estate.

Darwin discusses some of his friends in the *Autobiography,* but one must read his correspondence to know how closely he was involved with a number of his scientific contemporaries; at first with the geologist Sir Charles Lyell; then with the four principal supporters of natural selection: the zoologist and anatomist Huxley, the botanist Hooker, the naturalist Alfred Russell Wallace, and the American botanist Asa Gray. (Letters to Lyell and to Hooker are included below, on pp. 431–435.) The filial relationship with Lyell, who was the major source of Darwin's evolutionary thinking, gradually cooled when the cautious Lyell refused to make public acknowledgment of his belief in evolution by natural selection. Darwin wrote to him bitterly in 1860: "You cut my throat, and your own throat; and I believe will live to be sorry for it." Hooker then became Darwin's closest friend; as Darwin's daughter Henrietta later wrote: "My father had been more attached to him than to anyone outside his own family." The correspondence reveals not only the enormous scope of Darwin's operations—exchanging specimens and information with hundreds of amateur and professional scientists —but the warmth of his attachment to many correspondents he never met. (In his later years, Darwin was too sick to see any-one but relatives and a few intimate friends.)

Darwin goes not at all into his appearance and not much into his habits in the *Autobiography*. He was of a ruddy com-plexion, about six feet tall, with long legs, and he habitually stooped. In 1863 he grew a beard to avoid the burden of shav-ing, and he wore it for the rest of his life, allowing it to get quite long in his last years and hardly trimming it. Darwin became almost entirely bald. In the time not spent on his scientific work, his investments, and his correspondence, Darwin walked on the Down grounds, played (as much as his health permitted) with the children and, later, with the grandchildren, read the news-paper, and indulged in two games of backgammon with his wife every night. He pronounced foreign languages as though they were English. Darwin had no respect for books as objects and would readily cut a heavy book in half, to make it more con-venient to hold. He took snuff, when the doctor let him.

The remaining supplement that must be added to the *Auto-biography* is an account of Darwin's death and burial. He had a series of heart attacks in 1881 and 1882. These culminated in a very severe one during the night of April 18, 1882, when Darwin could be brought back to consciousness only with great difficulty. "I am not the least afraid to die," he said when he recovered consciousness. Darwin died the next afternoon, at the age of seventy-three. It had been the intention of the family to bury him at Down, but a letter signed by twenty members of Parliament asked the Dean of Westminster if Darwin might be buried in Westminster Abbey, and the Dean and the family acquiesced. The pallbearers included Hooker, Huxley, Wallace, and James Russell Lowell, the American Minister. Darwin was buried in the Abbey, a few feet from the grave of Sir Isaac Newton.

In his lifetime Darwin published seven thousand pages, about three million words. To reduce those to 400 pages and one hundred and seventy thousand words, and still give an adequate sense of the breadth of Darwin's range, is not easy. The eight works included here in whole or in part seem to me to represent his writing in all its more important and more engaging aspects.

Some of *The Voyage of the Beagle* (1839) is essential. (It

seems simpler to use its popular title; Darwin published it as
Journal of Researches.) It is Darwin's most widely read book
and was always his own favorite. As he recognized, it was the
voyage that determined his life and made him a scientist. More
specifically, some of the things Darwin saw on the voyage, principally the geology of South America and what he calls the
"eminently curious" natural history of the Galapagos Islands,
turned his mind toward evolution. Darwin later wrote in *The
Origin of Species:* "Many years ago, when comparing, and seeing others compare, the birds from the closely neighbouring islands of the Galapagos archipelago, one with another, and with
those from the American mainland, I was much struck how
entirely vague and arbitrary is the distinction between species
and varieties." In the Galapagos chapter here reprinted, Darwin
remarks: "We seem to be brought somewhat near to that great
fact—that mystery of mysteries—the first appearance of new
beings on this earth." He goes on to warn himself: "No one
has a right to speculate without distinct facts." (In this warning
we can see the true germ of *The Origin of Species,* toward which
Darwin began keeping a notebook of facts not long after he
got back to England.) But he continues to speculate. A page or
two after the warning, Darwin writes: "Seeing this gradation
and diversity of structure in one small, intimately related group
of birds, one might really fancy that from an original paucity of
birds in this archipelago, one species had been taken and modified for different ends."

The other chapter included here, on Tierra del Fuego, dis-
covers a very important feature of Darwin's theory, the struggle
for existence (although he did not understand its implications
until later, when he read Malthus). Darwin writes of the
Fuegians: "It is certainly true, that when pressed in winter by
hunger, they kill and devour their old women before they kill
their dogs: the boy, being asked by Mr. Low why they did this,
answered, 'Doggies catch otters, old women no.' " At another
place in the Tierra del Fuego chapter Darwin observes of the
landscape: "Death, instead of Life, seemed the predominant
spirit." Eventually his theory would discover, and rejoice in,
improved life arising out of death; here he is still philosophizing
poetically. (It is amusing to note that one of the reasons Darwin's
uncle Josiah Wedgwood gave Darwin's father in favor of the

Beagle voyage was: "The pursuit of Natural History, though certainly not professional, is very suitable to a clergyman.")

Darwin never did become a clergyman. He realized quite soon on the *Beagle* voyage that he was a born scientist, and a reader of the book makes that discovery on every page. Darwin's mind is always questing for causes. "How can this faculty be explained?" is his typical question. He will taste anything, from the meat of Galapagos tortoises to the urine of a frog. He is a tireless observer and primitive experimenter; his pages on plaguing the Galapagos lizards are a foretaste of a life of ingenious experimentation with any organisms at hand. He is no anthropologist, surely, since he dismisses the Fuegians as subhuman and childlike; but this is the result of his immaturity on the *Beagle* voyage, not of any lack of humanity. His remarks about slavery in the book's final chapter are a powerful and ringing statement of extreme moral revulsion. "I thank God, I shall never again visit a slave-country," he writes.

Darwin's *Essay of 1844* is included in its entirety, as a substitute for *The Origin of Species,* which by itself would more than fill the pages of this present volume. Darwin outlined his views on the origin of species in a notebook in 1837, then in a condensed *Sketch* in 1842, then in the much fuller *Essay* in 1844. When he had finished this, he wrote a letter to his wife requesting that, in case of his sudden death, she do her best to get a competent scientist to enlarge and publish it. The *Essay* naturally lacks a number of things that appeared in the *Origin* fifteen years later. Darwin thought afterward that its only theoretical flaw was that it overlooked a "problem of great importance," the problem of divergence of character. (To a reader now, as to Francis Darwin, divergence seems clearly implied in the idea of descent with modification, but Darwin wanted it spelled out.) The *Essay* of course lacks the enormous body of evidence that makes the *Origin* almost four times as long and so convincing to the reader. It lacks the discussion of pigeons, in which Darwin is able to talk about his own pigeon-breeding experiences, carried on in the years between the appearance of the two works. Some of the 1844 terminology does not represent Darwin's later views; the most important difference is that, where in 1844 it is a "Being" who selects, it is "nature" that does the selecting in 1859. The *Essay* lacks much of the power of imaginative presentation of the *Origin,* principally its strong rhetorical questions and vivid metaphors. The key metaphoric sequence in the *Origin* is a progress in complexity from the great Chain of Life to the great Tree of Life to the great Tangled Bank of Life; none of these is explicit in the *Essay.* Where in the *Essay* Darwin says that "seeing the contented face of nature" we may doubt the bloody war for survival that it masks, in the *Origin* he cries out eloquently: "We behold the face of nature bright with gladness."

Yet in some respects the *Essay* of 1844 is a better work than its famous successor. One editor, Francis Darwin, notes in his 1909 edition that the *Essay* "has a certain freshness which gives it a character of its own . . . an air of freedom, as if the author were letting himself go, rather than applying the curb," as he does in the *Origin.* Another editor, Sir Gavin de Beer, remarks in his 1959 edition: "It is fresher, shorter, simpler, more direct, and spends less space in countering possible objections which are now known to have been groundless. As Sir Ronald Fisher has pointed out, the *Essay* shows the reasons which led Darwin to his conclusions, whereas the *Origin* and later works only give the evidence on which they were based." If, as Huxley thought, the *Essay* is more Buffonian and Lamarckian, and less Darwinian, than the first edition of the *Origin,* it is certainly much less so than the sixth edition, which is the one the modern reader is likely to see. The *Essay* has its own imaginative touches, such as the comparison of nature on p. 105 to a surface on which "ten thousand sharp wedges" rest. (This metaphor appears in the first edition of the *Origin* but was afterwards discarded.) In many cases Darwin's argument is clearer in the *Essay* than in the *Origin,* where it is often obscured by detail. Thus the *Essay* is probably the best introduction to the theory of the evolution of species by means of natural selection.

The pages from the Conclusion of *The Origin of Species* (1859) are included here as an example of the way in which the *Essay* was expanded and altered. As a comparison of the two endings makes clear, Darwin's views had become far more affirmative in the fifteen-year interval. The passage in the earlier work about the evil and cruelty in nature has been omitted, and some of the statements that replace it are optimistic, even triumphant. The complex imaginative vision of the great Tangled Bank of Life appears at the beginning of the last paragraph, followed by the capitalized, almost divine, laws underlying the tangle.

The chapter from *The Descent of Man* (1871) shows Darwin fulfilling his promise in the conclusion to the *Origin* that "light would be thrown on the origin of man and his history" (in later editions, "much light"). Darwin had written to Wallace in 1857 that the problem of "man" "is the highest and most interesting problem for the naturalist." Finally in the *Descent,* countering "that arrogance which made our forefathers declare that they were descended from demi-gods," he put man back into his place in nature, a mammal like any other. Every fool had greeted Darwin's theory, as earlier fools had greeted earlier evolutionary theories (including that of Darwin's grandfather Erasmus Darwin), with the statement that it would make man a descendant of a baboon or a chimpanzee. In the chapter here included, Darwin dismisses that nonsense. He writes: "But we must not fall into the error of supposing that the early progenitor of the whole Simian stock, including man, was identical with, or even closely resembled, any existing ape or monkey." Indeed, things are much worse; ultimately the ancestor of all the vertebrates is a tiny primitive fish, the amphioxus or lancelet. But Darwin assures us: "Nor need we feel ashamed of it." We may even be proud of this "pedigree of prodigious length." The last words of the chapter affirm what increasingly was to be the message of Darwin's works: the marvel of life. "No one with an unbiased mind," he writes, "can study any living creature, however humble, without being struck with enthusiasm at its marvellous structure and properties."

The pages here included from *The Expression of the Emotions in Man and Animals* (1872), on the behavior of monkeys, are an example of the kind of study and ingenious, simple experiment that Darwin could do with animals before he became a complete invalid. These monkeys, as Darwin notes, are like people, and he studies them in exactly the same fashion as he studied his children or his own face when retching. As far back as the 1837 notebook, Darwin had written of animals as "our fellow brethren." In an 1857 letter to Hooker, his comment on a scheme of mammalian classification that would put man in a unique sub-class is: "I wonder what a chimpanzee would say to this?" During the composition of *The Expression of the Emotions,* Darwin wrote to the superintendent of the London Zoo, A. D. Bartlett, that he could not get down to London until spring, but that meanwhile he would appreciate an observation of the face of a screaming monkey. He adds characteristically: "Could you make it scream without hurting it much?" By that time Darwin had more fellow-feeling for monkeys than he had had for Fuegians in 1834, and in the remaining decade of his life he was to embrace even lower brethren.

The two chapters included from *Insectivorous Plants* (1875) deal with one of those brethren, Drosera, the common sundew. Darwin wrote to Hooker in 1860: "By Jove, I sometimes think *Drosera* is a disguised animal!" Three or four years later he wrote to Gray: "Depend on it you are unjust on the merits of my beloved Drosera; it is a wonderful plant, or rather a most sagacious animal. I will stick up for Drosera to the day of my death." The first chapter shows Darwin discovering its animal habits, and the later chapter shows his remarkable discovery of its animal digestion. "I do not think any discovery gave me more pleasure than proving a true act of digestion in *Drosera,*" Darwin wrote to Hooker in 1873. Here Darwin most fully shows his wild ingenuity at experiment, and plays the mad scientist with gusto. We may be reminded of the witches in *Macbeth* when Darwin tests his sundews with tissue from the visceral cavity of a toad, cartilage from the leg bone of a sheep, bits of the skinned ear of a cat, slivers of the dried bone of a fowl moistened with saliva, and slices of the canine tooth of a dog.

The *Autobiography* (1876–1882) is included here complete as the best portrait—far better than any biography yet published—of the man himself. Darwin's extreme modesty and winning style strike the reader in the first sentence, which explains his motive for writing: "I have thought that the attempt would amuse me, and might possibly interest my children or their children." Darwin's truthfulness extends from the somewhat comic frankness with which he confesses that as a little boy he once beat a puppy, although not very severely, to the unspairing, shocking frankness with which he confesses his "curious and lamentable loss of the higher aesthetic tastes." Darwin's account of the passionate beetle collecting of his youth, and his story of popping that third beetle into his mouth, is a delight to read, as well as an insight into the passions that burned beneath his placid surface. It may have been more important to his career than he thought, however. After Darwin's death, Wallace was asked to explain how he and Darwin independently and simultaneously happened to hit on the theory of evolution of species by natural selection. "First (and most important, as I believe), in early life both Darwin and myself became ardent beetle-hunters," Wallace said. "Now there is certainly no group of organisms that so impresses the collector by the almost infinite number of its specific forms, the endless modifications of structure, shape, colour, and surface-markings that distinguish them from each other, and their innumerable adaptations to diverse environments."

Darwin's last book, *The Formation of Vegetable Mould through the Action of Worms* (1881), is represented by a chapter here because it is, with *The Voyage of the Beagle,* Darwin at his most delightful. He notes in the chapter that the earthworm shares extra-stomachal digestion with Drosera, as it shares with that favorite of his youth, the corals, the ability to change the world enormously by what a reviewer called "the cumulative importance of the infinitely little." The book displays the general tendency Darwin admits to in the *Autobiography,* the tendency to exalt the lowly in the organic scale. "The subject may appear an insignificant one," he writes of earthworms in the introduction, "but we shall see that it possesses some interest; and the maxim 'de minimis lex non curat' does not apply to science."

Here we see the familiar experimental ingenuity, the willingness to try anything, however absurd (from which so many great scientific discoveries have come). In this chapter we see Darwin test the hearing of worms by having his son Francis play the bassoon to them. (In later chapters he designs a superior sort of leaf for his worms made of writing paper smeared with raw fat, and creates an improved rain, watering their castings through a very fine rose.) At all times Darwin is trying to imagine what it is like to be a worm. While at work on the book in 1880, he wrote to G. J. Romanes: "I tried to observe what passed in my own mind when I did the work of a worm."

All of Darwin's writings are so interrelated that there is a sense in which he wrote only one book. Lyell, Murray, and Huxley all advised him to follow the *Origin* with "separate volumes in detail," and so he did, for the rest of his life. Even the eight years he spent classifying barnacles were spent classifying them in evolutionary sequences. Darwin explains in the *Autobiography* (pp. 392–97) how his botanical books relate to the general theory. What worms represented for him is shown in the introduction to *Vegetable Mould*. Replying to a reviewer's contention that worms are too weak and small to accomplish all that the book says they do, Darwin writes: "Here we have an instance of that inability to sum up the effects of a continually recurrent cause, which has often retarded the progress of science, as formerly in the case of geology, and more recently in that of the principle of evolution."

In this volume, the various pieces similarly interrelate. The discoveries made on the *Beagle* voyage underlie the *Essay of 1844* and *The Origin of Species*. The Fuegians' murder of their old women rather than their otter-dogs, which struck Darwin as "horrid" in *The Voyage of the Beagle,* by 1844 has become simply a neutral scientific fact. He writes in the *Essay:* "The value set upon animals by savages is shown by the inhabitants of Tierra del Fuego devouring their old women before their dogs, which as they asserted are useful in otter-hunting: who can doubt but that in every case of famine and war, the best otter-hunters would be preserved, and therefore in fact selected for breeding." Similarly the variety of species on the Galapagos Islands becomes an illustration for a general theory in the *Essay of 1844* (p. 155). The close observation of monkeys in *The Expression of Emotions* underlies the theorizing of *The Descent of Man. Vegetable Mould* and *Insectivorous Plants* are demonstrations of evolutionary adaptation, but they also supplement the *Autobiography* by showing us Darwin at work in his greenhouses and in his study.

These eight selections thus represent the range of Darwin's work about as well as one can in a collection of this size. The *Essay of 1844,* the pages from *The Origin of Species,* and the chapter from *The Descent of Man* show Darwin as a theorist: bold and vigorous in his ideas, scrupulously careful in his selection of evidence. The selections from *The Expression of the Emotions, Insectivorous Plants,* and *Vegetable Mould* show

Darwin at work as a practising scientist: a gifted observer and a tireless experimenter. (It is instructive to contrast the primitiveness of his experiments on the *Beagle* voyage—throwing one lizard into the sea or pulling another one's tail—with the sophistication of the experimentation on Drosera.) Finally, the chapters from *The Voyage of the Beagle* and the full *Autobiography* give us a picture of Darwin the man: an extraordinarily attractive and sympathetic figure.

A few facets of Darwin's work have had to remain unrepresented in this volume. The books on coral reefs and geology are not as interesting to the general reader as Darwin's work in other scientific fields. The books on the taxonomy of the Cirripedes are too technical for the layman. The other botanical volumes must be represented by the chapters from *Insectivorous Plants. Variation Under Domestication,* which has many things of interest, including an account of Darwin's fascinating genetic theory, Pangenesis, cannot profitably be excerpted.

As the list of his works suggests, Darwin had an amazing range among the sciences, with almost all of his knowledge self-taught. When he began he was an almost completely untrained naturalist. In his obituary on Darwin, Huxley somewhat unkindly characterized his work during the *Beagle* voyage: "But with no previous training in dissection, hardly any power of drawing, and next to no knowledge of comparative anatomy, his occupation with work of this kind—notwithstanding all his zeal and industry—resulted, for the most part, in a vast accumulation of useless manuscript." Francis Darwin remarks of his father: "He was a Naturalist in the old sense of the word, that is, a man who works at many branches of the science, not merely a specialist in one." Darwin nevertheless attained a remarkable proficiency in several of these branches. His first love was geology, and as he says in the *Autobiography,* he would certainly have become a professional geologist had his health been up to it. Restricted as he was much of the time to his house and grounds, he turned first to zoology and spent the eight years from 1846 to 1854 dissecting and classifying tiny barnacles, the Cirripedes. Darwin writes in the *Autobiography:* "I doubt whether the work was worth the consumption of so much time." In the opinion of Huxley and Hooker, it was; by the time he had finished the work Darwin had become a trained scientist. As for his skill as a taxonomist, the biologist George Romanes wrote in an obituary notice: "Had Mr. Darwin chosen to devote himself to a life of purely morphological work, his name would probably have been second to none in that department of biology."

After Darwin finished with the Cirripedes, he bred pigeons, to learn at first hand about variation under domestication. For the latter part of his life he worked principally at plant physiology, and he founded whole new divisions of the subject: the fertilization of flowers, insectivorous plants, dimorphism, and others. In the opinion of some, Darwin was the greatest botanist since Linnaeus. In 1958 a modern botanist, J. Heslop-Harrison, assessed Darwin's work in the field: "In his technique of experiment, Darwin was irreproachable within the limits of his faculties, and in powers of detailed observation it is doubtful whether any biologist has ever surpassed him."

Darwin's greatest contributions as a scientist were in none of these traditional branches of science, however, but in others that he founded or revived. *The Origin of Species,* with its vision of the great Tangled Bank of Life, founded the modern science of ecology, the study of the interrelationship of organisms and environment. *The Expression of the Emotions* credits the founding of the science of ethology, or behavioral psychology, to Sir Charles Bell, but its modern development stems largely from Darwin's own work. Margaret Mead has described Darwin's behavioral studies as pioneer work in "the new science of kinesics," the study of the nonverbal aspects of human communication. In 1872 Darwin wrote to a naturalist studying the habits of insects: "How incomparably more valuable are such researches than the mere description of a thousand species!" The extent to which the modern emphasis in biology has shifted away from the taxonomic is in considerable part due to Darwin's own shift and its influence.

Darwin's genetic theory, the "provisional hypothesis of Pangenesis," is a speculation that invisible "gemmules" thrown off by every cell in the organism combine to make up the sexual elements. It was derided by his contemporaries, including his scientific friends and supporters. Darwin wrote to Hooker in 1868: "It is a comfort to me to think that you will be surely haunted on your death-bed for not honouring the great god Pan." Two years later he insisted to another correspondent: "I fully believe Pangenesis will have its successful day." After the rediscovery of Mendel in 1900 and the general acceptance in our century of Weismann's theory of the continuity of the germ plasm, Darwin's faith seemed extremely ill-founded. But the great geneticist Hugo de Vries wrote of Darwin in 1909: "In his Pangenesis hypothesis he has given us the clue for a close study and ultimate elucidation of the subject under discussion," the subject of variation. And a present-day geneticist, Donald Michie, in 1958 described Pangenesis as "an arresting and prophetic hypothesis" that fits some present evidence as germ-line heredity does not, and concluded: "I estimate that genetics has about ten years to go before it can claim fully to have caught up with Darwin."

Beyond all of these, Darwin's contribution to science was as a compiler and theorist, in *The Origin of Species*. In a letter to Huxley he mocked himself for belonging to the "blessed gang of compilers," but it was in compiling facts and theorizing from them that his greatest scientific genius lay. The Darwinian theory of evolution has profoundly affected all the sciences, the social sciences, even the humanities and the arts.

Underlying Darwin the scientist is Darwin the man. The most mysterious thing about him was the quality of his mind, which displayed no visible brilliance, as he himself notes in the *Autobiography,* so that his genius seemed surprising to everyone who knew him. Rereading the *Origin* some years after Darwin's death, Huxley wrote to a friend: "There is a marvelous dumb sagacity about him—like that of a sort of miraculous dog—and he gets to the truth by ways as dark as those of the Heathen Chinese." Darwin's genius was a combination of lesser qualities best described by S. A. Barnett: "Darwin was primarily a naturalist of extraordinary gifts, imaginative, with insatiable curiosity and immense enthusiasm, thorough, critical and inexhaustible." His quest for causes and explanations was ceaseless. As Francis Darwin put it, "It was as though he were charged with theorizing power ready to flow into any channel on the slightest disturbance, so that no fact, however small, could avoid releasing a stream of theory." Wallace remarked that in Darwin's case "the restless curiosity of the child to know the 'what for?' the 'why?' and the 'how?' of everything" seems "never to have abated its forces."

Darwin also had the child's sense of awe and wonder at the marvels of the natural world. That most objective of works, the *Essay of 1844,* notes that the mistletoe is "wonderfully" related to other beings, that the modifications of the mammalian foot are "wonderful," the abortive organs are "justly considered wonderful"; and it finishes on "endless forms most beautiful and most wonderful." Words like "wonderful" and "marvelous" are proportionately more frequent in the other books.

If Darwin's mind was not intuitive and quick, it was dogged and patient beyond belief. His experiments in the fertilization of plants went on for eleven years and involved so much tedious counting of seeds that he did not believe that anyone would ever repeat them. A favorite quotation of Darwin's was from Trollope's Brickmaker: "It's dogged as does it." Collecting facts on the origin of species for twenty years, he was building, like his corals and earthworms, an enormous change at a pace almost imperceptible.

Darwin's character, Huxley wrote, was even nobler than his intellect. The modesty of Darwin's account of his mental powers in the *Autobiography* is enormously impressive. Huxley notes that he accepted "criticisms and suggestions from anybody and everybody, not only without impatience, but with expressions of gratitude sometimes almost comically in excess of their value." Darwin's correspondence is always flattering to the recipient, and usually deprecatory about Darwin. "I believe I am the slowest (perhaps the worst) thinker in England," he wrote, typically, to Hooker. There is nevertheless a good deal of evidence that Darwin fully appreciated the greatness of his achievement. The *Origin* makes an implied comparison of Darwin with Galileo, and a letter from Darwin to Lyell implies a comparison of himself with Newton. But Darwin has no sense at all of any personal greatness; it is as though he had won his success in a raffle.

Darwin's character had a quality of geniality, of simple goodness, that shines through the *Autobiography,* as when he credits all his critics (with one exception) with good faith. Related to this is what Jane Harrison called Darwin's *"piety*—in the beautiful Roman sense—towards tradition and association." Most important of all is his absolute honesty and integrity. In his scrupulous regard for negative evidence, Darwin anticipated the psychoanalytic principle of selective forgetting, and he discusses his precautions against such forgetting in the *Autobiography* (p. 390). His books emphasize evidence and arguments against his views, and freely admit errors. Darwin wrote to a correspondent: "It is a golden rule, which I try to follow, to put every fact which is opposed to one's preconceived opinion in the strongest light." Basil Willey refers to "that candour in meeting objections which perhaps won more converts than any of his other devices of persuasion." This is perhaps discounting as rhetoric what is surely a quality of Darwin's character, although it is certainly true that it does persuade. Possibly the most impressive single example of his integrity was his decision, made early in life, not to theorize after sixty, since he had seen so many scientists make old fools of themselves. Pangenesis, Darwin's last ambitious theory, was published when he was fifty-nine.

As an expression of his mind and character, Darwin's writing is always interesting and sometimes charming. He calls on us not only to think and understand but to imagine and wonder; and, beyond that, to metamorphose ourselves magically into an insect fertilizing an orchid, an earthworm tugging a leaf into its burrow. With Darwin we rejoice in life. *The Origin of Species* cries out: "There is so much beauty throughout nature"; in *The Descent of Man* even naked sea slugs are of "extreme beauty." In his introduction to *Orchids* Darwin says: "An examination of their many beautiful contrivances will exalt the whole vegetable kingdom in most persons' estimation"; midway he exclaims: "How numerous and beautiful are the contrivances"; he concludes that the orchids "exhibit an almost endless diversity of beautiful adaptation."

Sometimes Darwin writes with high style, as in this sentence on subsidence from *Coral Reefs:* "This, however, can hardly be expected, for it must ever be most difficult, excepting in country long civilized, to detect a movement the tendency of which is to conceal the part affected." At all times he writes very personally, as he himself discovered to his horror when someone pointed out forty-three uses of the first person singular in the first four paragraphs of the *Origin*'s introduction. He sometimes even manages a kind of wit, as when he writes in *The Descent of Man:* "Hardly any colour is finer than that of arterial blood; but there is no reason to suppose that the colour of the blood is in itself any advantage; and though it adds to the beauty of the maiden's cheek, no one will pretend that it has been acquired for this purpose." As Darwin disparaged his mind, so he disparaged his prose. "I do not believe any man in England naturally writes so vile a style as I do," he wrote to Hooker in 1867.

Darwin's place in history seems a high and assured one. In one sense it is the result of lucky accidents. As Darwin notes in the *Autobiography,* it depended on the *Beagle* voyage, which in turn "depended on so small a circumstance as my uncle offering to drive me thirty miles to Shrewsbury, which few uncles would have done, and on such a trifle as the shape of my nose." After that, it depended on his reading Malthus at the right time, and there is a wonderful understatement in the *Autobiography*'s bland remark: "I happened to read for amusement Malthus on *Population.*" In another sense Darwin's success was entirely deserved. In an unspectacular fashion he probably was, as Jane Harrison says, "intellectual king of his generation." In the qualified opinion of Julian Huxley, Darwin managed "to produce a scientific output which for quantity and quality has never been exceeded." Reading him is both a duty and a pleasure.

Chronology of the Life and Works of Charles Darwin

1809   Born at Shrewsbury, England, February 12.

1817   Death of his mother.

Attended a day school at Shrewsbury kept by the Rev. G. Case.

1818   Attended a school at Shrewsbury kept by Dr. Butler, father of Samuel Butler.

1825   In October, began two years at Edinburgh University, intending to become a doctor.

1828   Began residence at Christ's College, Cambridge, intending to become a clergyman.

1831   Passed the examination for the B.A. degree, without honors, in January.

Sailed from England as naturalist aboard H.M.S. *Beagle* on December 27.

1836   Returned to Shrewsbury on October 4, after five years spent sailing around the world.

Moved to Cambridge on December 13.

1837   Moved to lodgings in London on March 7, to be near the British Museum.

Began a notebook of facts relating to the origin of species in July.

1838   Read Malthus on *Population* in October and realized that the struggle for existence would have the effect of originating new species.

1839   Married at Maer in Staffordshire to his first cousin Emma Wedgwood. On the last day of the year they moved to a house in London.

Publication of *Journal of Researches* of the *Beagle* voyage.

1842   Wrote a brief abstract of his theory of the origin of species in June.

Settled at the village of Down in Kent, where he remained for the rest of his life.

Publication of *The Structure and Distribution of Coral Reefs*.

1844 Enlarged the 1842 *Abstract* into an *Essay* of 230 pages.
Publication of *Geological Observations on the Volcanic Islands*.

1846 Publication of *Geological Observations on South America*.

1849 Death of his father.

1851 Publication of two monographs on Cirripedes.

1854 Publication of two more monographs on Cirripedes.
Began working full time on the origin of species, compiling his notes and other material.

1856 At the suggestion of Sir Charles Lyell, began writing a large book on the origin of species.

1858 Received a paper from Alfred Russell Wallace stating Wallace's independent discovery, also based on a reading of Malthus, of Darwin's theory.
Lyell and Sir Joseph Hooker communicated a joint paper by Darwin and Wallace to the Linnaean Society.
Began abstracting his enormous manuscript on species on July 20.

1859 Publication of *The Origin of Species,* in an edition of 1250 copies, on November 24. The edition sold out that day.

1860 Second edition of the *Origin* (3000 copies).

1861 Third edition of the *Origin* (2000 copies).

1862 Publication of *On the various contrivances by which Orchids are fertilised by Insects.*

1866 Fourth edition of the *Origin* (1250 copies).

1868 Publication of *Variation of Animals and Plants under Domestication.*

1869 Fifth edition of the *Origin.*

1871 Publication of *The Descent of Man.*

1872 Sixth and final edition of the *Origin.*
Publication of *The Expression of the Emotions in Man and Animals.*

1874 Second edition of *Descent.*
Second edition of *Coral Reefs.*

1875 Publication of *Insectivorous Plants.*
Second edition of *Variation.*

Publication of *The Movements and Habits of Climbing Plants*.

1876 Wrote *Autobiography*.

Publication of *The Effects of Cross- and Self-Fertilisation*.

Second edition of *Volcanic Islands*.

1877 Publication of *The Different Forms of Flowers*.

Second edition of *Orchids*.

1878 Second edition of *Cross- and Self-Fertilisation*.

1880 Publication of *The Power of Movement in Plants*.

Second edition of *Forms of Flowers*.

1881 Wrote a continuation of the *Autobiography*.

Publication of *The Formation of Vegetable Mould through the Action of Worms*.

1883 Died at Down, April 19, and was buried in Westminster Abbey, April 26.

The Voyage of the Beagle

The Voyage of the Beagle

CHAPTER X

Tierra del Fuego

December 17, 1832. Having now finished with Patagonia and the Falkland Islands, I will describe our first arrival in Tierra del Fuego. A little after noon we doubled Cape St. Diego, and entered the famous strait of Le Maire. We kept close to the Fuegian shore, but the outline of the rugged, inhospitable Staten-land was visible amidst the clouds. In the afternoon we anchored in the Bay of Good Success. While entering we were saluted in a manner becoming the inhabitants of this savage land. A group of Fuegians partly concealed by the entangled forest were perched on a wild point overhanging the sea; and as we passed by, they sprang up and waving their tattered cloaks sent forth a loud and sonorous shout. The savages followed the ship, and just before dark we saw their fire, and again heard their wild cry. The harbour consists of a fine piece of water half surrounded by low rounded mountains of clay-slate, which are covered to the water's edge by one dense gloomy forest. A single glance at the landscape was sufficient to show me how widely different it was from any thing I had ever beheld. At night it blew a gale of wind, and heavy squalls from the mountains swept past us. It would have been a bad time out at sea, and we, as well as others, may call this Good Success Bay.

In the morning the Captain sent a party to communicate with the Fuegians. When we came within hail, one of the four natives who were present advanced to receive us, and began to shout most vehemently, wishing to direct us where to land. When we were on shore the party looked rather alarmed, but continued talking and making gestures with great rapidity. It was without exception the most curious and interesting spectacle I ever beheld: I could not have believed how wide was the difference between savage and civilised man. It is greater than between a wild and domesticated animal, inasmuch as in man there is a greater power of improvement. The chief spokesman was old, and appeared to be the head of the family; the three others were powerful young men, about six feet high. The women and children had been sent away. These Fuegians are a very different race from the stunted, miserable wretches farther westward; and they seem closely allied to the famous Patagonians of the Strait of Magellan. Their only garment consists of a mantle made of guanaco skin, with the wool outside; this they wear just thrown over their shoulders, leaving their persons as often exposed as covered. Their skin is of a dirty coppery red colour.

The old man had a fillet of white feathers tied round his head, which partly confined his black, coarse, and entangled hair. His face was crossed by two broad transverse bars; one, painted bright red, reached from ear to ear and included the upper lip; the other, white like chalk, extended above and parallel to the first, so that even his eyelids were thus coloured. The other two men were ornamented by streaks of black powder, made of charcoal. The party altogether closely resembled the devils which come on the stage in plays like *Der Freischutz*.

Their very attitudes were abject, and the expression of their countenances distrustful, surprised, and startled. After we had presented them with some scarlet cloth, which they immediately tied round their necks, they became good friends. This was shown by the old man patting our breasts, and making a chuckling kind of noise, as people do when feeding chickens. I walked with the old man, and this demonstration of friendship was repeated several times; it was concluded by three hard slaps, which were given me on the breast and back at the same time. He then bared his bosom for me to return the compliment, which being done, he seemed highly pleased. The language of these people, according to our notions, scarcely deserves to be called articulate. Captain Cook has compared it to a man clearing his throat, but certainly no European ever cleared his throat with so many hoarse, guttural, and clicking sounds.

They are excellent mimics: As often as we coughed or yawned, or made any odd motion, they immediately imitated us. Some of our party began to squint and look awry; but one of the young Fuegians (whose whole face was painted black, excepting a white band across his eyes) succeeded in making far more hideous grimaces. They could repeat with perfect correctness each word in any sentence we addressed them, and they remem-

bered such words for some time. Yet we Europeans all know
how difficult it is to distinguish apart the sounds in a foreign
language. Which of us, for instance, could follow an American
Indian through a sentence of more than three words? All
savages appear to possess, to an uncommon degree, this power
of mimicry. I was told, almost in the same words, of the same
ludicrous habit among the Caffres; the Australians, likewise,
have long been notorious for being able to imitate and describe
the gait of any man, so that he may be recognised. How can
this faculty be explained? Is it a consequence of the more practised habits of perception and keener senses, common to all men
in a savage state, as compared with those long civilised?

When a song was struck up by our party, I thought the
Fuegians would have fallen down with astonishment. With equal
surprise they viewed our dancing; but one of the young men,
when asked, had no objection to a little waltzing. Little accustomed to Europeans as they appeared to be, yet they knew
and dreaded our fire-arms; nothing would tempt them to take a
gun in their hands. They begged for knives, calling them by
the Spanish word *cuchilla*. They explained also what they
wanted by acting as if they had a piece of blubber in their mouth,
and then pretending to cut instead of tear it.

I have not as yet noticed the Fuegians whom we had on
board. During the former voyage of the *Adventure* and *Beagle*
in 1826 to 1830, Captain Fitz-Roy seized on a party of natives
as hostages for the loss of a boat, which had been stolen, to the
great jeopardy of a party employed on the survey; and some of
these natives, as well as a child whom he bought for a pearl-
button, he took with him to England, determining to educate
them and instruct them in religion at his own expense. To settle
these natives in their own country was one chief inducement to
Captain Fitz-Roy to undertake our present voyage; and before
the Admiralty had resolved to send out this expedition, Captain
Fitz-Roy had generously chartered a vessel, and would himself
have taken them back. The natives were accompanied by a missionary, R. Matthews, of whom and of the natives Captain Fitz-
Roy had published a full and excellent account. Two men, one
of whom died in England of the small-pox, a boy and a little
girl, were originally taken; and we had now on board York
Minster, Jemmy Button (whose name expresses his purchase-

money), and Fuegia Basket. York Minster was a full-grown, short, thick, powerful man; his disposition was reserved, taciturn, morose, and when excited violently passionate; his affections were very strong towards a few friends on board; his intellect good. Jemmy Button was a universal favourite, but likewise passionate; the expression of his face at once showed his nice disposition. He was merry and often laughed, and was remarkably sympathetic with any one in pain; when the water was rough, I was often a little sea-sick, and he used to come to me and say in a plaintive voice, "Poor, poor fellow!" But the notion, after his aquatic life, of a man being sea-sick was too ludicrous, and he was generally obliged to turn on one side to hide a smile or laugh, and then he would repeat his "Poor, poor fellow!" He was of a patriotic disposition; and he liked to praise his own tribe and country, in which he truly said there were "plenty of trees," and he abused all the other tribes. He stoutly declared that there was no devil in his land. Jemmy was short, thick, and fat, but vain of his personal appearance; he used always to wear gloves, his hair was neatly cut, and he was distressed if his well-polished shoes were dirtied. He was fond of admiring himself in a looking-glass; and a merry-faced little Indian boy from the Rio Negro, whom we had for some months on board, soon perceived this, and used to mock him. Jemmy, who was always rather jealous of the attention paid to this little boy, did not at all like this, and used to say, with rather a contemptuous twist of his head, "Too much skylark." It seems yet wonderful to me, when I think over all his many good qualities, that he should have been of the same race, and doubtless partaken of the same character, with the miserable, degraded savages whom we first met here. Lastly, Fuegia Basket was a nice, modest, reserved young girl, with a rather pleasing but sometimes sullen expression, and very quick in learning anything, especially languages. This she showed in picking up some Portuguese and Spanish, when left on shore for only a short time at Rio de Janeiro and Monte Video, and in her knowledge of English. York Minster was very jealous of any attention paid to her; for it was clear he determined to marry her as soon as they were settled on shore.

Although all three could both speak and understand a good deal of English, it was singularly difficult to obtain much information from them concerning the habits of their countrymen;

this was partly owing to their apparent difficulty in understanding the simplest alternative. Every one accustomed to very young children knows how seldom one can get an answer even to so simple a question as whether a thing is black *or* white; the idea of black or white seems alternately to fill their minds. So it was with these Fuegians, and hence it was generally impossible to find out, by cross-questioning, whether one had rightly understood anything which they had asserted. Their sight was remarkably acute: It is well known that sailors, from long practice, can make out a distant object much better than a landsman; but both York and Jemmy were much superior to any sailor on board. Several times they have declared what some distant object has been, and though doubted by every one, they have proved right, when it has been examined through a telescope. They were quite conscious of this power; and Jemmy, when he had any little quarrel with the officer on watch, would say, "Me see ship, me no tell."

It was interesting to watch the conduct of the savages, when we landed, towards Jemmy Button; they immediately perceived the difference between him and ourselves, and held much conversation one with another on the subject. The old man addressed a long harangue to Jemmy, which it seems was to invite him to stay with them. But Jemmy understood very little of their language, and was, moreover, thoroughly ashamed of his countrymen. When York Minster afterwards came on shore, they noticed him in the same way, and told him he ought to shave; yet he had not twenty dwarf hairs on his face, whilst we all wore our untrimmed beards. They examined the colour of his skin, and compared it with ours. One of our arms being bared, they expressed the liveliest surprise and admiration at its whiteness, just in the same way in which I have seen the ourang-outang do at the Zoological Gardens. We thought that they mistook two or three of the officers, who were rather shorter and fairer, though adorned with large beards, for the ladies of our party. The tallest amongst the Fuegians was evidently much pleased at his height being noticed. When placed back to back with the tallest of the boat's crew, he tried his best to edge on higher ground, and to stand on tiptoe. He opened his mouth to show his teeth, and turned his face for a side view; and all this was done with such alacrity that I dare say he thought himself

the handsomest man in Tierra del Fuego. After our first feeling of grave astonishment was over, nothing could be more
ludicrous than the odd mixture of surprise and imitation which
these savages every moment exhibited.

The next day I attempted to penetrate some way into the
country. Tierra del Fuego may be described as a mountainous
land, partly submerged in the sea, so that deep inlets and bays
occupy the place where valleys should exist. The mountain
sides, except on the exposed western coast, are covered from
the water's edge upwards by one great forest. The trees reach
to an elevation of between a thousand and fifteen hundred feet,
and are succeeded by a band of peat, with minute alpine plants;
and this again is succeeded by the line of perpetual snow, which,
according to Captain King, in the Strait of Magellan descends to
between three thousand and four thousand feet. To find an acre
of level land in any part of the country is most rare. I recollect
only one little flat piece near Port Famine, and another of rather
larger extent near Goeree Road. In both places, and everywhere else, the surface is covered by a thick bed of swampy
peat. Even within the forest, the ground is concealed by a mass
of slowly putrefying vegetable matter, which, from being soaked
with water, yields to the foot.
    Finding it nearly hopeless to push my way through the wood,
I followed the course of a mountain torrent. At first, from the
waterfalls and number of dead trees, I could hardly crawl along;
but the bed of the stream soon became a little more open, from
the floods having swept the sides. I continued slowly to advance
for an hour along the broken and rocky banks, and was amply
repaid by the grandeur of the scene. The gloomy depth of the
ravine well accorded with the universal signs of violence. On
every side were lying irregular masses of rock and torn-up trees;
other trees, though still erect, were decayed to the heart and
ready to fall. The entangled mass of the thriving and the fallen
reminded me of the forests within the tropics—yet there was a
difference: For in these still solitudes, Death, instead of Life,
seemed the predominant spirit. I followed the watercourse till
I came to a spot where a great slip had cleared a straight space
down the mountain side. By this road I ascended to a considerable elevation, and obtained a good view of the surrounding

woods. The trees all belong to one kind, the Fagus betuloides; for the number of the other species of Fagus and of the Winter's Bark is quite inconsiderable. This beech keeps its leaves throughout the year; but its foliage is of a peculiar brownish-green colour, with a tingle of yellow. As the whole landscape is thus coloured, it has a sombre, dull appearance; nor is it often enlivened by the rays of the sun.

December 20. One side of the harbour is formed by a hill about fifteen hundred feet high, which Captain Fitz-Roy has called after Sir J. Banks, in commemoration of his disastrous excursion, which proved fatal to two men of his party and nearly so to Dr. Solander. The snow-storm, which was the cause of their misfortune, happened in the middle of January, corresponding to our July, and in the latitude of Durham! I was anxious to reach the summit of this mountain to collect alpine plants; for flowers of any kind in the lower parts are few in number. We followed the same watercourse as on the previous day, till it dwindled away, and we were then compelled to crawl blindly among the trees. These, from the effects of the elevation and of the impetuous winds, were low, thick, and crooked. At length we reached that which from a distance appeared like a carpet of fine green turf, but which, to our vexation, turned out to be a compact mass of little beech-trees about four or five feet high. They were as thick together as box in the border of a garden, and we were obliged to struggle over the flat but treacherous surface. After a little more trouble we gained the peat, and then the bare slate rock.

A ridge connected this hill with another, distant some miles, and more lofty, so that patches of snow were lying on it. As the day was not far advanced, I determined to walk there and collect plants along the road. It would have been very hard work had it not been for a well-beaten and straight path made by the guanacos; for these animals, like sheep, always follow the same line. When we reached the hill we found it the highest in the immediate neighbourhood, and the waters flowed to the sea in opposite directions. We obtained a wide view over the surrounding country: To the north a swampy moorland extended, but to the south we had a scene of savage magnificence, well becoming Tierra del Fuego. There was a degree of mysterious grandeur in mountain behind mountain, with the deep intervening valleys, all covered by one thick, dusky mass of forest. The atmosphere, likewise, in this climate, where gale succeeds gale, with rain, hail, and sleet, seems blacker than anywhere else. In the Strait of Magellan, looking due southward from Port Famine, the distant channels between the mountains appeared from their gloominess to lead beyond the confines of this world.

December 21. The *Beagle* got under way; and on the succeeding day, favoured to an uncommon degree by a fine easterly breeze, we closed in with the Barnevelts, and running past Cape Deceit with its stony peaks, about three o'clock doubled the weather-beaten Cape Horn. The evening was calm and bright, and we enjoyed a fine view of the surrounding isles. Cape Horn, however, demanded his tribute, and before night sent us a gale of wind directly in our teeth. We stood out to sea, and on the second day again made the land, when we saw on our weather-bow this notorious promontory in its proper form—veiled in a mist, and its dim outline surrounded by a storm of wind and water. Great black clouds were rolling across the heavens, and squalls of rain, with hail, swept by us with such extreme violence that the Captain determined to run into Wigwam Cove. This is a snug little harbour, not far from Cape Horn; and here, at Christmas-eve, we anchored in smooth water. The only thing which reminded us of the gale outside was every now and then a puff from the mountains, which made the ship surge at her anchors.

December 25. Close by the cove, a pointed hill, called Kater's Peak, rises to the height of seventeen hundred feet. The surrounding islands all consist of conical masses of greenstone, associated sometimes with less regular hills of baked and altered clay-slate. This part of Tierra del Fuego may be considered as the extremity of the submerged chain of mountains already alluded to. The cove takes its name of "Wigwam" from some of the Fuegian habitations; but every bay in the neighbourhood might be so called with equal propriety. The inhabitants, living chiefly upon shell-fish, are obliged constantly to change their place of residence; but they return at intervals to the same spots, as is evident from the piles of old shells, which must often amount to many tons in weight. These heaps can be distinguished at a long distance by the bright green colour of certain plants which invariably grow on them. Among these may be enumerated the wild celery and scurvy grass, two very serviceable plants, the use of which has not been discovered by the natives.

The Fuegian wigwam resembles, in size and dimensions, a haycock. It merely consists of a few broken branches stuck in the ground, and very imperfectly thatched on one side with a few tufts of grass and rushes. The whole cannot be the work of an hour, and it is only used for a few days. At Goeree Roads I saw a place where one of these naked men had slept which absolutely offered no more cover than the form of a hare. The man was evidently living by himself, and York Minster said he was "very bad man," and that probably he had stolen something. On the west coast, however, the wigwams are rather better, for they are covered with seal-skins. We were detained here several days by the bad weather. The climate is certainly wretched: The summer solstice was now passed, yet every day snow fell on the hills, and in the valleys there was rain, accompanied by sleet. The thermometer generally stood about 45°, but in the night fell to 38° or 40°. From the damp and boisterous state of the atmosphere, not cheered by a gleam of sunshine, one fancied the climate even worse than it really was.

While going one day on shore near Wollaston Island, we pulled alongside a canoe with six Fuegians. These were the most abject and miserable creatures I anywhere beheld. On the east coast the natives, as we have seen, have guanaco cloaks, and on the west, they possess seal-skins. Amongst these central tribes the men generally have an otter-skin, or some small scrap about as large as a pocket-handkerchief, which is barely sufficient to cover their backs as low down as their loins. It is laced across the breast by strings, and according as the wind blows, it is shifted from side to side. But these Fuegians in the canoe were quite naked, and even one full-grown woman was absolutely so. It was raining heavily, and the fresh water, together with the spray, trickled down her body. In another harbour not far distant, a woman, who was suckling a recently born child, came one day alongside the vessel, and remained there out of mere curiosity, whilst the sleet fell and thawed on her naked bosom, and on the skin of her naked baby! These poor wretches were stunted in their growth, their hideous faces bedaubed with white paint, their skins filthy and greasy, their hair entangled, their voices discordant, and their gestures violent. Viewing such men, one can hardly make oneself believe that they are fellow-creatures, and inhabitants of the same world. It is a common subject of conjecture what pleasure in life some of the lower animals can enjoy; how much more reasonably the same question may be asked with respect to these barbarians! At night, five or six human beings, naked and scarcely protected from the wind and rain of this tempestuous climate, sleep on the wet ground coiled up like animals. Whenever it is low water, winter or summer, night or day, they must rise to pick shell-fish from the rocks; and the women either dive to collect sea-eggs, or sit patiently in their canoes and, with a baited hair-line without any hook, jerk out little fish. If a seal is killed, or the floating carcass of a putrid whale discovered, it is a feast; and such miserable food is assisted by a few tasteless berries and fungi.

They often suffer from famine. I heard Mr. Low, a sealing-master intimately acquainted with the natives of this country, give a curious account of the state of a party of one hundred and fifty natives on the west coast, who were very thin and in great distress. A succession of gales prevented the women from getting shell-fish on the rocks, and they could not go out in their canoes to catch seal. A small party of these men one morning set out, and the other Indians explained to him that they were going a four days' journey for food. On their return, Low went to meet them, and he found them excessively tired, each man carrying a great square piece of putrid whales-blubber with a hole in the middle, through which they put their heads, like the Gauchos do through their ponchos or cloaks. As soon as the blubber was brought into a wigwam, an old man cut off thin slices, and muttering over them, broiled them for a minute, and distributed them to the famished party, who during this time preserved a profound silence. Mr. Low believes that whenever a whale is cast on shore, the natives bury large pieces of it in the sand, as a resource in time of famine; and a native boy, whom he had on board, once found a stock thus buried. The different tribes when at war are cannibals. From the concurrent, but quite independent, evidence of the boy taken by Mr. Low and of Jemmy Button, it is certainly true that when pressed in winter by hunger, they kill and devour their old women before they kill their dogs; the boy, being asked by Mr. Low why they did this, answered, "Doggies catch otters, old women no." This boy described the manner in which they are killed by being held over smoke and thus choked; he imitated their screams as a joke, and described the parts of their bodies which are considered best to eat. Horrid as such a death by the hands of their friends and relatives must be, the fears of the old women, when hunger begins to press, are more painful to think of; we were told that they then often run away into the mountains, but that they are pursued by the men and brought back to the slaughter-house at their own fire-sides!

Captain Fitz-Roy could never ascertain that the Fuegians have any distinct belief in a future life. They sometimes bury their dead in caves, and sometimes in the mountain forests; we do not know what ceremonies they perform. Jemmy Button would not eat land-birds, because they "eat dead men": they are unwilling even to mention their dead friends. We have no reason to believe that they perform any sort of religious worship, though perhaps the muttering of the old man before he distributed the putrid blubber to his famished party may be of this nature. Each family or tribe has a wizard or conjuring doctor, whose office we could never clearly ascertain. Jemmy believed in dreams, though not, as I have said, in the devil. I do not think that our Fuegians were much more superstitious than some of the sailors; for an old quarter-master firmly believed that the successive heavy gales, which we encountered off Cape Horn, were caused by our having the Fuegians on board. The nearest approach to a religious feeling which I heard of was shown by York Minster, who, when Mr. Bynoe shot some very young ducklings as specimens, declared in the most solemn manner, "Oh, Mr. Bynoe, much rain, snow, blow much." This was evidently a retributive punishment for wasting human food. In a wild and excited manner he also related that his brother, one day whilst returning to pick up some dead birds which he had left on the coast, observed some feathers blown by the wind. His brother said (York imitating his manner), "What that?"; and crawling onwards, he peeped over the cliff, and saw "wild man" picking his birds; he crawled a little nearer, and then hurled down a great stone and killed him. York declared for a long time afterwards storms raged, and much rain and snow fell. As far as we could make out, he seemed to consider the elements themselves as the avenging agents; it is evident in this case, how naturally, in a race a little more advanced in culture, the elements would become personified. What the "bad wild men" were has always appeared to me most mysterious. From what York said, when we found the place like the form of a hare, where a single man had slept the night before, I should have thought that they were thieves who had been driven from their tribes; but other obscure speeches made me doubt this. I have sometimes imagined that the most probable explanation was that they were insane.

The different tribes have no government or chief; yet each is surrounded by other hostile tribes, speaking different dialects, and separated from each other only by a deserted border or neutral territory. The cause of their warfare appears to be the means of subsistence. Their country is a broken mass of wild rocks, lofty hills, and useless forests; and these are viewed through mists and endless storms. The habitable land is reduced to the stones on the beach; in search of food they are compelled unceasingly to wander from spot to spot, and so steep is the coast that they can only move about in their wretched canoes. They cannot know the feeling of having a home, and still less that of domestic affection; for the husband is to the wife a brutal master to a laborious slave. Was a more horrid deed ever perpetrated than that witnessed on the west coast by Byron, who saw a wretched mother pick up her bleeding dying infant-boy, whom her husband had mercilessly dashed on the stones for dropping a basket of sea-eggs! How little can the higher powers of the mind be brought into play; what is there for imagination to picture, for reason to compare, for judgment to decide upon? To knock a limpet from the rock does not require even cunning, that lowest power of the mind. Their skill in some respects may be compared to the instinct of animals; for it is not improved by experience: The canoe, their most ingenious work, poor as it is, has remained the same, as we know from Drake, for the last two hundred and fifty years.

Whilst beholding these savages, one asks, whence have they

ed, or what change compelled a
e regions of the north, to travel
one of America, to invent and
sed by the tribes of Chile, Peru,
on one of the most inhospitable
the globe? Although such reflec-
mind, yet we may feel sure that
e is no reason to believe that the
therefore we must suppose that
f happiness, of whatever kind it
having. Nature by making habit
editary, has fitted the Fuegian to
of his miserable country.

ix days in Wigwam Cove by very
the 30th of December. Captain
ard to land York and Fuegia in
we had a constant succession of
ainst us; we drifted to 57° 23′
1833, by carrying a press of sail,
of the great rugged mountain of
York Minster (so called by Captain Cook, and the origin of the
name of the elder Fuegian), when a violent squall compelled
us to shorten sail and stand out to sea. The surf was breaking
fearfully on the coast, and the spray was carried over a cliff
estimated at two hundred feet in height. On the 12th the gale
was very heavy, and we did not know exactly where we were; it
was a most unpleasant sound to hear constantly repeated, "Keep
a good look-out to leeward." On the 13th the storm raged with
its full fury; our horizon was narrowly limited by the sheets of
spray borne by the wind. The sea looked ominous, like a dreary
waving plain with patches of drifted snow; whilst the ship
laboured heavily, the albatross glided with its expanded wings
right up the wind. At noon a great sea broke over us, and filled
one of the whale-boats, which was obliged to be instantly cut
away. The poor *Beagle* trembled at the shock, and for a few
minutes would not obey her helm; but soon, like a good ship
that she was, she righted and came up to the wind again. Had
another sea followed the first, our fate would have been decided soon, and forever. We had now been twenty-four days

trying in vain to get westward; the men were worn out with
fatigue, and they had not had for many nights or days a dry thing
to put on. Captain Fitz-Roy gave up the attempt to get westward
by the outside coast. In the evening we ran in behind False Cape
Horn, and dropped our anchor in forty-seven fathoms, fire
flashing from the windlass as the chain rushed round it. How
delightful was that still night, after having been so long involved in the din of the warring elements!

January 15, 1833. The *Beagle* anchored in Goeree Roads.
Captain Fitz-Roy having resolved to settle the Fuegians, according to their wishes, in Ponsonby Sound, four boats were
equipped to carry them there through the Beagle Channel. This
channel, which was discovered by Captain Fitz-Roy during the
last voyage, is a most remarkable feature in the geography of
this, or indeed of any other, country; it may be compared to the
valley of Lochness in Scotland, with its chain of lakes and friths.
It is about one hundred and twenty miles long, with an average
breadth, not subject to any very great variation, of about two
miles, and is throughout the greater part so perfectly straight
that the view, bounded on each side by a line of mountains,
gradually becomes indistinct in the long distance. It crosses the
southern part of Tierra del Fuego in an east and west line, and
in the middle is joined at right angles on the south side by an
irregular channel, which has been called Ponsonby Sound. This
is the residence of Jemmy Button's tribe and family.

January 19. Three whale-boats and the yawl, with a party of
twenty-eight, started under the command of Captain Fitz-Roy.
In the afternoon we entered the eastern mouth of the channel,
and shortly afterwards found a snug little cove concealed by
some surrounding islets. Here we pitched our tents and lighted
our fires. Nothing could look more comfortable than this scene.
The glassy water of the little harbour, with the branches of the
trees hanging over the rocky beach, the boats at anchor, the
tents supported by the crossed oars, and the smoke curling up
the wooded valley, formed a picture of quiet retirement. The
next day (the 20th) we smoothly glided onwards in our little
fleet, and came to a more inhabited district. Few if any of these
natives could ever have seen a white man; certainly nothing

could exceed their astonishment at the apparition of the four boats. Fires were lighted on every point (hence the name of Tierra del Fuego, or the land of fire), both to attract our attention and to spread far and wide the news. Some of the men ran for miles along the shore. I shall never forget how wild and savage one group appeared; suddenly four or five men came to the edge of an overhanging cliff. They were absolutely naked, and their long hair streamed about their faces; they held rugged staffs in their hands, and, springing from the ground, they waved their arms round their heads, and sent forth the most hideous yells.

At dinner-time we landed among a party of Fuegians. At first they were not inclined to be friendly; for until the Captain pulled in ahead of the other boats, they kept their slings in their hands. We soon, however, delighted them by trifling presents, such as tying red tape round their heads. They liked our biscuit: but one of the savages touched with his finger some of the meat preserved in tin cases which I was eating and, feeling it soft and cold, showed as much disgust at it as I should have done at putrid blubber. Jemmy was thoroughly ashamed of his country-men, and declared his own tribe were quite different, in which he was wofully mistaken. It was as easy to please as it was difficult to satisfy these savages. Young and old, men and children, never ceased repeating the word "yammerschooner," which means "give me." After pointing to almost every object, one after the other, even to the buttons on our coats, and saying their favourite word in as many intonations as possible, they would then use it in a neuter sense, and vacantly repeat "yam-merschooner." After yammerschoonering for any article very eagerly, they would by a simple artifice point to their young women or little children, as much as to say, "If you will not give it me, surely you will to such as these."

At night we endeavoured in vain to find an uninhabited cove and at last were obliged to bivouac not far from a party of natives. They were very inoffensive as long as they were few in numbers, but in the morning (the 21st) being joined by others they showed symptoms of hostility, and we thought that we should have come to a skirmish. An European labours under great disadvantages when treating with savages like these, who have not the least idea of the power of fire-arms. In the very act of levelling his musket he appears to the savage far inferior to a man armed with a bow and arrow, a spear, or even a sling. Nor is it easy to teach them our superiority except by striking a fatal blow. Like wild beasts, they do not appear to compare numbers; for each individual, if attacked, instead of retiring, will endeavour to dash your brains out with a stone, as certainly as a tiger under similar circumstances would tear you. Captain Fitz-Roy on one occasion being very anxious, from good reasons, to frighten away a small party first flourished a cutlass near them, at which they only laughed; he then twice fired his pistol close to a native. The man both times looked astounded, and carefully but quickly rubbed his head; he then stared awhile, and gabbled to his companions, but he never seemed to think of running away. We can hardly put ourselves in the position of these savages, and understand their actions. In the case of this Fuegian, the possibility of such a sound as the report of a gun close to his ear could never have entered his mind. He perhaps literally did not for a second know whether it was a sound or a blow, and therefore very naturally rubbed his head. In a similar manner, when a savage sees a mark struck by a bullet, it may be some time before he is able at all to understand how it is effected; for the fact of a body being invisible from its velocity would perhaps be to him an idea totally inconceivable. Moreover, the extreme force of a bullet, that penetrates a hard substance without tearing it, may convince the savage that it has no force at all. Certainly I believe that many savages of the lowest grade, such as these of Tierra del Fuego, have seen objects struck, and even small animals killed, by the musket, without being in the least aware how deadly an instrument it is.

January 22. After having passed an unmolested night, in what would appear to be neutral territory between Jemmy's tribe and the people whom we saw yesterday, we sailed pleasantly along. I do not know anything which shows more clearly the hostile state of the different tribes than these wide border or neutral tracts. Although Jemmy Button well knew the force of our party, he was, at first, unwilling to land amidst the hostile tribe nearest to his own. He often told us how the savage Oens men, "when the leaf red," crossed the mountains from the eastern coast of Tierra del Fuego, and made inroads on the natives of this part of the country. It was most curious to watch him when thus talking, and see his eyes gleaming and his whole face assume a new and wild expression. As we proceeded along the Beagle Channel, the scenery assumed a peculiar and very magnificent character; but the effect was much lessened from the lowness of the point of view in a boat, and from looking along the valley, and thus losing all the beauty of a succession of ridges. The mountains were here about three thousand feet high, and terminated in sharp and jagged points. They rose in one unbroken sweep from the water's edge, and were covered to the height of fourteen or fifteen hundred feet by the dusky-coloured forest. It was most curious to observe, as far as the eye could range, how level and truly horizontal the line on the mountain side was, at which trees ceased to grow; it precisely resembled the high-water mark of drift-weed on a sea-beach.

At night we slept close to the junction of Ponsonby Sound with the Beagle Channel. A small family of Fuegians, who were living in the cove, were quiet and inoffensive, and soon joined our party round a blazing fire. We were well clothed, and though sitting close to the fire were far from too warm; yet these naked savages, though further off, were observed, to our great surprise, to be streaming with perspiration at undergoing such a roasting. They seemed, however, very well pleased, and all joined in the chorus of the seamen's songs; but the manner in which they were invariably a little behindhand was quite ludicrous.

During the night the news had spread, and early in the morning (23d) a fresh party arrived, belonging to the Tekenika, or Jemmy's tribe. Several of them had run so fast that their noses were bleeding, and their mouths frothed from the rapidity with which they talked; and with their naked bodies all bedaubed with black, white,[1] and red, they looked like so many demoniacs

[1] This substance, when dry, is tolerably compact, and of little specific gravity: Prof. Ehrenberg has examined it; he states (*König Akad, der Wissen,* Berlin, Feb. 1845) that it is composed of infusoria, including fourteen polygastrica, and four phytolitharia. He says that they are all inhabitants of fresh-water; this is a beautiful example of the results obtainable through Professor Ehrenberg's microscopic researches; for Jemmy Button told me that it is always collected at the bottoms of

who had been fighting. We then proceeded (accompanied by twelve canoes, each holding four or five people) down Ponsonby Sound to the spot where poor Jemmy expected to find his mother and relatives. He had already heard that his father was dead; but as he had had a "dream in his head" to that effect, he did not seem to care much about it, and repeatedly comforted himself with the very natural reflection, "Me no help it." He was not able to learn any particulars regarding his father's death, as his relations would not speak about it.

Jemmy was now in a district well known to him, and guided the boats to a quiet pretty cove named Woollya, surrounded by islets, everyone of which and every point had its proper native name. We found here a family of Jemmy's tribe, but not his relations. We made friends with them; and in the evening they sent a canoe to inform Jemmy's mother and brothers. The cove was bordered by some acres of good sloping land, not covered (as elsewhere) either by peat or by forest-trees. Captain Fitz-Roy originally intended, as before stated, to have taken York Minster and Fuegia to their own tribe on the west coast; but as they expressed a wish to remain here, and as the spot was singularly favourable, Captain Fitz-Roy determined to settle here the whole party, including Matthews, the missionary. Five days were spent in building for them three large wigwams, in landing their goods, in digging two gardens, and sowing seeds.

The next morning after our arrival (the 24th) the Fuegians began to pour in, and Jemmy's mother and brothers arrived. Jemmy recognised the stentorian voice of one of his brothers at a prodigious distance. The meeting was less interesting than that between a horse, turned out into a field, when he joins an old companion. There was no demonstration of affection; they simply stared for a short time at each other, and the mother immediately went to look after her canoe. We heard, however, through York that the mother had been inconsolable for the loss of Jemmy, and had searched everywhere for him, thinking that he might have been left after having been taken in the boat. The women took much notice of and were very kind to Fuegia.

---

mountain-brooks. It is, moreover, a striking fact in the geographical distribution of the infusoria, which are well known to have very wide ranges, that all the species in this substance, although brought from the extreme southern point of Tierra del Fuego, are old, known forms.

We had already perceived that Jemmy had almost forgotten his own language. I should think there was scarcely another human being with so small a stock of language, for his English was very imperfect. It was laughable, but almost pitiable, to hear him speak to his wild brother in English, and then ask him in Spanish (*"No sabe?"*) whether he did not understand him.

Everything went on peaceably during the three next days, whilst the gardens were digging and wigwams building. We estimated the number of natives at about one hundred and twenty. The women worked hard, whilst the men lounged about all day long, watching us. They asked for everything they saw, and stole what they could. They were delighted at our dancing and singing, and were particularly interested at seeing us wash in a neighbouring brook; they did not pay much attention to anything else, not even to our boats. Of all the things which York saw, during his absence from his country, nothing seems more to have astonished him than an ostrich, near Maldonado; breathless with astonishment he came running to Mr. Bynoe, with whom he was out walking: "Oh, Mr. Bynoe, oh, bird all same horse!" Much as our white skins surprised the natives, by Mr. Low's account a Negro cook to a sealing vessel did so more effectually; and the poor fellow was so mobbed and shouted at that he would never go on shore again. Everything went on so quietly that some of the officers and myself took long walks in the surrounding hills and woods. Suddenly, however, on the 27th, every woman and child disappeared. We were all uneasy at this, as neither York nor Jemmy could make out the cause. It was thought by some that they had been frightened by our cleaning and firing off our muskets on the previous evening, by others that it was owing to offence taken by an old savage, who, when told to keep further off, had coolly spit in the sentry's face, and had then, by gestures acted over a sleeping Fuegian, plainly showed, as it was said, that he should like to cut up and eat our man. Captain Fitz-Roy, to avoid the chance of an encounter, which would have been fatal to so many of the Fuegians, thought it advisable for us to sleep at a cove a few miles distant. Matthews, with his usual quiet fortitude (remarkable in a man apparently possessing little energy of character), determined to stay with the Fuegians, who evinced no alarm for themselves; and so we left them to pass their first awful night.

On our return in the morning (the 28th) we were delighted to find all quiet, and the men employed in their canoes spearing fish. Captain Fitz-Roy determined to send the yawl and one whale-boat back to the ship, and to proceed with the two other boats, one under his own command (in which he most kindly allowed me to accompany him), and one under Mr. Hammond, to survey the western parts of the Beagle Channel, and afterwards to return and visit the settlement. The day to our astonishment was overpoweringly hot, so that our skins were scorched; with this beautiful weather, the view in the middle of the Beagle Channel was very remarkable. Looking towards either hand, no object intercepted the vanishing points of this long canal between the mountains. The circumstance of its being an arm of the sea was rendered very evident by several huge whales [2] spouting in different directions. On one occasion I saw two of these monsters, probably male and female, slowly swimming one after the other, within less than a stone's throw of the shore, over which the beech-tree extended its branches.

We sailed on till it was dark, and then pitched our tents in a quiet creek. The greatest luxury was to find for our beds a beach of pebbles, for they were dry and yielded to the body. Peaty soil is damp; rock is uneven and hard; sand gets into one's meat, when cooked and eaten boat-fashion; but when lying in our blanket-bags, on a good bed of smooth pebbles, we passed most comfortable nights.

It was my watch till one o'clock. There is something very solemn in these scenes. At no time does the consciousness in what a remote corner of the world you are then standing come so strongly before the mind. Everything tends to this effect; the stillness of the night is interrupted only by the heavy breathing of the seamen beneath the tents, and sometimes by the cry of a night-bird. The occasional barking of a dog, heard in the distance, reminds one that it is the land of the savage.

January 29. Early in the morning we arrived at the point where the Beagle Channel divides into two arms; and we en-

---

[2] One day, off the east coast of Tierra del Fuego, we saw a grand sight in several spermaceti whales jumping upright quite out of the water, with the exception of their tail-fins. As they fell down sideways, they splashed the water high up, and the sound reverberated like a distant broadside.

tered the northern one. The scenery here becomes even grander than before. The lofty mountains on the north side compose the granitic axis, or backbone of the country, and boldly rise to a height of between three and four thousand feet, with one peak above six thousand feet. They are covered by a wide mantle of perpetual snow, and numerous cascades pour their waters, through the woods, into the narrow channel below. In many parts, magnificent glaciers extend from the mountain side to the water's edge. It is scarcely possible to imagine anything more beautiful than the beryl-like blue of these glaciers, and especially as contrasted with the dead white of the upper expanse of snow. The fragments which had fallen from the glacier into the water were floating away, and the channel with its icebergs presented, for the space of a mile, a miniature likeness of the Polar Sea. The boats being hauled on shore at our dinner-hour, we were admiring from the distance of half a mile a perpendicular cliff of ice, and were wishing that some more fragments would fall. At last, down came a mass with a roaring noise, and immediately we saw the smooth outline of a wave travelling towards us. The men ran down as quickly as they could to the boats; for the chance of their being dashed to pieces was evident. One of the seamen just caught hold of the bows, as the curling breaker reached it. He was knocked over and over, but not hurt; and the boats, though thrice lifted on high and let fall again, received no damage. This was most fortunate for us, for we were a hundred miles distant from the ship, and we should have been left without provisions or fire-arms. I had previously observed that some large fragments of rock on the beach had been lately displaced; but until seeing this wave, I did not understand the cause. One side of the creek was formed by a spur of mica-slate; the head by a cliff of ice about forty feet high; and the other side by a promontory fifty feet high, built up of huge rounded fragments of granite and mica-slate, out of which old trees were growing. This promontory was evidently a moraine, heaped up at a period when the glacier had greater dimensions.

When we reached the western mouth of this northern branch of the Beagle Channel, we sailed amongst many unknown desolate islands, and the weather was wretchedly bad. We met with no natives. The coast was almost everywhere so steep that we had several times to pull many times before we could find space enough to pitch our two tents; one night we slept on large round boulders, with putrefying sea-weed between them; and when the tide rose, we had to get up and move our blanket-bags. The farthest point westward which we reached was Stewart Island, a distance of about one hundred and fifty miles from our ship. We returned into the Beagle Channel by the southern arm, and thence proceeded, with no adventure, back to Ponsonby Sound.

February 6. We arrived at Woollya. Matthews gave so bad an account of the conduct of the Fuegians that Captain Fitz-Roy determined to take him back to the *Beagle;* and ultimately he was left at New Zealand, where his brother was a missionary. From the time of our leaving, a regular system of plunder commenced. Fresh parties of the natives kept arriving; York and Jemmy lost many things, and Matthews almost every thing which had not been concealed underground. Every article seemed to have been torn up and divided by the natives. Matthews described the watch he was obliged always to keep as most harassing; night and day he was surrounded by the natives, who tried to tire him out by making an incessant noise close to his head. One day an old man, whom Matthews asked to leave his wigwam, immediately returned with a large stone in his hand; another day a whole party came armed with stones and stakes, and some of the younger men and Jemmy's brother were crying. Matthews met them with presents. Another party showed by signs that they wished to strip him naked and pluck all the hairs out of his face and body. I think we arrived just in time to save his life. Jemmy's relatives had been so vain and foolish that they had showed to strangers their plunder, and their manner of obtaining it. It was quite melancholy leaving the three Fuegians with their savage countrymen; but it was a great comfort that they had no personal fears. York, being a powerful resolute man, was pretty sure to get on well, together with his wife Fuegia. Poor Jemmy looked rather disconsolate, and would then, I have little doubt, have been glad to have returned with us. His own brother had stolen many things from him; and as he remarked, "What fashion call that?"; he abused his countrymen, "All bad men, *no sabe* (know nothing)," and, though I never heard him swear before, "Damned fools." Our three Fuegians, though they had been only three years with civilized men, would, I am sure, have been glad to have retained their new habits; but this was obviously impossible. I fear it is more than doubtful whether their visit will have been of any use to them.

In the evening, with Matthews on board, we made sail back to the ship, not by the Beagle Channel, but by the southern coast. The boats were heavily laden and the sea rough, and we had a dangerous passage. By the evening of the 7th we were on board the *Beagle* after an absence of twenty days, during which time we had gone three hundred miles in the open boats. On the 11th, Captain Fitz-Roy paid a visit by himself to the Fuegians and found them going on well; and that they had lost very few more things.

On the last day of February in the succeeding year (1834), the *Beagle* anchored in a beautiful little cove at the eastern entrance of the Beagle Channel. Captain Fitz-Roy determined on the bold, and as it proved successful, attempt to beat against the westerly winds by the same route which we had followed in the boats to the settlement at Woollya. We did not see many natives until we were near Ponsonby Sound, where we were followed by ten or twelve canoes. The natives did not at all understand the reason of our tacking and, instead of meeting us at each tack, vainly strove to follow us in our zig-zag course. I was amused at finding what a difference the circumstance of being quite superior in force made, in the interest of beholding these savages. While in the boats I got to hate the very sound of their voices, so much trouble did they give us. The first and last word was "yammerschooner." When, entering some quiet little cove, we have looked round and thought to pass a quiet night, the odious word "yammerschooner" has shrilly sounded from some gloomy nook, and then the little signal-smoke has curled up to spread the news far and wide. On leaving some place we have said to each other, "Thank Heaven, we have at last fairly left these wretches!," when one more faint halloo from an all-powerful voice, heard at a prodigious distance, would reach our ears, and clearly could we distinguish "yammerschooner." But now, the more Fuegians the merrier; and very merry work it was. Both parties laughing, wondering, gaping at each other; we pitying them, for giving us good fish and crabs for rags, &c.; they grasping at the chance of finding people so foolish as to exchange such splendid ornaments for a good supper. It was most amusing to see the undisguised smile of satisfaction with which one young woman, with her face painted black, tied several bits of scarlet cloth round her head with rushes. Her husband, who enjoyed the very universal privilege in this country of possessing two wives, evidently became jealous of all the attention paid to his young wife; and, after a consultation with his naked beauties, was paddled away by them.

Some of the Fuegians plainly showed that they had a fair notion of barter. I gave one man a large nail (a most valuable present) without making any signs for a return; but he immediately picked out two fish, and handed them up on the point of his spear. If any present was designed for one canoe, and it fell near another, it was invariably given to the right owner. The Fuegian boy, whom Mr. Low had on board, showed, by going into the most violent passion, that he quite understood the reproach of being called a liar, which in truth he was. We were this time, as on all former occasions, much surprised at the little notice, or rather none whatever, which was taken of many things, the use of which must have been evident to the natives. Simple circumstances—such as the beauty of scarlet cloth or blue beads, the absence of women, our care in washing ourselves—excited their admiration far more than any grand or complicated object, such as our ship. Bougainville has well remarked concerning these people that they treat the "chef-d'oeuvres de l'industrie humaine, comme ils traitent les loix de la nature et ses phénomènes."

On the 5th of March, we anchored in the cove at Woollya, but we saw not a soul there. We were alarmed at this, for the natives in Ponsonby Sound showed by gestures that there had been fighting; and we afterwards heard that the dreaded Oens men had made a descent. Soon a canoe, with a little flag flying, was seen approaching, with one of the men in it washing the paint off his face. This man was poor Jemmy—now a thin haggard savage, with long disordered hair, and naked, except a bit of a blanket round his waist. We did not recognise him till he was close to us; for he was ashamed of himself, and turned his back to the ship. We had left him plump, fat, clean, and well dressed; I never saw so complete and grievous a change. As soon however as he was clothed, and the first flurry was over, things wore a good appearance. He dined with Captain Fitz-Roy, and ate his dinner as tidily as formerly. He told us he had "too much" (meaning enough) to eat, that he was not cold, that his relations were very good people, and that he did not wish to go back to England; in the evening we found out the cause of this great change in Jemmy's feelings, in the arrival of his young and nice-looking wife. With his usual good feeling, he brought two beautiful otter-skins for two of his best friends, and some spear-heads and arrows made with his own hands for the Captain. He said he had built a canoe for himself, and he boasted that he could talk a little of his own language! But it is a most singular fact that he appears to have taught all his tribe some English: An old man spontaneously announced, "Jemmy Button's wife." Jemmy had lost all his property. He told us that York Minster had built a large canoe, and with his wife Fuegia [3] had several months since gone to his own country, and had taken farewell by an act of consummate villainy; he persuaded Jemmy and his mother to come with him, and then on the way deserted them by night, stealing every article of their property.

Jemmy went to sleep on shore, and in the morning returned, and remained on board till the ship got under weigh, which frightened his wife, who continued crying violently till he got into his canoe. He returned loaded with valuable property. Every soul on board was heartily sorry to shake hands with him for the last time. I do not now doubt that he will be as happy as, perhaps happier than, if he had never left his own country. Every one must sincerely hope that Captain Fitz-Roy's noble hope may be fulfilled, of being rewarded for the many generous sacrifices which he made for these Fuegians, by some ship-wrecked sailor being protected by the descendants of Jemmy Button and his tribe! When Jemmy reached the shore, he lighted a signal fire, and the smoke curled up, bidding us a last and long farewell, as the ship stood on her course into the open sea.

[3] Captain Sulivan, who, since his voyage in the *Beagle,* has been employed on the survey of the Falkland Islands, heard from a sealer in 1842 (?), that when in the western part of the Strait of Magellan he was astonished by a native woman coming on board who could talk some English. Without doubt this was Fuegia Basket. She lived (I fear the term probably bears a double interpretation) some days on board.

The perfect equality among the individuals composing the Fuegian tribes must for a long time retard their civilisation. As we see those animals whose instinct compels them to live in society and obey a chief are most capable of improvement, so is it with the races of mankind. Whether we look at it as a cause or a consequence, the more civilised always have the most artificial governments. For instance, the inhabitants of Otaheite, who, when first discovered, were governed by hereditary kings, had arrived at a far higher grade than another branch of the same people, the New Zealanders—who, although benefited by being compelled to turn their attention to agriculture, were republicans in the most absolute sense. In Tierra del Fuego, until some chief shall arise with power sufficient to secure any acquired advantage, such as the domesticated animals, it seems scarcely possible that the political state of the country can be improved. At present, even a piece of cloth given to one is torn into shreds and distributed; and no one individual becomes richer than another. On the other hand, it is difficult to understand how a chief can arise till there is property of some sort by which he might manifest his superiority and increase his power.

I believe, in this extreme part of South America, man exists in a lower state of improvement than in any other part of the world. The South Sea Islanders of the two races inhabiting the Pacific are comparatively civilised. The Esquimaux, in his subterranean hut, enjoys some of the comforts of life, and in his canoe, when fully equipped, manifests much skill. Some of the tribes of Southern Africa, prowling about in search of roots, and living concealed on the wild and arid plains, are sufficiently wretched. The Australian, in the simplicity of the arts of life, comes nearest the Fuegian; he can, however, boast of his boomerang, his spear, and throwing-stick, his method of climbing trees, of tracking animals, and of hunting. Although the Australian may be superior in acquirements, it by no means follows that he is likewise superior in mental capacity; indeed, from what I saw of the Fuegians when on board, and from what I have read of the Australians, I should think the case was exactly the reverse.

CHAPTER XVII

Galapagos Archipelago

September 15. This archipelago consists of ten principal islands, of which five exceed the others in size. They are situated under the equator, and between five and six hundred miles westward of the coast of America. They are all formed of volcanic rocks; a few fragments of granite, curiously glazed and altered by the heat, can hardly be considered as an exception. Some of the craters surmounting the larger islands are of immense size, and they rise to a height of between three and four thousand feet. Their flanks are studded by innumerable smaller orifices. I scarcely hesitate to affirm that there must be in the whole archipelago at least two thousand craters. These consist either of lava and scoriae, or of finely stratified, sandstone-like tuff. Most of the latter are beautifully symmetrical; they owe their origin to eruptions of volcanic mud without any lava. It is a remarkable circumstance that every one of the twenty-eight tuff-craters which were examined had their southern sides either much lower than the other sides, or quite broken down and removed. As all these craters apparently have been formed when standing in the sea, and as the waves from the trade wind and the swell from the open Pacific here unite their forces on the southern coasts of all the islands, this singular uniformity in the broken state of the craters, composed of the soft and yielding tuff, is easily explained.

Considering that these islands are placed directly under the equator, the climate is far from being excessively hot; this seems chiefly caused by the singularly low temperature of the surrounding water, brought here by the great southern polar current. Excepting during one short season, very little rain falls, and even then it is irregular; but the clouds generally hang low. Hence, whilst the lower parts of the islands are very sterile, the upper parts, at a height of a thousand feet and upwards, possess a damp climate and a tolerably luxuriant vegetation. This is especially the case on the windward sides of the islands, which first receive and condense the moisture from the atmosphere.

In the morning (the 17th) we landed on Chatham Island,

which, like the others, rises with a tame and rounded outline, broken here and there by scattered hillocks, the remains of former craters. Nothing could be less inviting than the first appearance. A broken field of black basaltic lava, thrown into the most rugged waves, and crossed by great fissures, is everywhere covered by stunted, sun-burnt brushwood, which shows little signs of life. The dry and parched surface, being heated by the noon-day sun, gave to the air a close and sultry feeling, like that from a stove; we fancied even that the bushes smelt unpleasantly. Although I diligently tried to collect as many plants as possible, I succeeded in getting very few; and such wretched-looking little weeds would have better become an arctic than an equatorial Flora. The brushwood appears, from a short distance, as leafless as our trees during winter; and it was some time before I discovered that not only almost every plant was now in full leaf but that the greater number were in flower. The commonest bush is one of the Euphorbiaceae; in acacia and a great odd-looking cactus are the only trees which afford any shade. After the season of heavy rains, the islands are said to appear for a short time partially green. The volcanic island of Fernando Noronha, placed in many respects under nearly similar conditions, is the only other country where I have seen a vegetation at all like this of the Galapagos Islands.

The *Beagle* sailed round Chatham Island, and anchored in several bays. One night I slept on shore on a part of the island, where black truncated cones were extraordinarily numerous; from one small eminence I counted sixty of them, all surmounted by craters more or less perfect. The greater number consisted merely of a ring of red scoriae or slags, cemented together; and their height above the plain of lava was not more than from fifty to a hundred feet—none had been very lately active. The entire surface of this part of the island seems to have been permeated, like a sieve, by the subterranean vapours: Here and there the lava, whilst soft, has been blown into great bubbles; and in other parts, the tops of caverns similarly formed have fallen in, leaving circular pits with steep sides. From the regular form of the many craters, they gave to the country an artificial appearance, which vividly reminded me of those parts of Staffordshire, where the great iron-foundries are most numerous. The day was glowing hot, and the scrambling over the rough surface and through the intricate thickets was very fatiguing; but I was well repaid by the strange Cyclopean scene. As I was walking along I met two large tortoises, each of which must have weighed at least two hundred pounds. One was eating a piece of cactus, and as I approached, it stared at me and slowly stalked away; the other gave a deep hiss, and drew in its head. These huge reptiles, surrounded by the black lava, the leafless shrubs, and large cacti, seemed to my fancy like some antediluvian animals. The few dull-coloured birds cared no more for me than they did for the great tortoises.

September 23. The *Beagle* proceeded to Charles Island. This archipelago has long been frequented, first by the buccaneers, and latterly by whalers, but it is only within the last six years that a small colony has been established here. The inhabitants are between two and three hundred in number; they are nearly all people of colour, who have been banished for political crimes from the Republic of the Equator, of which Quito is the capital. The settlement is placed about four and a half miles inland, and at a height probably of a thousand feet. In the first part of the road we passed through leafless thickets, as in Chatham Island. Higher up, the woods gradually became greener; and as soon as we crossed the ridge of the island, we were cooled by a fine southerly breeze, and our sight refreshed by a green and thriving vegetation. In this upper region coarse grasses and ferns abound; but there are no tree-ferns. I saw nowhere any member of the palm family, which is the more singular as, three hundred and sixty miles northward, Cocos Island takes its name from the number of cocoa-nuts. The houses are irregularly scattered over a flat space of ground, which is cultivated with sweet potatoes and bananas. It will not easily be imagined how pleasant the sight of black mud was to us, after having been so long accustomed to the parched soil of Peru and northern Chile. The inhabitants, although complaining of poverty, obtain, without much trouble, the means of subsistence. In the woods there are many wild pigs and goats; but the staple article of animal food is supplied by the tortoises. Their numbers have of course been greatly reduced in this island, but the people yet count on two days' hunting giving them food for the rest of the week. It is said that formerly single vessels have taken away as many as seven hundred, and that the ship's company of a frigate some years since brought down in one day two hundred tortoises to the beach.

September 29. We doubled the south-west extremity of Albemarle Island, and the next day were nearly becalmed between it and Narborough Island. Both are covered with immense deluges of black naked lava, which have flowed either over the rims of the great caldrons, like pitch over the rim of a pot in which it has been boiled, or have burst forth from smaller orifices on the flanks; in their descent they have spread over miles of the sea-coast. On both of these islands, eruptions are known to have taken place; and in Albemarle, we saw a small jet of smoke curling from the summit of one of the great craters. In the evening we anchored in Bank's Cove, in Albemarle Island. The next morning I went out walking. To the south of the broken tuff-crater, in which the *Beagle* was anchored, there was another beautifully symmetrical one of an elliptic form; its longer axis was a little less than a mile, and its depth about five hundred feet. At its bottom there was a shallow lake, in the middle of which a tiny crater formed an islet. The day was overpoweringly hot, and the lake looked clear and blue. I hurried down the cindery slope, and choked with dust eagerly tasted the water—but, to my sorrow, I found it salt as brine.

The rocks on the coast abounded with great black lizards, between three and four feet long; and on the hills an ugly yellowish-brown species was equally common. We saw many of this latter kind, some clumsily running out of our way, and others shuffling into their burrows. I shall presently describe in more detail the habits of both these reptiles. The whole of this northern part of Albemarle Island is miserably sterile.

October 8. We arrived at James Island; this island, as well as Charles Island, were long since thus named after our kings of the Stuart line. Mr. Bynoe, myself, and our servants were left here for a week, with provisions and a tent, whilst the *Beagle* went for water. We found here a party of Spaniards, who had been sent from Charles Island to dry fish, and to salt tortoise-meat. About six miles inland, and at the height of nearly two thousand feet, a hovel had been built in which two men lived, who were employed in catching tortoises, whilst the others were fishing on the coast. I paid this party two visits, and slept there one night. As in the other islands, the lower region was covered by nearly leafless bushes, but the trees were here of a larger growth than elsewhere, several being two feet and some even two feet nine inches in diameter. The upper region, being kept damp by the clouds, supports a green and flourishing vegetation. So damp was the ground that there were large beds of a coarse cyperus, in which great numbers of a very small water-rail lived and bred. While staying in this upper region, we lived entirely upon tortoise-meat; the breast-plate roasted (as the Gauchos do, *carne con cuero*), with the flesh on it, is very good; and the young tortoises make excellent soup; but otherwise the meat to my taste is indifferent.

One day we accompanied a party of the Spaniards in their whale-boat to a salina, or lake from which salt is procured. After landing, we had a very rough walk over a rugged field of recent lava, which has almost surrounded a tuff-crater, at the bottom of which the salt-lake lies. The water is only three or four inches deep, and rests on a layer of beautifully crystallised white salt. The lake is quite circular, and is fringed with a border of bright green succulent plants; the almost precipitous walls of the crater are clothed with wood, so that the scene was altogether both picturesque and curious. A few years since, the sailors belonging to a sealing-vessel murdered their captain in this quiet spot; and we saw his skull lying among the bushes.

During the greater part of our stay of a week, the sky was cloudless, and if the trade-wind failed for an hour, the heat became very oppressive. On two days, the thermometer within the tent stood for some hours at 93°; but in the open air, in the wind and sun, at only 85°. The sand was extremely hot; the thermometer placed in some of a brown colour immediately rose to 137°, and how much above that it would have risen I do not know, for it was not graduated any higher. The black sand felt much hotter, so that even in thick boots it was quite disagreeable to walk over it.

The natural history of these islands is eminently curious, and well deserves attention. Most of the organic productions are aboriginal creations, found nowhere else; there is even a difference between the inhabitants of the different islands, yet all show a marked relationship with those of America, though separated from that continent by an open space of ocean, between five hundred and six hundred miles in width. The archipelago is a little world within itself, or rather a satelite attached to America, whence it has derived a few stray colonists, and has received the general character of its indigenous productions. Considering the small size of these islands, we feel the more astonished at the number of their aboriginal beings, and at their confined range. Seeing every height crowned with its crater, and the boundaries of most of the lava-streams still distinct, we are led to believe that within a period, geologically recent, the unbroken ocean was here spread out. Hence, both in space and time, we seem to be brought somewhat near to that great fact, that mystery of mysteries—the first appearance of new beings on this earth.

Of terrestrial mammals, there is only one which must be considered as indigenous, namely, a mouse (Mus Galapagoensis), and this is confined, as far as I could ascertain, to Chatham Island, the most easterly island of the group. It belongs, as I am informed by Mr. Waterhouse, to a division of the family of mice characteristic of America. At James Island, there is a rat sufficiently distinct from the common kind to have been named and described by Mr. Waterhouse; but as it belongs to the old-world division of the family, and as this island has been frequented by ships for the last hundred and fifty years, I can hardly doubt that this rat is merely a variety, produced by the new and peculiar climate, food, and soil to which it has been subjected. Although no one has a right to speculate without distinct facts, yet even with respect to the Chatham Island mouse, it should be borne in mind that it may possibly be an American species imported here; for I have seen, in a most unfrequented part of the Pampas, a native mouse living in the roof of a newly built hovel, and therefore its transportation in a vessel is not improbable. Analogous facts have been observed by Dr. Richardson in North America.

Of land-birds I obtained twenty-six kinds, all peculiar to the group and found nowhere else, with the except of one lark-like finch from North America (Dolichonyx oryzivorus), which ranges on that continent as far north as 54°, and generally frequents marshes. The other twenty-five birds consist, firstly, of a hawk, curiously intermediate in structure between a buzzard and the American group of carrion-feeding Polybori; and with these latter birds it agrees most closely in every habit and even tone of voice. Secondly, there are two owls, representing the short-eared and white barn-owls of Europe. Thirdly, a wren, three tyrant fly-catchers (two of them species of Pyrocephalus, one or both of which would be ranked by some ornithologists as only varieties), and a dove—all analogous to, but distinct from, American species. Fourthly, a swallow, which though differing from the Progne purpurea of both Americas only in being rather duller coloured, smaller, and slenderer, is considered by Mr. Gould as specifically distinct. Fifthly, there are three species of mocking-thrush—a form highly characteristic of America. The remaining land-birds form a most singular group of finches, related to each other in the structure of their beaks, short tails, form of body, and plumage: There are thirteen species, which Mr. Gould has divided into four sub-groups. All these species are peculiar to this archipelago; and so is the whole group, with the exception of one species of the sub-group Cactornis, lately brought from Bow Island, in the Low Archipelago. Of Cactornis, the two species may be often seen climbing about the flowers of the great cactus-trees; but all the other species of this group of finches, mingled together in flocks, feed on the dry and sterile ground of the lower districts. The males of all, or certainly of the greater number, are jet black; and the females (with perhaps one or two exceptions) are brown. The most curious fact is the perfect gradation in the size of the beaks in the different species of Geospiza, from one as large as that of a hawfinch to that of a chaffinch, and (if Mr. Gould is right in including his sub-group, Certhidea, in the main group), even to that of a warbler. The largest beak in the genus Geospiza is shown in fig. 1,* and the smallest in fig. 3; but instead of there being only one inter-mediate species, with a beak of the size shown in fig. 2, there are no less than six species with insensibly graduated beaks. The beak of the sub-group Certhidea is shown in fig. 4. The beak of Cactornis is somewhat like that of a starling, and that of the fourth sub-group, Camarhynchus, is slightly parrot-shaped. See-ing this gradation and diversity of structure in one small, in-timately related group of birds, one might really fancy that from

* In the original edition.—S.E.H.

an original paucity of birds in this archipelago one species had been taken and modified for different ends. In a like manner it might be fancied that a bird originally a buzzard had been induced here to undertake the office of the carrion-feeding Polybori of the American continent.

Of waders and water-birds I was able to get only eleven kinds, and of these only three (including a rail confined to the damp summits of the islands) are new species. Considering the wandering habits of the gulls, I was surprised to find that the species inhabiting these islands is peculiar, but allied to one from the southern parts of South America. The far greater peculiarity of the land-birds, namely, twenty-five out of twenty-six being new species or at least new races, compared with the waders and web-footed birds, is in accordance with the greater range which these latter orders have in all parts of the world. We shall hereafter see this law of aquatic forms, whether marine or freshwater, being less peculiar at any given point of the earth's surface than the terrestrial forms of the same classes strikingly illustrated in the shells, and in a lesser degree in the insects, of this archipelago.

Two of the waders are rather smaller than the same species brought from other places; the swallow is also smaller, though it is doubtful whether or not it is distinct from its analogue. The two owls, the two tyrant fly-catchers (Pyrocephalus), and the dove are also smaller than the analogous but distinct species to which they are most nearly related; on the other hand, the gull is rather larger. The two owls, the swallow, all three species of mocking-thrush, the dove in its separate colours though not in its whole plumage, the Totanus, and the gull are likewise duskier coloured than their analogous species; and in the case of the mocking-thrush and Totanus, than any other species of the two genera. With the exception of a wren with a fine yellow breast, and of a tyrant fly-catcher with a scarlet tuft and breast, none of the birds are brilliantly coloured, as might have been expected in an equatorial district. Hence it would appear probable that the same causes which here make the immigrants of some species smaller make most of the peculiar Galapageian species also smaller, as well as very generally more dusky coloured. All the plants have a wretched, weedy appearance, and I did not see one beautiful flower. The insects, again, are small sized and dull coloured, and, as Mr. Waterhouse informs me, there is nothing in their general appearance which would have led him to imagine that they had come from under the equator. The birds, plants, and insects have a desert character, and are not more brilliantly coloured than those from southern Patagonia; we may, therefore, conclude that the usual gaudy colouring of the intertropical productions is not related either to the heat or light of those zones, but to some other cause, perhaps to the conditions of existence being generally favourable to life.

We will now turn to the order of reptiles, which gives the most striking character to the zoology of these islands. The species are not numerous, but the numbers of individuals of each species are extraordinarily great. There is one small lizard belonging to a South American genus, and two species (and probably more) of the Amblyrhynchus—a genus confined to the Galapagos Islands. There is one snake which is numerous; it is identical, as I am informed by M. Bibron, with the Psammophis Temminckii from Chile. Of sea-turtle I believe there is more than one species; and of tortoises there are, as we shall presently show, two or three species or races. Of toads and frogs there are none. I was surprised at this, considering how well suited for them the temperate and damp upper woods appeared to be. It recalled to my mind the remark made by Bory St. Vincent,[1] namely, that none of this family are found on any of the volcanic islands in the great oceans. As far as I can ascertain from various works, this seems to hold good throughout the Pacific, and even in the large islands of the Sandwich archipelago. Mauritius offers an apparent exception, where I saw the Rana Mascariensis in abundance; this frog is said now to inhabit the Seychelles, Madagascar, and Bourbon. But on the other hand, Du Bois, in his voyage in 1669, states that there were no reptiles in Bourbon except tortoises; and the Officier du Roi asserts that before 1768 it had been attempted, without success, to introduce frogs into Mauritius—I presume, for the purpose of eating. Hence it may be well doubted whether

[1] *Voyage aux Quatre Iles d'Afrique.* With respect to the Sandwich Islands, see Tyerman and Bennett's *Journal*, Vol. I, p. 434. For Mauritius, see *Voyage par un Officier, &c.*, Part I, p. 170. There are no frogs in the Canary Islands (Webb *et* Berthelot, *Hist. Nat. des Iles Canaries*). I saw none at St. Jago in the Cape Verdes. There are none at St. Helena.

this frog is an aboriginal of these islands. The absence of the frog family in the oceanic islands is the more remarkable when contrasted with the case of lizards, which swarm on most of the smallest islands. May this difference not be caused by the greater facility with which the eggs of lizards, protected by calcareous shells, might be transported through salt-water, than could the slimy spawn of frogs?

I will first describe the habits of the tortoise (Testudo nigra, formerly called Indica) which has been so frequently alluded to. These animals are found, I believe, on all the islands of the Archipelago; certainly on the greater number. They frequent in preference the high damp parts, but they likewise live in the lower and arid districts. I have already shown, from the numbers which have been caught in a single day, how very numerous they must be. Some grow to an immense size: Mr. Lawson, an Englishman and vice-governor of the colony, told us that he had seen several so large that it required six or eight men to lift them from the ground, and that some had afforded as much as two hundred pounds of meat. The old males are the largest, the females rarely growing to so great a size; the male can readily be distinguished from the female by the greater length of its tail. The tortoises which live on those islands where there is no water, or in the lower and arid parts of the others, feed chiefly on the succulent cactus. Those which frequent the higher and damp regions eat the leaves of various trees, a kind of berry (called guayavita) which is acid and austere, and likewise a pale green filamentous lichen (Usnera plicata) that hangs in tresses from the boughs of the trees.

The tortoise is very fond of water, drinking large quantities and wallowing in the mud. The larger islands alone possess springs, and these are always situated towards the central parts, and at a considerable height. The tortoises, therefore, which frequent the lower districts, when thirsty, are obliged to travel from a long distance. Hence broad and well-beaten paths branch off in every direction from the wells down to the sea-coast; and the Spaniards by following them up first discovered the watering-places. When I landed at Chatham Island, I could not imagine what animal travelled so methodically along well-chosen tracks. Near the springs it was a curious spectacle to behold many of these huge creatures, one set eagerly travelling onwards with outstretched necks, and another set returning, after having drunk their fill. When the tortoise arrives at the spring, quite regardless of any spectator, he buries his head in the water above his eyes, and greedily swallows great mouthfuls, at the rate of about ten in a minute. The inhabitants say each animal stays three or four days in the neighbourhood of the water, and then returns to the lower country; but they differed respecting the frequency of these visits. The animal probably regulates them according to the nature of the food on which it has lived. It is, however, certain that tortoises can subsist even on those islands where there is no other water than what falls during a few rainy days in the year.

I believe it is well ascertained that the bladder of the frog acts as a reservoir for the moisture necessary to its existence; such seems to be the case with the tortoise. For some time after a visit to the springs, their urinary bladders are distended with fluid, which is said gradually to decrease in volume, and to become less pure. The inhabitants, when walking in the lower district, and overcome with thirst, often take advantage of this circumstance, and drink the contents of the bladder if full; in one I saw killed, the fluid was quite limpid, and had only a very slightly bitter taste. The inhabitants, however, always first drink the water in the pericardium, which is described as being best.

The tortoises, when purposely moving towards any point, travel by night and day, and arrive at their journey's end much sooner than would be expected. The inhabitants, from observing marked individuals, consider that they travel a distance of about eight miles in two or three days. One large tortoise, which I watched, walked at the rate of sixty yards in ten minutes, that is, three hundred and sixty yards in the hour, or four miles a day—allowing a little time for it to eat on the road. During the breeding season, when the male and female are together, the male utters a hoarse roar or bellowing, which, it is said, can be heard at the distance of more than a hundred yards. The female never uses her voice, and the male only at these times; so that when the people hear this noise, they know that the two are together. They were at this time (October) laying their eggs. The female, where the soil is sandy, deposits them together, and covers them up with sand; but where the ground is rocky she drops them indiscriminately in any hole. Mr. Bynoe found seven placed in a fissure. The egg is white and spherical; one which I measured was seven inches and three-eights in circumference, and therefore larger than a hen's egg. The young tortoises, as soon as they are hatched, fall a prey in great numbers to the carrion-feeding buzzard. The old ones seem generally to die from accidents, as from falling down precipices; at least, several of the inhabitants told me that they had never found one dead without some evident cause.

The inhabitants believe that these animals are absolutely deaf; certainly they do not overhear a person walking close behind them. I was always amused when overtaking one of these great monsters, as it was quietly pacing along, to see how suddenly, the instant I passed, it would draw in its head and legs, and uttering a deep hiss fall to the ground with a heavy sound, as if struck dead. I frequently got on their backs, and then giving a few raps on the hinder part of their shells, they would rise up and walk away; but I found it very difficult to keep my balance. The flesh of this animal is largely employed, both fresh and salted; and a beautifully clear oil is prepared from the fat. When a tortoise is caught, the man makes a slit in the skin near its tail, so as to see inside its body whether the fat under the dorsal plate is thick. If it is not, the animal is liberated; and it is said to recover soon from this strange operation. In order to secure the tortoises, it is not sufficient to turn them like turtle, for they are often able to get on their legs again.

There can be little doubt that this tortoise is an aboriginal inhabitant of the Galapagos; for it is found on all, or nearly all, the islands, even on some of the smaller ones where there is no water. Had it been an imported species, this would hardly have been the case in a group which has been so little frequented. Moreover, the old buccaneers found this tortoise in greater numbers even than at present; Wood and Rogers also, in 1708, say that it is the opinion of the Spaniards that it is found nowhere else in this quarter of the world. It is now widely distributed; but it may be questioned whether it is in any other place an aboriginal. The bones of a tortoise at Mauritius, associated with those of the extinct dodo, have generally been considered as belonging to this tortoise; if this had been so, undoubtedly it must have been there indigenous. But

M. Bibron informs me that he believes that it was distinct, as the species now living there certainly is.

The Amblyrhynchus, a remarkable genus of lizards, is confined to this archipelago: There are two species, resembling each other in general form, one being terrestrial and the other aquatic. This latter species (A. cristatus) was first characterised by Mr. Bell, who well foresaw, from its short, broad head, and strong claws of equal length, that its habits of life would turn out very peculiar, and different from those of its nearest ally, the Iguana. It is extremely common on all the islands throughout the group, and lives exclusively on the rocky sea-beaches, being never found, at least I never saw one, even ten yards inshore. It is a hideous-looking creature, of a dirty black colour, stupid, and sluggish in its movements. The usual length of a full-grown one is about a yard, but there are some even four feet long; a large one weighed twenty pounds. On the island of Albemarle they seem to grow to a greater size than elsewhere. Their tails are flattened sideways, and all four feet partially webbed. They are occasionally seen some hundred yards from the shore, swimming about; and Captain Collnett, in his *Voyage*, says, "They go to sea in herds a-fishing, and sun themselves on the rocks; and may be called alligators in miniature." It must not, however, be supposed that they live on fish. When in the water this lizard swims with perfect ease and quickness, by a serpentine movement of its body and flattened tail—the legs being motionless and closely collapsed on its sides. A seaman on board sank one, with a heavy weight attached to it, thinking thus to kill it directly; but when, an hour afterwards, he drew up the line, it was quite active. Their limbs and strong claws are admirably adapted for crawling over the rugged and fissured masses of lava, which everywhere form the coast. In such situations, a group of six or seven of these hideous reptiles may oftentimes be seen on the black rocks, a few feet above the surf, basking in the sun with outstretched legs.

I opened the stomachs of several, and found them largely distended with minced sea-weed (Ulvae), which grows in thin foliaceous expansions of a bright green or a dull red colour. I do not recollect having observed this sea-weed in any quantity on the tidal rocks; and I have reason to believe it grows at the bottom of the sea, at some little distance from the coast. If such be the case, the object of these animals occasionally going out to sea is explained. The stomach contained nothing but the sea-weed. Mr. Bynoe, however, found a piece of a crab in one; but this might have got in accidentally, in the same manner as I have seen a caterpillar, in the midst of some lichen, in the paunch of a tortoise. The intestines were large, as in other herbivorous animals. The nature of this lizard's food, as well as the structure of its tail and feet, and the fact of its having been seen voluntarily swimming out at sea, absolutely prove its aquatic habits; yet there is in this respect one strange anomaly, namely, that when frightened it will not enter the water. Hence it is easy to drive these lizards down to any little point overhanging the sea, where they will sooner allow a person to catch hold of their tails than jump into the water. They do not seem to have any notion of biting; but when much frightened they squirt a drop of fluid from each nostril. I threw one several times as far as I could, into a deep pool left by the retiring tide; but it invariably returned in a direct line to the spot where I stood. It swam near the bottom, with a very graceful and rapid movement, and occasionally aided itself over the uneven ground with its feet. As soon as it arrived near the edge, but still being under water, it tried to conceal itself in the tufts of sea-weed, or it entered some crevice. As soon as it thought the danger was past, it crawled out on the dry rocks, and shuffled away as quickly as it could. I several times caught this same lizard, by driving it down to a point, and though possessed of such perfect powers of diving and swimming, nothing would induce it to enter the water; and as often as I threw it in, it returned in the manner above described. Perhaps this singular piece of apparent stupidity may be accounted for by the circumstance that this reptile has no enemy whatever on shore, whereas at sea it must often fall a prey to the numerous sharks. Hence, probably, urged by a fixed and hereditary instinct that the shore is its place of safety, whatever the emergency may be, it there takes refuge.

During our visit (in October), I saw extremely few small individuals of this species, and none I should think under a year old. From this circumstance it seems probable that the breeding season had not then commenced. I asked several of the inhabitants if they knew where it laid its eggs; they said that they knew nothing of its propagation, although well acquainted with the eggs of the land kind—a fact, considering how very common this lizard is, not a little extraordinary.

We will now turn to the terrestrial species (A. Demarlii), with a round tail, and toes without webs. This lizard, instead of being found like the other on all the islands, is confined to the central part of the archipelago, namely to Albermarle, James, Barrington, and Indefatigable Islands. To the southward, in Charles, Hood, and Chatham Islands, and to the northward, in Towers, Bindloes, and Abingdon, I neither saw nor heard of any. It would appear as if it had been created in the centre of the archipelago, and thence had been dispersed only to a certain distance. Some of these lizards inhabit the high and damp parts of the islands, but they are much more numerous in the lower and sterile districts near the coast. I cannot give a more forcible proof of their numbers than by stating that when we were left at James Island we could not for some time find a spot free from their burrows on which to pitch our single tent. Like their brothers the sea-kind, they are ugly animals, of a yellowish orange beneath, and of a brownish red colour above; from their low facial angle they have a singularly stupid appearance. They are, perhaps, of a rather less size than the marine species; but several of them weighed between ten and fifteen pounds. In their movements they are lazy and half torpid. When not frightened, they slowly crawl along with their tails and bellies dragging on the ground. They often stop, and doze for a minute or two, with closed eyes and hind legs spread out on the parched soil.

They inhabit burrows, which they sometimes make between fragments of lava, but more generally on level patches of the soft sandstone-like tuff. The holes do not appear to be very deep, and they enter the ground at a small angle; so that when walking over these lizard-warrens, the soil is constantly giving way, much to the annoyance of the tired walker. This animal, when making its burrow, works alternately the opposite sides of its body. One front leg for a short time scratches up the soil, and throws it towards the hind foot, which is well placed so as to heave it beyond the mouth of the hole. That side of the body being tired, the other takes up the task, and so on alternately. I watched one for a long time, till half its body was buried. I then walked up and pulled it by the tail; at this it was greatly astonished, and soon shuffled up to see what was the matter, and then stared me in the face, as much as to say, "What made you pull my tail?"

They feed by day, and do not wander far from their burrows; if frightened, they rush to them with a most awkward gait. Except when running down hill, they cannot move very fast, apparently from the lateral position of their legs. They are not at all timorous: When attentively watching anyone, they curl their tails, and, raising themselves on their front legs, nod their heads vertically, with a quick movement, and try to look very fierce. But in reality they are not at all so; if one just stamps on the ground, down go their tails, and off they shuffle as quickly as they can. I have frequently observed small fly-eating lizards, when watching anything, nod their heads in precisely the same manner; but I do not at all know for what purpose. If this Amblyrhynchus is held and plagued with a stick, it will bite it very severely; but I caught many by the tail, and they never tried to bite me. If two are placed on the ground and held together, they will fight, and bite each other till blood is drawn.

The individuals, and they are the greater number, which inhabit the lower country, can scarcely taste a drop of water throughout the year; but they consume much of the succulent cactus, the branches of which are occasionally broken off by the wind. I several times threw a piece to two or three of them when together; and it was amusing enough to see them trying to seize and carry it away in their mouths, like so many hungry dogs with a bone. They eat very deliberately, but do not chew their food. The little birds are aware how harmless these creatures are: I have seen one of the thick-billed finches picking at one end of a piece of cactus (which is much relished by all the animals of the lower region), whilst a lizard was eating at the other end; and afterwards the little bird with the utmost indifference hopped on the back of the reptile.

I opened the stomachs of several, and found them full of vegetable fibres and leaves of different trees, especially of an acacia. In the upper region they live chiefly on the acid and astringent berries of the guayavita, under which trees I have seen these lizards and the huge tortoises feeding together. To obtain the acacia-leaves they crawl up the low stunted trees; and it is not uncommon to see a pair quietly browsing, whilst seated on a branch several feet above the ground. These lizards, when cooked, yield a white meat, which is liked by those whose stomachs soar above all prejudices. Humboldt has remarked that in intertropical South America all lizards which inhabit dry regions are esteemed delicacies for the table. The inhabitants state that those which inhabit the upper damp parts drink water, but that the others do not, like the tortoises, travel up for it from the lower sterile country. At the time of our visit, the females had within their bodies numerous, large, elongated eggs, which they lay in their burrows; the inhabitants seek them for food.

These two species of Amblyrhynchus agree, as I have already stated, in their general structure, and in many of their habits. Neither have that rapid movement, so characteristic of the genera Lacerta and Iguana. They are both herbivorous, although the kind of vegetation on which they feed is so very different. Mr. Bell has given the name to the genus from the shortness of the snout. Indeed, the form of the mouth may almost be compared to that of the tortoise; one is led to suppose that this is an adaptation to their herbivorous appetites. It is very interesting thus to find a well-characterised genus, having its marine and terrestrial species, belonging to so confined a portion of the world. The aquatic species is by far the most remarkable, because it is the only existing lizard which lives on marine vegetable productions. As I at first observed, these islands are not so remarkable for the number of the species of reptiles, as for that of the individuals; when we remember the well-beaten paths made by the thousands of huge tortoises, the many turtles, the great warrens of the terrestrial Amblyrhynchus, and the groups of the marine species basking on the coast-rocks of every island, we must admit that there is no other quarter of the world where this order replaces the herbivorous mammalia in so extraordinary a manner. The geologist on hearing this will probably refer back in his mind to the Secondary epochs, when lizards, some herbivorous, some carnivorous, and of dimensions comparable only with our existing whales, swarmed on the land and in the sea. It is, therefore, worthy of

his observation that this archipelago, instead of possessing a
humid climate and rank vegetation, cannot be considered otherwise than extremely arid, and, for an equatorial region, remarkably temperate.

To finish with the zoology: The fifteen kinds of sea-fish which
I procured here are all new species; they belong to twelve
genera, all widely distributed, with the exception of Prionotus,
of which the four previously known species live on the eastern
side of America. Of land-shells I collected sixteen kinds (and
two marked varieties), of which, with the exception of one
Helix found at Tahiti, all are peculiar to this achipelago; a
single fresh-water shell (Paludina) is common to Tahiti and
Van Diemen's Land. Mr. Cuming, before our voyage, procured
here ninety species of sea-shells, and this does not include several species not yet specifically examined, of Trochus, Turbo,
Monodanta, and Nassa. He has been kind enough to give me
the following interesting results: Of the ninety shells, no less
than forty-seven are unknown elsewhere—a wonderful fact, considering how widely distributed sea-shells generally are. Of the
forty-three shells found in other parts of the world, twenty-five
inhabit the western coast of America, and of these eight are distinguishable as varieties; the remaining eighteen (including one
variety) were found by Mr. Cuming in the Low Archipelago,
and some of them also at the Philippines. This fact of shells
from islands in the central parts of the Pacific occurring here
deserves notice, for not one single sea-shell is known to be common to the islands of that ocean and to the west coast of America. The space of open sea running north and south off the west
coast separates two quite distinct conchological provinces; but
at the Galapagos Archipelago we have a halting-place, where
many new forms have been created, and whither these two great
conchological provinces have each sent several colonists. The
American province has also sent here representative species;
for there is a Galapageian species of Monoceros, a genus only
found on the west coast of America; and there are Galapageian
species of Fissurella and Cancellaria, genera common on the
west coast, but not found (as I am informed by Mr. Cuming)
in the central islands of the Pacific. On the other hand, there are
Galapageian species of Oniscia and Stylifer, genera common to
the West Indies and to the Chinese and Indian seas, but not

found either on the west coast of America or in the central Pacific. I may here add that after the comparison by Messrs. Cuming and Hinds of about two thousand shells from the eastern and western coasts of America, only one single shell was found in common, namely, the Pupura patula, which inhabits the West Indies, the coast of Panama, and the Galapagos. We have, therefore, in this quarter of the world, three great conchological sea-provinces, quite distinct, though surprisingly near each other, being separated by long north and south spaces either of land or of open sea.

I took great pains in collecting the insects, but, excepting Tierra del Fuego, I never saw in this respect so poor a country. Even in the upper and damp region I procured very few, excepting some minute Diptera and Hymenoptera, mostly of common mundane forms. As before remarked, the insects, for a tropical region, are of very small size and dull colours. Of beetles I collected twenty-five species (excluding a Dermestes and Corynetes imported wherever a ship touches); of these, two belong to the Harpalidæ, two to the Hydrophilidæ, nine to three families of the Heteromera, and the remaining twelve to as many different families. This circumstance of insects (and I may add plants), where few in number, belonging to many different families is, I believe, very general. Mr. Waterhouse, who has published [2] an account of the insects of this archipelago, and to whom I am indebted for the above details, informs me that there are several new genera; and that of the genera not new, one or two are American, and the rest of mundane distribution. With the exception of a wood-feeding Apate, and of one or probably two water-beetles from the American continent, all the species appear to be new.

The botany of this group is fully as interesting as the zoology. Dr. J. Hooker will soon publish in the *Linnaean Transactions* a full account of the flora, and I am much indebted to him for the following details. Of flowering plants there are, as far as at present is known, one hundred and eighty-five species, and forty cryptogamic species, making together two hundred and twenty-five; of this number I was fortunate enough to bring home one hundred and ninety-three. Of the flowering plants, one hundred are new species, and are probably confined to

[2] *Ann. and Mag. of Nat. Hist.,* Vol. XVI, p. 19.

this archipelago. Dr. Hooker conceives that, of the plants not so confined, at least ten species found near the cultivated ground at Charles Island have been imported. It is, I think, surprising that more American species have not been introduced naturally, considering that the distance is only between five hundred and six hundred miles from the continent; and that (according to Collnett, p. 58) drift-wood, bamboos, canes, and the nuts of a palm are often washed on the south-eastern shores. The proportion of one hundred flowering plants out of one hundred eighty-five (or one hundred seventy-five excluding the imported weeds) being new is sufficient, I conceive, to make the Galapagos Archipelago a distinct botanical province; but this flora is not nearly so peculiar as that of St. Helena, nor, as I am informed by Dr. Hooker, of Juan Fernandez. The peculiarity of the Galapageian flora is best shown in certain families: Thus there are twenty-one species of Compositæ, of which twenty are peculiar to this archipelago; these belong to twelve genera, and of these genera no less than ten are confined to the archipelago! Dr. Hooker informs me that the flora has an undoubted western American character; nor can he detect in it any affinity with that of the Pacific. If, therefore, we except the eighteen marine, the one fresh-water, and one land-shell, which have apparently come here as colonists from the central islands of the Pacific, and likewise the one distinct Pacific species of the Galapageian group of finches, we see that this archipelago, though standing in the Pacific Ocean, is zoologically part of America.

If this character were owing merely to immigrants from America, there would be little remarkable in it; but we see that a vast majority of all the land animals, and that more than half of the flowering plants, are aboriginal productions. It was most striking to be surrounded by new birds, new reptiles, new shells, new insects, new plants, and yet by innumerable trifling details of structure, and even by the tones of voice and plumage of the birds, to have the temperate plains of Patagonia, or the hot dry deserts of Northern Chile, vividly brought before my eyes. Why —on these small points of land, which within a late geological period must have been covered by the ocean, which are formed of basaltic lava, and therefore differ in geological character from the American continent, and which are placed under a peculiar climate—why were their aboriginal inhabitants, associated, I may add, in different proportions both in kind and number from those on the continent, and therefore acting on each other in a different manner—why were they created on American types of organization? It is probable that the islands of the Cape Verde group resemble, in all their physical conditions, far more closely the Galapagos Islands than these latter physically resemble the coast of America; yet the aboriginal inhabitants of the two groups are totally unlike, those of the Cape Verde Islands bearing the impress of Africa, as the inhabitants of the Galapagos Archipelago are stamped with that of America.

I have not as yet noticed by far the most remarkable feature in the natural history of this archipelago; it is that the different islands to a considerable extent are inhabited by a different set of beings. My attention was first called to this fact by the vice-governor, Mr. Lawson, declaring that the tortoises differed from the different islands, and that he could with certainty tell from which island any one was brought. I did not for some time pay sufficient attention to this statement, and I had already partially mingled together the collections from two of the islands. I never dreamed that islands about fifty or sixty miles apart, and most of them in sight of each other, formed of precisely the same rocks, placed under a quite similar climate, rising to a nearly equal height, would have been differently tenanted; but we shall soon see that this is the case. It is the fate of most voyagers no sooner to discover what is most interesting in any locality than they are hurried from it; but I ought, perhaps, to be thankful that I obtained sufficient materials to establish this most remarkable fact in the distribution of organic beings.

The inhabitants, as I have said, state that they can distinguish the tortoises from the different islands; and that they differ not only in size, but in other characters. Captain Porter has described [3] those from Charles and from the nearest island to it, namely, Hood Island, as having their shells in front thick and turned up like a Spanish saddle, whilst the tortoises from James Island are rounder, blacker, and have a better taste when cooked. M. Bibron, moreover, informs me that he has seen what

[3] *Voyage in the U.S. Ship Essex,* Vol. I, p. 215.

he considers two distinct species of tortoise from the Galapagos, but he does not know from which islands. The specimens that I brought from three islands were young ones; and probably owing to this cause, neither Mr. Gray nor myself could find in them any specific differences. I have remarked that the marine Amblyrhynchus was larger at Albemarle Island than elsewhere; and M. Bibron informs me that he has seen two distinct aquatic species of this genus; so that the different islands probably have their representative species or races of the Amblyrhynchus, as well as of the tortoise. My attention was first thoroughly aroused by comparing together the numerous specimens, shot by myself and several other parties on board, of the mocking-thrushes, when, to my astonishment, I discovered that all those from Charles Island belonged to one species (Mimus trifasciatus); all from Albemarle Island to M. parvulus; and all from James and Chatham Islands (between which two other islands are situated, as connecting links) belonged to M. melanotis. These two latter species are closely allied, and would by some ornithologists be considered as only well-marked races or varieties; but the Mimus trifasciatus is very distinct. Unfortunately most of the specimens of the finch tribe were mingled together; but I have strong reasons to suspect that some of the species of the sub-group Geospiza are confined to separate islands. If the different islands have their representatives of Geospiza, it may help to explain the singularly large number of the species of this sub-group in this one small archipelago and, as a probable consequence of their numbers, the perfectly graduated series in the size of their beaks. Two species of the sub-group Cactornis, and two of Camarhynchus, were procured in the archipelago; and of the numerous specimens of these two sub-groups shot by four collectors at James Island all were found to belong to one species of each; whereas the numerous specimens shot either on Chatham or Charles Island (for the two sets were mingled together) all belonged to the two other species. Hence we may feel almost sure that these islands possess their representative species of these two sub-groups. In land-shells this law of distribution does not appear to hold good. In my very small collection of insects, Mr. Waterhouse remarks, that of those which were ticketed with their locality, not one was common to any two of the islands.

If we now turn to the flora, we shall find the aboriginal plants of the different islands wonderfully different. I give all the following results on the high authority of my friend Dr. J. Hooker. I may premise that I indiscriminately collected everything in flower on the different islands, and fortunately kept my collections separate. Too much confidence, however, must not be placed in the proportional results, as the small collections brought home by some other naturalists, though in some respects confirming the results, plainly show that much remains to be done in the botany of this group; the Leguminosae, moreover, have as yet been only approximately worked out:

Name of island	Total no. of species	No. of species found in other parts of the world	No. of species confined to the Galapagos Archipelago	No. confined to the one island	No. of species confined to the Galapagos Archipelago, but found on more than the one island
James Island	71	33	38	30	8
Albemarle Island	46	18	26	22	4
Chatham Island	32	16	16	12	4
Charles Island	68	39 (or 29, if the probably imported plants be subtracted)	29	21	8

Hence we have the truly wonderful fact that, in James Island, of the thirty-eight Galapageian plants, or those found in no other part of the world, thirty are exclusively confined to this one island; and in Albemarle Island, of the twenty-six aboriginal Galapageian plants, twenty-two are confined to this one island, that is, only four are at present known to grow in the other islands of the archipelago; and so on, as shown in the above table, with the plants from Chatham and Charles Islands. This fact will, perhaps, be rendered even more striking, by giving a few illustrations: Thus, Scalesia, a remarkable arborescent genus of the Compositae, is confined to the archipelago; it has six species, one from Chatham, one from Albemarle, one from Charles Island, two from James Island, and the sixth from one of the three latter islands, but it is not known from which. Not one of these six species grows on any two islands. Again, Euphorbia, a mundane or widely distributed genus, has here eight species, of which seven are confined to the archipelago, and not one found on any two islands; Acalypha and Borreria, both mundane genera, have respectively six and seven species, none of which have the same species on two islands, with the exception of one Borreria, which does occur on two islands. The species of the Compositae are particularly local; and Dr. Hooker has furnished me with several other most striking illustrations of the difference of the species on the different islands. He remarks that this law of distribution holds good both with those genera confined to the archipelago, and those distributed in other quarters of the world; in like manner we have seen that the different islands have their proper species of the mundane genus of tortoise, and of the widely distributed American genus of the mocking-thrush, as well as of two of the Galapageian sub-groups of finches, and almost certainly of the Galapageian genus Amblyrhynchus.

The distribution of the tenants of this archipelago would not be nearly so wonderful if, for instance, one island had a mocking-thrush, and a second island some other quite distinct genus; if one island had its genus of lizard, and a second island another distinct genus, or none whatever; or if the different islands were inhabited not by representative species of the same genera of plants but by totally different genera, as does to a certain extent hold good. For, to give one instance, a large berry-bearing tree at James Island has no representative species in Charles Island. But it is the circumstance that several of the islands possess their own species of the tortoise, mocking-thrush, finches, and numerous plants, these species having the same general habits, occupying analogous situations, and obviously filling the same place in the natural economy of this archipelago, that strikes me with wonder. It may be suspected that some of these representative species, at least in the case of the tortoise and of some of the birds, may hereafter prove to be only well-marked races; but this would be of equally great interest to the philosophical naturalist. I have said that most of the islands are in sight of each other: I may specify that Charles Island is fifty miles from the nearest part of Chatham Island, and thirty-three miles from the nearest part of Albemarle Island. Chatham Island is sixty miles from the nearest part of James Island, but there are two intermediate islands between them which were not visited by me. James Island is only ten miles from the nearest part of Albemarle Island, but the two points where the collections were made are thirty-two miles apart. I must repeat that neither the nature of the soil, nor height of the land, nor the climate, nor the general character of the associated beings, and therefore their action one on another, can differ much in the different islands. If there be any sensible difference in their climates, it must be between the windward group (namely Charles and Chatham Islands), and that to leeward; but there seems to be no corresponding difference in the productions of these two halves of the archipelago.

The only light which I can throw on this remarkable difference in the inhabitants of the different islands, is that very strong currents of the sea running in a westerly and W.N.W. direction must separate, as far as transportal by the sea is concerned, the southern islands from the northern ones; and between these northern islands a strong N.W. current was observed, which must effectually separate James and Albemarle Islands. As the archipelago is free to a most remarkable degree from gales of wind, neither the birds, insects, nor lighter seeds would be blown from island to island. And lastly, the profound depth of the ocean between the islands, and their apparently recent (in a geological sense) volcanic origin, render it highly unlikely that they were ever united; and this, probably, is a far more important consideration than any other, with respect to the geographical distribution of their inhabitants. Reviewing the facts here given, one is astonished at the amount of creative force, if such an expression may be used, displayed on these small, barren, and rocky islands; and still more so, at its diverse yet analogous action on points so near each other. I have said that

the Galapagos Archipelago might be called a satellite attached
to America, but it should rather be called a group of satellites,
physically similar, organically distinct, yet intimately related
to each other, and all related in a marked, though much lesser
degree, to the great American continent.

I will conclude my description of the natural history of these
islands by giving an account of the extreme tameness of the
birds.

This disposition is common to all the terrestrial species:
namely, to the mocking-thrushes, the finches, wrens, tyrant-fly-
catchers, the dove, and carrion-buzzard. All of them often approached sufficiently near to be killed with a switch, and sometimes, as I myself tried, with a cap or hat. A gun is here almost
superfluous, for with the muzzle I pushed a hawk off the branch
of a tree. One day, whilst lying down, a mocking-thrush alighted
on the edge of a pitcher, made of the shell of a tortoise, which
I held in my hand, and began very quietly to sip the water; it
allowed me to lift it from the ground whilst seated on the
vessel. I often tried, and very nearly succeeded, in catching
these birds by their legs. Formerly the birds appear to have
been even tamer than at present. Cowley (in the year 1684)
says that the "turtle-doves were so tame, that they would often
alight upon our hats and arms, so as that we could take them
alive: they not fearing man, until such time as some of our company did fire at them, whereby they were rendered more shy."
Dampier also, in the same year, says that a man in a morning's
walk might kill six or seven dozen of these doves. At present,
although certainly very tame, they do not alight on people's
arms, nor do they suffer themselves to be killed in such large
numbers. It is surprising that they have not become wilder;
for these islands during the last hundred and fifty years have
been frequently visited by buccaneers and whalers; and the sailors, wandering through the woods in search of tortoises, always
take cruel delight in knocking down the little birds.

These birds, although now still more persecuted, do not
readily become wild; in Charles Island, which had then been
colonized about six years, I saw a boy sitting by a well with a
switch in his hand, with which he killed the doves and finches
as they came to drink He had already procured a little heap of

them for his dinner; and he said that he had constantly been in the habit of waiting by this well for the same purpose. It would appear that the birds of this archipelago, not having as yet learnt that man is a more dangerous animal than the tortoise or the Amblyrhynchus, disregard him, in the same manner as in England shy birds, such as magpies, disregard the cows and horses grazing in our fields.

The Falkland Islands offer a second instance of birds with a similar disposition. The extraordinary tameness of the little Opetiorhynchus has been remarked by Pernety, Lesson, and other voyagers. It is not, however, peculiar to that bird; the Polyborus, snipe, upland and lowland goose, thrush, bunting, and even some true hawks are all more or less tame. As the birds are so tame there, where foxes, hawks, and owls occur, we may infer that the absence of all rapacious animals at the Galapagos is not the cause of their tameness here. The upland geese at the Falklands show, by the precaution they take in building on the islets, that they are aware of their danger from the foxes; but they are not by this rendered wild towards man. This tameness of the birds, especially of the waterfowl, is strongly contrasted with the habits of the same species in Tierra del Fuego, where for ages past they have been persecuted by the wild inhabitants. In the Falklands, the sportsman may sometimes kill more of the upland geese in one day than he can carry home, whereas in Tierra del Fuego it is nearly as difficult to kill one as it is in England to shoot the common wild goose.

In the time of Pernety (1763) all the birds there appear to have been much tamer than at present; he states that the Opetiorhynchus would almost perch on his finger, and that with a wand he killed ten in half an hour. At that period the birds must have been about as tame as they now are at the Galapagos. They appear to have learnt caution more slowly at these latter islands than at the Falklands, where they have had proportionate means of experience; for besides frequent visits from vessels, those islands have been at intervals colonized during the entire period. Even formerly, when all the birds were so tame, it was impossible by Pernety's account to kill the black-necked swan—a bird of passage, which probably brought with it the wisdom learnt in foreign countries.

I may add that, according to Du Bois, all the birds at Bourbon in 1571–72, with the exception of the flamingoes and geese, were so extremely tame that they could be caught by the hand, or killed in any number with a stick. Again, at Tristan d'Acunha in the Atlantic, Carmichael [4] states that the only two land-birds, a thrush and a bunting, were "so tame as to suffer themselves to be caught with a hand-net." From these several facts we may, I think, conclude, first, that the wildness of birds with regard to man is a particular instinct directed against *him,* and not dependent on any general degree of caution arising from other sources of danger; secondly, that it is not acquired by individual birds in a short time, even when much persecuted, but that in the course of successive generations it becomes hereditary. With domesticated animals we are accustomed to see new mental habits or instincts acquired and rendered hereditary; but with animals in a state of nature, it must always be most difficult to discover instances of acquired hereditary knowledge. In regard to the wildness of birds towards man, there is no way of accounting for it, except as an inherited habit; comparatively few young birds, in any one year, have been injured by man in England, yet almost all, even nestlings, are afraid of him. Many individuals, on the other hand, both at the Galapagos and at the Falklands, have been pursued and injured by man, but yet have not learned a salutary dread of him. We may infer from these facts what havoc the introduction of any new beast of prey must cause in a country, before the instincts of the indigenous inhabitants have become adapted to the stranger's craft or power.

[4] *Linn. Trans.,* Vol. XII, p. 496. The most anomalous fact on this subject which I have met with is the wildness of the small birds in the arctic parts of North America (as described by Richardson, *Fauna Bor.,* Vol. II, p. 332), where they are said never to be persecuted. This case is the more strange because it is asserted that some of the same species in their winter-quarters in the United States are tame. There is much, as Dr. Richardson well remarks, utterly inexplicable connected with the different degrees of shyness and care with which birds conceal their nests. How strange it is that the English wood-pigeon, generally so wild a bird, should very frequently rear its young in shrubberies close to houses!

The Essay of 1844

PART I

CHAPTER I

On the Variation of Organic Beings under Domestication; and on the Principles of Selection

The most favourable conditions for variation seem to be when organic beings are bred for many generations under domestication: one may infer this from the simple fact of the vast number of races and breeds of almost every plant and animal, which has long been domesticated. Under certain conditions organic beings even during their individual lives become slightly altered from their usual form, size, or other characters; and many of the peculiarities thus acquired are transmitted to their offspring. Thus in animals, the size and vigour of body, fatness, period of maturity, habits of body or consensual movements, habits of mind and temper, are modified or acquired during the life of the individual, and become inherited. There is reason to believe that when long exercise has given to certain muscles great development, or disuse has lessened them, that such development is also inherited. Food and climate will occasionally produce changes in the colour and texture of the external coverings of animals; and certain unknown conditions affect the horns of cattle in parts of Abyssinia; but whether these peculiarities, thus acquired during individual lives, have been inherited, I do not know. It appears certain that malconformation and lameness in horses, produced by too much work on hard roads— that affections of the eyes in this animal probably caused by bad ventilation—that tendencies towards many diseases in man, such as gout, caused by the course of life and ultimately producing changes of structure, and that many other diseases produced by unknown agencies, such as goitre, and the idiocy resulting from it, all become hereditary.

It is very doubtful whether the flowers and leaf-buds annually produced from the same bulb, root, or tree can properly be considered as parts of the same individual, though in some respects they certainly seem to be so. If they are parts of an individual, plants also are subject to considerable changes during their *individual* lives. Most florist-flowers if neglected degenerate, that is, they lose some of their characters; so common is this, that trueness is often stated as greatly enhancing the value of a variety.[1] Tulips break their colours only after some years' culture; some plants become double and others single, by neglect or care; these characters can be transmitted by cuttings or grafts, and in some cases by true or seminal propagation. Occasionally a single bud on a plant assumes at once a new and widely different character: thus it is certain that nectarines have been produced on peach trees and moss roses on provence roses; white currants on red currant bushes; flowers of a different colour from that of the stock, in chrysanthemums, dahlias, sweet-williams, azaleas, etc.; variegated leaf-buds on many trees, and other similar cases. These new characters appearing in single buds, can, like those lesser changes affecting the whole plant, be multiplied not only by cuttings and such means, but often likewise by true seminal generation.

The changes thus appearing during the lives of individual animals and plants are extremely rare compared with those which are congenital or which appear soon after birth. Slight differences thus arising are infinitely numerous: the proportions and forms of every part of the frame, inside and outside, appear to vary in very slight degrees: anatomists dispute what is the "beau ideal" of the bones, the liver and kidneys, like painters do of the proportions of the face: the proverbial expression that no two animals or plants are born absolutely alike is much truer when applied to those under domestication, than to those in a state of nature. Besides these slight differences, single individuals are occasionally born considerably unlike in certain parts or in their whole structure to their parents: these are called by horticulturalists and breeders

---

[1] It is not clear where the following note is meant to come: "Case of *Orchis*—most remarkable as not long cultivated by seminal propagation. Case of varieties which soon acquire, like *Ægilops* and carrot (and maize), *a certain general character* and then go on varying." [All the footnotes in this *Essay* are by Francis Darwin.—S.E.H.]

"sports," and are not uncommon except when very strongly marked. Such sports are known in some cases to have been parents of some of our domestic races; and such probably have been the parents of many other races, especially of those which in some senses may be called hereditary monsters; for instance where there is an additional limb, or where all the limbs are stunted (as in the Ancon sheep), or where a part is wanting, as in rumpless fowls and tailless dogs or cats. The effects of external conditions on the size, colour and form, which can rarely and obscurely be detected during one individual life, become apparent after several generations: the slight differences, often hardly describable, which characterise the stock of different countries, and even of districts in the same country, seem to be due to such continued action.

On the Hereditary Tendency

A volume might be filled with facts showing what a strong tendency there is to inheritance, in almost every case of the most trifling, as well as of the most remarkable congenital peculiarities. The term congenital peculiarity, I may remark, is a loose expression and can only mean a peculiarity apparent when the part affected is nearly or fully developed: in Part II, I shall have to discuss at what period of the embryonic life connatal peculiarities probably first appear; and I shall then be able to show from some evidence, that at whatever period of life a new peculiarity first appears, it tends hereditarily to appear at a corresponding period. Numerous though slight changes, slowly supervening in animals during mature life (often, though by no means always, taking the form of disease), are, as stated in the first paragraphs, very often hereditary. In plants, again, the buds which assume a different character from their stock likewise tend to transmit their new peculiarities. There is not sufficient reason to believe that either mutilations or changes of form produced by mechanical pressure, even if continued for hundreds of generations, or that any changes of structure quickly produced by disease, are inherited; it would appear as if the tissue of the part affected must slowly and freely grow into the new form, in order to be inheritable. There is a very great difference in the hereditary tendency of different peculiarities, and of the same peculiarity, in different individuals and species; thus twenty thousand seeds of the weeping ash have been sown and not one came up true; out of seventeen seeds of the weeping yew, nearly all came up true. The ill-formed and almost monstrous "Niata" cattle of South America and Ancon sheep, both when bred together and when crossed with other breeds, seem to transmit their peculiarities to their offspring as truly as the ordinary breeds. I can throw no light on these differences in the power of hereditary transmission. Breeders believe, and apparently with good cause, that a peculiarity generally becomes more firmly implanted after having passed through several generations; that is, if one offspring out of twenty inherits a peculiarity from its parents, then its descendants will tend to transmit this peculiarity to a larger proportion than one in twenty; and so on in succeeding generations. I have said nothing about mental peculiarities being inheritable for I reserve this subject for a separate chapter.

Causes of Variation

Attention must here be drawn to an important distinction in the first origin or appearance of varieties: when we see an animal highly kept producing offspring with an hereditary tendency to early maturity and fatness; when we see the wild-duck and Australian dog always becoming, when bred for one or a few generations in confinement, mottled in their colours; when we see people living in certain districts or circumstances becoming subject to an hereditary taint to certain organic diseases, as consumption or plica polonica—we naturally attribute such changes to the direct effect of known or unknown agencies acting for one or more generations on the parents. It is probable that a multitude of peculiarities may be thus directly caused by unknown external agencies. But in breeds, characterized by an extra limb or claw, as in certain fowls and dogs; by an extra joint in the vertebrae; by the loss of a part, as the tail; by the substitution of a tuft of feathers for a comb in certain poultry; and in a multitude of other cases, we can hardly attribute these peculiarities directly to external influences, but indirectly to the laws of embryonic growth and of reproduction. When we see a multitude of varieties (as has often been the case, where a cross has been carefully guarded against) produced from seeds matured in the very same capsule, with the male and female principle nourished from the same roots and necessarily exposed to the same external influences; we cannot believe that the endless slight differences between seedling varieties thus produced can be the effect of any corresponding difference in their exposure. We are led (as Müller has remarked) to the same conclusion, when we see in the same litter, produced by the same act of conception, animals considerably different.

As variation to the degree here alluded to has been observed only in organic beings under domestication, and in plants amongst those most highly and long cultivated, we must attribute, in such cases, the varieties (although the difference between each variety cannot possibly be attributed to any corresponding difference of exposure in the parents) to the indirect effects of domestication on the action of the reproductive system. It would appear as if the reproductive powers failed in their ordinary function of producing new organic beings closely like their parents; and as if the entire organisation of embryo, under domestication, became in a slight degree plastic. We shall hereafter have occasion to show, that in organic beings, a considerable change from the natural conditions of life, affects, independently of their general state of health, in another and remarkable manner the reproductive system. I may add, judging from the vast number of new varieties of plants which have been produced in the same districts and under nearly the same routine of culture, that probably the indirect effects of domestication in making the organisation plastic, is a much more efficient source of variation than any direct effect which external causes may have on the colour, texture, or form of each part. In the few instances in which, as in the dahlia, the course of variation has been recorded, it appears that domestication produces little effect for several generations in rendering the organisation plastic; but afterwards, as if by an accumulated effect, the original character of the species suddenly gives way or breaks.

On Selection

We have hitherto only referred to the first appearance in individuals of new peculiarities; but to make a race or breed,
something more is generally requisite than such peculiarities
(except in the case of the peculiarities being the direct effect
of constantly surrounding conditions) should be inheritable—
namely the principle of selection, implying separation. Even
in the rare instances of sports, with the hereditary tendency
very strongly implanted, crossing must be prevented with other
breeds, or if not prevented the best characterised of the half-
bred offspring must be carefully selected. Where the external
conditions are constantly tending to give some character, a
race possessing this character will be formed with far greater
ease by selecting and breeding together the individuals most
affected. In the case of the endless slight variations produced
by the indirect effects of domestication on the action of the
reproductive system, selection is indispensable to form races;
and when carefully applied, wonderfully numerous and diverse
races can be formed. Selection, though so simple in theory, is
and has been important to a degree which can hardly be
overrated. It requires extreme skill, the result of long practice,
in detecting the slightest difference in the forms of animals,
and it implies some distinct object in view; with these req-
uisites and patience, the breeder has simply to watch for every,
even the smallest, approach to the desired end, to select such
individuals and pair them with the most suitable forms, and
so continue with succeeding generations. In most cases careful
selection and the prevention of accidental crosses will be necessary for several generations, for in new breeds there is a strong
tendency to vary and especially to revert to ancestral forms:
but in every succeeding generation less care will be requisite
for the breed will become truer; until ultimately only an
occasional individual will require to be separated or destroyed.
Horticulturalists in raising seeds regularly practise this, and
call it "roguing," or destroying the "rogues" or false varieties.
There is another and less efficient means of selection amongst
animals: namely, repeatedly procuring males with some de-

sirable qualities, and allowing them and their offspring to breed freely together; and this in the course of time will affect the whole lot. These principles of selection have been *methodically* followed for scarcely a century; but their high importance is shown by the practical results, and is admitted in the writings of the most celebrated agriculturalists and horticulturalists; I need only name Anderson, Marshall, Bakewell, Coke, Western, Sebright, and Knight.

Even in well-established breeds the individuals of which to an unpractised eye would appear absolutely similar, which would give, it might have been thought, no scope to selection, the whole appearance of the animal has been changed in a few years (as in the case of Lord Western's sheep), so that practised agriculturalists could scarcely credit that a change had not been effected by a cross with other breeds. Breeders both of plants and animals frequently give their means of selection greater scope, by crossing different breeds and selecting the offspring; but we shall have to recur to this subject again.

The external conditions will doubtless influence and modify the results of the most careful selection; it has been found impossible to prevent certain breeds of cattle from degenerating on mountain pastures; it would probably be impossible to keep the plumage of the wild-duck in the domesticated race; in certain soils, no care has been sufficient to raise cauliflower seed true to its character; and so in many other cases. But with patience it is wonderful what man has effected. He has selected and therefore in one sense made one breed of horses to race and another to pull; he has made sheep with fleeces good for carpets and other sheep good for broadcloth; he has, in the same sense, made one dog to find game and give him notice when found, and another dog to fetch him the game when killed; he has made by selection the fat to lie mixed with the meat in one breed and in another to accumulate in the bowels for the tallow-chandler; he has made the legs of one breed of pigeons long, and the beak of another so short, that it can hardly feed itself; he has previously determined how the feathers on a bird's body shall be coloured, and how the petals of many flowers shall be streaked or fringed, and has given prizes for complete success; by selection, he has made the leaves of one variety and the flower-buds of another variety of the

cabbage good to eat, at different seasons of the year; and thus
has he acted on endless varieties. I do not wish to affirm that
the long- and short-woolled sheep, or that the pointer and retriever, or that the cabbage and cauliflower have certainly descended from one and the same aboriginal wild stock; if they
have not so descended, though it lessens what man has effected,
a large result must be left unquestioned.

In saying as I have done that man makes a breed, let it not
be confounded with saying that man makes the individuals,
which are given by nature with certain desirable qualities;
man only adds together and makes a permanent gift of nature's
bounties. In several cases, indeed, for instance in the Ancon
sheep, valuable from not getting over fences, and in the turnspit dog, man has probably only prevented crossing; but in
many cases we positively know that he has gone on selecting,
and taking advantage of successive small variations.

Selection has been *methodically* followed, as I have said, for
barely a century; but it cannot be doubted that occasionally it
has been practised from the remotest ages, in those animals
completely under the dominion of man. In the earliest chapters of the Bible there are rules given for influencing the colours
of breeds, and black and white sheep are spoken of as separated. In the time of Pliny the barbarians of Europe and Asia
endeavoured by cross-breeding with a wild stock to improve
the races of their dogs and horses. The savages of Guiana now
do so with their dogs: such care shows at least that the characters of individual animals were attended to. In the rudest
times of English history, there were laws to prevent the ex-
portation of fine animals of established breeds, and in the case
of horses, in Henry VIII's time, laws for the destruction of all
horses under a certain size. In one of the oldest numbers of
the *Philosophical Transactions,* there are rules for selecting
and improving the breeds of sheep. Sir H. Bunbury, in 1660,
has given rules for selecting the finest seedling plants, with as
much precision as the best recent horticulturalist could. Even
in the most savage and rude nations, in the wars and famines
which so frequently occur, the most useful of their animals
would be preserved: the value set upon animals by savages is
shown by the inhabitants of Tierra del Fuego devouring their
old women before their dogs, which as they asserted are useful

in otter-hunting: who can doubt but that in every case of famine and war, the best otter-hunters would be preserved, and therefore in fact selected for breeding. As the offspring so obviously take after their parents, and as we have seen that savages take pains in crossing their dogs and horses with wild stocks, we may even conclude as probable that they would sometimes pair the most useful of their animals and keep their offspring separate. As different races of men require and admire different qualities in their domesticated animals, each would thus slowly, though unconsciously, be selecting a different breed. As Pallas has remarked, who can doubt but that the ancient Russian would esteem and endeavour to preserve those sheep in his flocks which had the thickest coats. This kind of insensible selection by which new breeds are not selected and kept separate, but a peculiar character is slowly given to the whole mass of the breed, by often saving the life of animals with certain characteristics, we may feel nearly sure, from what we see has been done by the more direct method of separate selection within the last fifty years in England, would in the course of some thousand years produce a marked effect.

Crossing Breeds

When once two or more races are formed, or if more than one race, or species fertile *inter se,* originally existed in a wild state, their crossing becomes a most copious source of new races. When two well-marked races are crossed the offspring in the first generation take more or less after either parent or are quite intermediate between them, or rarely assume characters in some degree new. In the second and several succeeding generations, the offspring are generally found to vary exceedingly, one compared with another, and many revert nearly to their ancestral forms. This greater variability in succeeding generations seems analogous to the breaking or variability of organic beings after having been bred for some generations under domestication. So marked is this variability in crossbred descendants, that Pallas and some other naturalists have supposed that all variation is due to an original cross; but I conceive that the history of the potato, dahlia, Scotch rose, the guinea-pig, and of many trees in this country, where only one species of the genus exists, clearly shows that a species may vary where there can have been no crossing. Owing to this variability and tendency to reversion in cross-bred beings, much careful selection is requisite to make intermediate or new permanent races: nevertheless crossing has been a most powerful engine, especially with plants, where means of propagation exist by which the cross-bred varieties can be secured without incurring the risk of fresh variation from seminal propagation: with animals the most skilful agriculturalists now greatly prefer careful selection from a well-established breed, rather than from uncertain cross-bred stocks.

Although intermediate and new races may be formed by the mingling of others, yet if the two races are allowed to mingle quite freely, so that none of either parent race remain pure, then, especially if the parent races are not widely different, they will slowly blend together, and the two races will be destroyed, and one mongrel race left in its place. This will of course happen in a shorter time, if one of the parent races exists in greater number than the other. We see the effect of this mingling, in the manner in which the aboriginal breeds of dogs and pigs in the Oceanic Islands and the many breeds of our domestic animals introduced into South America have all been lost and absorbed in a mongrel race. It is probably owing to the freedom of crossing, that, in uncivilised countries, where enclosures do not exist, we seldom meet with more than one race of a species: it is only in enclosed countries where the inhabitants do not migrate, and have conveniences for separating the several kinds of domestic animals, that we meet with a multitude of races. Even in civilised countries, want of care for a few years has been found to destroy the good results of far longer periods of selection and separation.

This power of crossing will affect the races of all *terrestrial* animals; for all terrestrial animals require for their reproduction the union of two individuals. Amongst plants, races will not cross and blend together with so much freedom as in terrestrial animals; but this crossing takes place through various curious contrivances to a surprising extent. In fact such contrivances exist in so very many hermaphrodite flowers by which an occasional cross may take place, that I cannot avoid suspecting (with Mr. Knight) that the reproductive action requires at *intervals,* the concurrence of distinct individuals. Most breeders of plants and animals are firmly convinced that benefit is derived from an occasional cross, not with another race, but with another family of the same race; and that, on the other hand, injurious consequences follow from long-continued close interbreeding in the same family. Of marine animals, many more, than was till lately believed, have their sexes on separate individuals; and where they are hermaphrodite, there seems very generally to be means through the water of one individual occasionally impregnating another: if individual animals can singly propagate themselves for perpetuity, it is unaccountable that no terrestrial animal, where the means of observation are more obvious, should be in this predicament of singly perpetuating its kind. I conclude, then, that races of most animals and plants, when unconfined in the same country, would tend to blend together.

Whether Our Domestic Races Have Descended from One or More Wild Stocks

Several naturalists, of whom Pallas regarding animals, and Humboldt regarding certain plants, were the first, believe that the breeds of many of our domestic animals such as of the horse, pig, dog, sheep, pigeon, and poultry, and of our plants have descended from more than one aboriginal form. They leave it doubtful, whether such forms are to be considered wild races, or true species, whose offspring are fertile when crossed *inter se.* The main arguments for this view consist, firstly, of the great difference between such breeds, as the race- and cart-horse, or the greyhound and bull-dog, and of our ignorance of the steps or stages through which these could have passed from a common parent; and secondly that in the most ancient historical periods, breeds resembling some of those at present most different, existed in different countries. The wolves of North America and of Siberia are thought to be different species; and it has been remarked that the dogs belonging to the savages in these two countries resemble the wolves of the same country; and therefore that they have probably descended from two different wild stocks. In the same manner, these naturalists believe that the horse of Arabia and of Europe have probably descended from two wild stocks both apparently now extinct. I do not think the assumed fertility of these wild stocks any very great difficulty on this view; for although in animals the offspring of most cross-bred species are infertile, it is not always remembered that the experiment is very seldom fairly tried, except when two near species *both* breed freely (which does not readily happen, as we shall hereafter see) when under the dominion of man. Moreover in the case of the China and common goose, the canary and siskin, the hybrids breed freely; in other cases the offspring from hybrids crossed with either pure parent are fertile, as is practically taken advantage of with the yak and cow; as far as the analogy of plants serves, it is impossible to deny that some species are quite fertile *inter se;* but to this subject we shall recur.

On the other hand, the upholders of the view that the several breeds of dogs, horses, etc., have descended each from one stock may aver that their view removes all *difficulty about fertility,* and that the main argument from the high antiquity of different breeds, somewhat similar to the present breeds, is worth little without knowing the date of the domestication of such animals, which is far from being the case. They may also with more weight aver that, knowing that organic beings under domestication do vary in some degree, the argument from the great difference between certain breeds is worth nothing, without we know the limits of variation during a long course of time, which is far from the case. They may argue that in almost every county in England, and in many districts of other countries, for instance in India, there are slightly different breeds of the domestic animals; and that it is opposed to all that we know of the distribution of wild animals to suppose that these have descended from so many different wild races or species: if so, they may argue, is it not probable that countries quite separate and exposed to different climates would have breeds not slightly, but considerably, different? Taking the most favourable case, on both sides, namely that of the dog; they might urge that such breeds as the bull-dog and turnspit have been reared by man, from the ascertained fact that strictly analogous breeds (namely the Niata ox and Ancon sheep) in other quadrupeds have thus originated. Again they may say, seeing what training and careful selection has effected for the greyhound, and seeing how absolutely unfit the Italian greyhound is to maintain itself in a state of nature, is it not probable that at least all greyhounds—from the rough deerhound, the smooth Persian, the common English, to the Italian—have descended from one stock? If so, is it so improbable that the deerhound and long-legged shepherd dog have so descended? If we admit this, and give up the bull-dog, we can hardly dispute the probable common descent of the other breeds.

The evidence is so conjectural and balanced on both sides that at present I conceive that no one can decide: for my own part, I lean to the probability of most of our domestic animals having descended from more than one wild stock; though from the arguments last advanced and from reflecting on the slow though inevitable effect of different races of mankind, under different circumstances, saving the lives of and therefore selecting the individuals most useful to them, I cannot doubt but that one class of naturalists have much overrated the probable number of the aboriginal wild stocks. As far as we admit the difference of our races to be due to the differences of their original stocks, so much must we give up of the amount of variation produced under domestication. But this appears to me unimportant, for we certainly know in some few cases, for instance in the dahlia, and potato, and rabbit, that a great number of varieties have proceeded from one stock; and, in many of our domestic races, we know that man, by slowly selecting and by taking advantage of sudden sports, has considerably modified old races and produced new ones. Whether we consider our races as the descendants of one or several wild stocks, we are in far the greater number of cases equally ignorant what these stocks were.

Limits to Variation in Degree and Kind

Man's power in making races depends, in the first instance, on the stock on which he works being variable; but his labours are modified and limited, as we have seen, by the direct effects of the external conditions—by the deficient or imperfect hereditariness of new peculiarities—and by the tendency to continual variation and especially to reversion to ancestral forms. If the stock is not variable under domestication, of course he can do nothing; and it appears that species differ considerably in this tendency to variation, in the same way as even sub-varieties from the same variety differ greatly in this respect, and transmit to their offspring this difference in tendency. Whether the absence of a tendency to vary is an unalterable quality in certain species, or depends on some deficient condition of the particular state of domestication to which they are exposed, there is no evidence. When the organisation is rendered variable, or plastic, as I have expressed it, under domestication, different parts of the frame vary more or less in different species: thus in the breeds of cattle it has been remarked that the horns are the most constant or least variable character, for these often remain constant, whilst the colour, size, proportions of the body, tendency to fatten, etc., vary; in sheep, I believe, the horns are much more variable. As a general rule the less important parts of the organisation seem to vary most, but I think there is sufficient evidence that every part occasionally varies in a slight degree. Even when man has the primary requisite variability he is necessarily checked by the health and life of the stock he is working on: thus he has already made pigeons with such small beaks that they can hardly eat and will not rear their own young; he has made families of sheep with so strong a tendency to early maturity and to fatten, that in certain pastures they cannot live from their extreme liability to inflammation; he has made (i.e. selected) sub-varieties of plants with a tendency to such early growth that they are frequently killed by the spring frosts; he has made a breed of cows having calves with such large hind quarters that they are born with great difficulty, often to the death of their mothers; the breeders were compelled to remedy this by the selection of a breeding stock with smaller hind quarters; in such a case, however, it is possible by long patience and great loss, a remedy might have been found in selecting cows capable of giving birth to calves with large hind quarters, for in human kind there are no doubt hereditary bad and good confinements. Besides the limits already specified, there can be little doubt that the variation of different parts of the frame are connected together by many laws: thus the two sides of the body, in health and disease, seem almost always to vary together: it has been asserted by breeders that if the head is much elongated, the bones of the extremities will likewise be so; in seedling-apples large leaves and fruit generally go together, and serve the horticulturalist as some guide in his selection; we can here see the reason, as the fruit is only a metamorphosed leaf. In animals the teeth and hair seem connected, for the hairless Chinese dog is almost toothless. Breeders believe that one part of the frame or function being increased causes other parts to decrease: they dislike great horns and great bones as so much flesh lost; in hornless breeds of cattle certain bones of the head become more developed: it is said that fat accumulating in one part checks its accumulation in another, and likewise checks the action of the udder. The whole organisation is so connected that it is probable there are many conditions determining the variation of each part, and causing other parts to vary with it; and man in making new races must be limited and ruled by all such laws.

In What Consists Domestication

In this chapter we have treated of variation under domestication, and it now remains to consider in what does this power of domestication consist, a subject of considerable difficulty. Observing that organic beings of almost every class, in all climates, countries, and times, have varied when long bred under domestication, we must conclude that the influence is of some very general nature.[2] Mr. Knight alone, as far as I know, has tried to define it; he believes it consists of an excess of food, together with transport to a more genial climate, or protection from its severities. I think we cannot admit this latter proposition, for we know how many vegetable products, aborigines of this country, here vary, when cultivated without any protection from the weather; and some of our variable trees, as apricots, peaches, have undoubtedly been derived from a more genial climate. There appears to be much more truth in the doctrine

[2] Note in the original: "Isidore G. St. Hilaire insists that breeding in captivity essential element. Schleiden on alkalies. What is it in domestication which causes variation?"

of excess of food being the cause, though I much doubt whether this is the sole cause, although it may well be requisite for the kind of variation desired by man, namely, increase of size and vigour. No doubt horticulturalists, when they wish to raise new seedlings, often pluck off all the flower-buds, except a few, or remove the whole during one season, so that a great stock of nutriment may be thrown into the flowers which are to seed. When plants are transported from high-lands, forests, marshes, heaths, into our gardens and greenhouses, there must be a considerable change of food, but it would be hard to prove that there was in every case an excess of the kind proper to the plant. If it be an excess of food, compared with that which the being obtained in its natural state,[3] the effects continue for an improbably long time; during how many ages has wheat been cultivated, and cattle and sheep reclaimed, and we cannot suppose their *amount* of food has gone on increasing, nevertheless these are amongst the most variable of our domestic productions. It has been remarked (Marshall) that some of the most highly kept breeds of sheep and cattle are truer or less variable than the straggling animals of the poor, which subsist on commons, and pick up a bare subsistence. In the case of forest-trees raised in nurseries, which vary more than the same trees do in their aboriginal forest, the cause would seem simply to lie in their not having to struggle against other trees and weeds, which in their natural state doubtless would limit the conditions of their existence. It appears to me that the power of domestication resolves itself into the accumulated effects of a change of all or some of the natural conditions of the life of the species, often associated with excess of food. These conditions moreover, I may add, can seldom remain, owing to the mutability of the affairs, migrations, and knowledge of man, for very long periods the same. I am the more inclined to come to this conclusion from finding, as we shall hereafter show, that changes of the natural conditions of existence seem peculiarly to affect the action of the reproductive system. As we see that hybrids and mongrels, after the first generation, are apt to

[3] Note in the original: "It appears that slight changes of condition are good for health; that more change affects the generative system, so that variation results in the offspring; that still more change checks or destroys fertility not of the offspring." What the meaning of "not of the offspring" may be is not clear.

vary much, we may at least conclude that variability does not altogether depend on excess of food.

After these views, it may be asked how it comes that certain animals and plants, which have been domesticated for a considerable length of time, and transported from very different conditions of existence, have not varied much, or scarcely at all; for instance, the ass, peacock, guinea-fowl, asparagus, Jerusalem artichoke. I have already said that probably different species, like different sub-varieties, possess different degrees of tendency to vary; but I am inclined to attribute in these cases the want of numerous races less to want of variability than to selection not having been practised on them. No one will take the pains to select without some corresponding object, either of use or amusement; the individuals raised must be tolerably numerous, and not so precious, but that he may freely destroy those not answering to his wishes. If guinea-fowls or peacocks [4] became "fancy" birds, I cannot doubt that after some generations several breeds would be raised. Asses have not been worked on from mere neglect; but they differ in *some* degree in different countries. The insensible selection, due to different races of mankind preserving those individuals most useful to them in their different circumstances, will apply only to the oldest and most widely domesticated animals. In the case of plants, we must put entirely out of the case those exclusively (or almost so) propagated by cuttings, layers, or tubers, such as the Jerusalem artichoke and laurel; and if we put on one side plants of little ornament or use, and those which are used at so early a period of their growth that no especial characters signify, as asparagus [5] and seakale, I can think of none long cultivated which have not varied. In no case ought we to expect to find as much variation in a race when it alone has been formed, as when several have been formed, for their crossing and recrossing will greatly increase their variability.

Summary of First Chapter

To sum up this chapter. Races are made under domestication: first, by the direct effects of the external conditions to which

[4] Note in the original: "There are white peacocks."
[5] Note in the original: "There are varieties of asparagus."

the species is exposed: secondly, by the indirect effects of the exposure to new conditions, often aided by excess of food, rendering the organisation plastic, and by man's selecting and separately breeding certain individuals, or introducing to his stock selected males, or often preserving with care the life of the individuals best adapted to his purposes: thirdly, by crossing and recrossing races already made, and selecting their offspring. After some generations man may relax his care in selection: for the tendency to vary and to revert to ancestral forms will decrease, so that he will have only occasionally to remove or destroy one of the yearly offspring which departs from its type. Ultimately, with a large stock, the effects of free crossing would keep, even without this care, his breed true. By these means man can produce infinitely numerous races, curiously adapted to ends, both most important and most frivolous; at the same time that the effects of the surrounding conditions, the laws of inheritance, of growth, and of variation, will modify and limit his labours.

CHAPTER II

On the Variation of Organic Beings in a Wild State; on the Natural Means of Selection; and on the Comparison of Domestic Races and True Species

Having treated of variation under domestication, we now come to it in a *state of nature*.

Most organic beings in a state of nature vary exceedingly little: I put out of the case variations (as stunted plants, etc., and sea-shells in brackish water) which are directly the effect of external agencies and which we do not *know are in the breed*,[1] or are *hereditary*. The amount of hereditary variation

[1] Note in the original: "Here discuss *what is a species,* sterility can most rarely be told when crossed. Descent from common stock."

is very difficult to ascertain, because naturalists (partly from the want of knowledge, and partly from the inherent difficulty of the subject) do not all agree whether certain forms are species or races.[2] Some strongly marked races of plants, comparable with the decided sports of horticulturalists, undoubtedly exist in a state of nature, as is actually known by experiment, for instance in the primrose and cowslip, in two so-called species of dandelion, in two of foxglove,[3] and I believe in some pines. Lamarck has observed that, as long as we confine our attention to one limited country, there is seldom much difficulty in deciding what forms to call species and what varieties; and that it is when collections flow in from all parts of the world that naturalists often feel at a loss to decide the limit of variation. Undoubtedly so it is, yet amongst British plants (and I may add land shells), which are probably better known than any in the world, the best naturalists differ very greatly in the relative proportions of what they call species and what varieties. In many genera of insects, and shells, and plants, it seems almost hopeless to establish which are which. In the higher classes there are less doubts; though we find considerable difficulty in ascertaining what deserve to be called species amongst foxes and wolves, and in some birds, for instance in the case of the white barn-owl. When specimens are brought from different parts of the world, how often do naturalists dispute this same question, as I found with respect to the birds brought from the Galapagos Islands. Yarrell has remarked that the individuals of the same undoubted species of birds, from Europe and North America, usually present slight, indefinable though perceptible differences. The recognition indeed of one animal by another of its kind seems to imply some difference. The disposition of wild animals undoubtedly differs. The variation, such as it is, chiefly affects the same parts in wild organisms as in domestic breeds; for instance, the

[2] Note in the original: "Give only rule: chain of intermediate forms, and *analogy;* this important. Every naturalist at first when he gets hold of new variable type is *quite puzzled* to know what to think species and what variations."

[3] Notes in original: "Compare feathered heads in very different birds with spines in *Echidna* and hedgehog. Plants under very different climate not varying. *Digitalis* shows jumps in variation, like *Laburnum* and *Orchis* case—in fact hostile cases. Variability of sexual characters alike in domestic and wild."

size, colour, and the external and less important parts. In many species the variability of certain organs or qualities is even stated as one of the specific characters; thus, in plants, colour, size, hairiness, the number of the stamens and pistils, and even their presence, the form of the leaves; the size and form of the mandibles of the males of some insects; the length and curvature of the beak in some birds (as in *Opetiorhynchus*) are variable characters in some species and quite fixed in others. I do not perceive that any just distinction can be drawn between this recognised variability of certain parts in many species and the more general variability of the whole frame in domestic races.

Although the amount of variation be exceedingly small in most organic beings in a state of nature, and probably quite wanting (as far as our senses serve) in the majority of cases; yet considering how many animals and plants, taken by mankind from different quarters of the world for the most diverse purposes, have varied under domestication in every country and in every age, I think we may safely conclude that all organic beings with few exceptions, if capable of being domesticated and bred for long periods, would vary. Domestication seems to resolve itself into a change from the natural conditions of the species (generally perhaps including an increase of food); if this be so, organisms in a state of nature must *occasionally,* in the course of ages, be exposed to analogous influences; for geology clearly shows that many places must, in the course of time, become exposed to the widest range of climatic and other influences; and if such places be isolated, so that new and better adapted organic beings cannot freely emigrate, the old inhabitants will be exposed to new influences, probably far more varied than man applies under the form of domestication. Although every species no doubt will soon breed up to the full number which the country will support, yet it is easy to conceive that on an average some species may receive an increase of food; for the times of dearth may be short, yet enough to kill, and recurrent only at long intervals. All such changes of conditions from geological causes would be exceedingly slow; what effect the slowness might have we are ignorant; under domestication it appears that the effects of change of conditions accumulate, and then break out. Whatever might be the result of these slow geological changes, we may feel sure, from the means of dissemination common in a lesser or greater degree to every organism taken conjointly with the changes of geology, which are steadily (and sometimes suddenly, as when an isthmus at last separates) in progress, that occasionally organisms must suddenly be introduced into new regions, where, if the conditions of existence are not so foreign as to cause its extermination, it will often be propagated under circumstances still more closely analogous to those of domestication; and therefore we expect will evince a tendency to vary. It appears to me quite *inexplicable* if this has never happened; but it can happen very rarely. Let us then suppose that an organism by some chance (which might be hardly repeated in a thousand years) arrives at a modern volcanic island in process of formation and not fully stocked with the most appropriate organisms; the new organism might readily gain a footing, although the external conditions were considerably different from its native ones. The effect of this we might expect would influence in some small degree the size, colour, nature of covering, etc., and from inexplicable influences even special parts and organs of the body. But we might further (and this is far more important) expect that the reproductive system would be affected, as under domesticity, and the structure of the offspring rendered in some degree plastic. Hence almost every part of the body would tend to vary from the typical form in slight degrees, and in no determinate way, and therefore *without selection* the free crossing of these small variations (together with the tendency to reversion to the original form) would constantly be counteracting this unsettling effect of the extraneous conditions on the reproductive system. Such, I conceive, would be the unimportant result without selection. And here I must observe that the foregoing remarks are equally applicable to that small and admitted amount of variation which has been observed in some organisms in a state of nature; as well as to the above hypothetical variation consequent on changes of condition.

Let us now suppose a Being with penetration sufficient to perceive differences in the outer and innermost organisation quite imperceptible to man, and with forethought extending over future centuries to watch with unerring care and select

for any object the offspring of an organism produced under the
foregoing circumstances; I can see no conceivable reason why
he could not form a new race (or several, were he to separate
the stock of the original organism and work on several islands)
adapted to new ends. As we assume his discrimination, and his
forethought, and his steadiness of object, to be incomparably
greater than those qualities in man, so we may suppose the
beauty and complications of the adaptations of the new races
and their differences from the original stock to be greater than
in the domestic races produced my man's agency: the groundwork of his labours we may aid by supposing that the external
conditions of the volcanic island, from its continued emergence
and the occasional introduction of new immigrants, vary; and
thus to act on the reproductive system of the organism, on
which he is at work, and so keep its organisation somewhat
plastic. With time enough, such a Being might rationally (without some unknown law opposed him) aim at almost any result.

For instance, let this imaginary Being wish, from seeing a
plant growing on the decaying matter in a forest and choked
by other plants, to give it power of growing on the rotten
stems of trees, he would commence selecting every seedling
whose berries were in the smallest degree more attractive to
tree-frequenting birds, so as to cause a proper dissemination of
the seeds, and at the same time he would select those plants
which had in the slightest degree more and more power of
drawing nutriment from rotten wood; and he would destroy
all other seedlings with less of this power. He might thus, in
the course of century after century, hope to make the plant by
degrees grow on rotten wood, even high up on trees, wherever
birds dropped the non-digested seeds. He might then, if the
organisation of the plant was plastic, attempt by continued selection of chance seedlings to make it grow on less and less
rotten wood, till it would grow on sound wood. Supposing
again, during these changes the plant failed to seed quite
freely from non-impregnation, he might begin selecting seedlings with a little sweeter, differently tasted honey or pollen,
to tempt insects to visit the flowers regularly: having effected
this, he might wish, if it profited the plant, to render abortive
the stamens and pistils in different flowers, which he could do
by continued selection. By such steps he might aim at making

a plant as wonderfully related to other organic beings as is the mistletoe, whose existence absolutely depends on certain insects for impregnation, certain birds for transportal, and certain trees for growth. Furthermore, if the insect which had been induced regularly to visit this hypothetical plant profited much by it, our same Being might wish by selection to modify by gradual selection the insect's structure, so as to facilitate its obtaining the honey or pollen: in this manner he might adapt the insect (always presupposing its organisation to be in some degree plastic) to the flower, and the impregnation of the flower to the insect; as is the case with many bees and many plants.

Seeing what blind capricious man has actually effected by selection during the few last years, and what in a ruder state he has probably effected without any systematic plan during the last few thousand years, he will be a bold person who will positively put limits to what the supposed Being could effect during whole geological periods. In accordance with the plan by which this universe seems governed by the Creator, let us consider whether there exists any *secondary* means in the economy of nature by which the process of selection could go on adapting, nicely and wonderfully, organisms, if in ever so small a degree plastic, to diverse ends. I believe such secondary means to exist.[4]

Natural Means of Selection

De Candolle, in an eloquent passage, has declared that all nature is at war, one organism with another, or with external nature. Seeing the contented face of nature, this may at first be well doubted; but reflection will inevitably prove it is too true. The war, however, is not constant, but only recurrent in a slight degree at short periods and more severely at occasional more distant periods; and hence its effects are easily overlooked. It is the doctrine of Malthus applied in most cases with ten-fold force. As in every climate there are seasons for

[4] Note in original: "The selection, in cases where adult lives only few hours as *Ephemera*, must fall on larva—curious speculation of the effect which changes in it would bring in parent."

each of its inhabitants of greater and less abundance, so all annually breed; and the moral restraint, which in some small degree checks the increase of mankind, is entirely lost. Even slow-breeding mankind has doubled in twenty-five years, and if he could increase his food with greater ease, he would double in less time. But for animals, without artificial means, *on an average* the amount of food for each species must be constant; whereas the increase of all organisms tends to be geometrical, and in a vast majority of cases at an enormous ratio. Suppose in a certain spot there are eight pairs of birds [robins], and that *only* four pairs of them annually (including double hatches) rear only four young; and that these go on rearing their young at the same rate: then at the end of seven years (a short life, excluding violent deaths, for any birds) there will be two thousand and forty-eight robins, instead of the original sixteen; as this increase is quite impossible, so we must conclude either that robins do not rear nearly half their young or that the average life of a robin when reared is from accident not nearly seven years. Both checks probably concur. The same kind of calculation applied to all vegetables and animals produces results either more or less striking, but in scarcely a single instance less striking than in man.

Many practical illustrations of this rapid tendency to increase are on record, namely, during peculiar seasons, in the extraordinary increase of certain animals, for instance during the years 1826 to 1828, in La Plata, when from drought, some millions of cattle perished, the whole country *swarmed* with innumerable mice: now I think it cannot be doubted that during the breeding season all the mice (with the exception of a few males or females in excess) ordinarily pair; and therefore that this astounding increase during three years must be attributed to a greater than usual number surviving the first year, and then breeding, and so on, till the third year, when their numbers were brought down to their usual limits on the return of wet weather. Where man has introduced plants and animals into a new country favourable to them, there are many accounts in how surprisingly few years the whole country has become stocked with them. This increase would necessarily stop as soon as the country was fully stocked; and yet we have every reason to believe from what is known of wild animals that *all* would pair in the spring. In the majority of cases it is most difficult to imagine where the check falls, generally no doubt on the seeds, eggs, and young; but when we remember how impossible even in mankind (so much better known than any other animal) it is to infer from repeated casual observations what the average of life is, or to discover how different the percentage of deaths to the births in different countries, we ought to feel no legitimate surprise at not seeing where the check falls in animals and plants. It should always be remembered that in most cases the checks are yearly recurrent in a small regular degree, and in an extreme degree during occasionally unusually cold, hot, dry, or wet years, according to the constitution of the being in question. Lighten any check in the smallest degree, and the geometrical power of increase in every organism will instantly increase the average numbers of the favoured species. Nature may be compared to a surface on which rest ten thousand sharp wedges touching each other and driven inwards by incessant blows. Fully to realise these views much reflection is requisite; Malthus on man should be studied; and all such cases as those of the mice in La Plata, of the cattle and horses when first turned out in South America, of the robins by our calculation, etc., should be well considered: reflect on the enormous multiplying power *inherent and annually in action* in all animals; reflect on the countless seed scattered by a hundred ingenious contrivances, year after year, over the whole face of the land; and yet we have every reason to suppose that the average percentage of every one of the inhabitants of a country will *ordinarily* remain constant. Finally, let it be borne in mind that this average number of individuals (the external conditions remaining the same) in each country is kept up by recurrent struggles against other species or against external nature (as on the borders of the arctic regions,[5] where the cold checks life); and that ordinarily each individual of each species holds its place either by its own struggle and capacity of acquiring nourishment in some period (from the egg upwards) of its life, or by the struggle of its parents (in short lived organisms,

[5] Note in the original: "In case like mistletoe, it may be asked why not more species, no other species interferes; answer almost sufficient, same causes which check the multiplication of individuals."

when the main check occurs at long intervals) against and compared with other individuals of the *same* or *different* species.

But let the external conditions of a country change; if in a small degree, the relative proportions of the inhabitants will in most cases simply be slightly changed; but let the number of inhabitants be small, as in an island, and free access to it from other countries be circumscribed; and let the change of condition continue progressing (forming new stations); in such case the original inhabitants must cease to be so perfectly adapted to the changed conditions as they originally were. It has been shown that probably such changes of external conditions would, from acting on the reproductive system, cause the organisation of the beings most affected to become, as under domestication, plastic. Now can it be doubted from the struggle each individual (or its parents) has to obtain subsistence that any minute variation in structure, habits, or instincts, adapting that individual better to the new conditions, would tell upon its vigour and health? In the struggle it would have a better *chance* of surviving, and those of its offspring which inherited the variation, let it be ever so slight, would have a better *chance* to survive. Yearly more are bred than can survive; the smallest grain in the balance, in the long run must tell on which death shall fall, and which shall survive. Let this work of selection, on the one hand, and death on the other, go on for a thousand generations; who would pretend to affirm that it would produce no effect, when we remember what in a few years Bakewell effected in cattle and Western in sheep, by this identical principle of selection.

To give an imaginary example, from changes in progress on an island, let the organisation of a canine animal become slightly plastic, which animal preyed chiefly on rabbits, but sometimes on hares; let these same changes cause the number of rabbits very slowly to decrease and the number of hares to increase; the effect of this would be that the fox or dog would be driven to try to catch more hares, and his numbers would tend to decrease; his organisation, however, being slightly plastic, those individuals with the lightest forms, longest limbs, and best eyesight (though perhaps with less cunning or scent) would be slightly favoured, let the difference be ever so small.

and would tend to live longer and to survive during that time of the year when food was shortest; they would also rear more young, which young would tend to inherit these slight peculiarities. The less fleet ones would be rigidly destroyed. I can see no more reason to doubt but that these causes in a thousand generations would produce a marked effect, and adapt the form of the fox to catching hares instead of rabbits, than that greyhounds can be improved by selection and careful breeding. So would it be with plants under similar circumstances; if the number of individuals of a species with plumed seeds could be increased by greater powers of dissemination within its own area (that is if the check to increase fell chiefly on the seeds), those seeds which were provided with ever so little more down, or with a plume placed so as to be slightly more acted on by the winds, would in the long run tend to be most disseminated; and hence a greater number of seeds thus formed would germinate, and would tend to produce plants inheriting this slightly better adapted down.

Besides this natural means of selection, by which those individuals are preserved, whether in their egg or seed or in their mature state, which are best adapted to the place they fill in nature, there is a second agency at work in most bisexual animals tending to produce the same effect, namely the struggle of the males for the females. These struggles are generally decided by the law of battle; but in the case of birds, apparently, by the charms of their song, by their beauty or their power of courtship, as in the dancing rock-thrush of Guiana. Even in the animals which pair there seems to be an excess of males which would aid in causing a struggle: in the polygamous animals,[6] however, as in deer, oxen, poultry, we might expect there would be severest struggle: is it not in the polygamous animals that the males are best formed for mutual war? The most vigorous males, implying perfect adaptation, must generally gain the victory in their several contests. This kind of selection, however, is less rigorous than the other; it does not require the death of the less successful, but gives to them fewer descendants. This struggle falls, moreover, at a time of year when food is generally abundant, and perhaps the effect chiefly produced would be the alteration of sexual characters, and the

* Note in original: "Seals– Pennant about battles of seals."

selection of individual forms, no way related to their power
of obtaining food, or of defending themselves from their natural enemies, but of fighting one with another. This natural
struggle amongst the males may be compared in effect, but in
a less degree, to that produced by those agriculturalists who
pay less attention to the careful selection of all the young animals which they breed and more to the occasional use of a
choice male.

Differences Between "Races" and "Species": First, in Their Trueness or Variability

Races produced by these natural means of selection [7] we may
expect would differ in some respects from those produced by
man. Man selects chiefly by the eye, and is not able to perceive the course of every vessel and nerve, or the form of the
bones, or whether the internal structure corresponds to the
outside shape. He is unable to select shades of constitutional
differences, and by the protection he affords and his endeavours to keep his property alive, in whatever country he lives,
he checks, as much as lies in his power, the selecting action of
nature which will, however, go on to a lesser degree with all
living things, even if their length of life is not determined by
their own powers of endurance. He has bad judgment, is capricious, he does not, or his successors do not, wish to select
for the same exact end for hundreds of generations. He cannot
always suit the selected form to the properest conditions; nor
does he keep those conditions uniform: he selects that which is
useful to him, not that best adapted to those conditions in
which each variety is placed by him: he selects a small dog,
but feeds it highly; he selects a long-backed dog, but does not
exercise it in any peculiar manner, at least not during every
generation. He seldom allows the most vigorous males to struggle for themselves and propagate, but picks out such as he possesses, or such as he prefers, and not necessarily those best

[7] The last 23 lines of p. 108 and 14 lines of p. 109 are, in the MS.,
marked through in pencil with vertical lines, beginning at "Races produced, etc.," and ending with "to these conditions."

adapted to the existing conditions. Every agriculturalist and breeder knows how difficult it is to prevent an occasional cross with another breed. He often grudges to destroy an individual which departs considerably from the required type. He often begins his selection by a form or sport considerably departing from the parent form. Very differently does the natural law of selection act; the varieties selected differ only slightly from the parent forms; the conditions are constant for long periods and change slowly; rarely can there be a cross; the selection is rigid and unfailing, and continued through many generations; a selection can *never be made* without the form be *better* adapted to the conditions than the parent form; the selecting power goes on without caprice, and steadily, for thousands of years adapting the form to these conditions. The selecting power is not deceived by external appearances, it tries the being during its whole life; and if less well adapted than its *congeners,* without fail it is destroyed; every part of its structure is thus scrutinised and proved good towards the place in nature which it occupies.

We have every reason to believe that in proportion to the number of generations that a domestic race is kept free from crosses, and to the care employed in continued steady selection with one end in view, and to the care in not placing the variety in conditions unsuited to it; in such proportion does the new race become "true" or subject to little variation. How incomparably "truer" then would a race produced by the above rigid, steady, natural means of selection, excellently trained and perfectly adapted to its conditions, free from stains of blood or crosses, and continued during thousands of years, be compared with one produced by the feeble, capricious, mis-directed, and ill-adapted selection of man. Those races of domestic animals produced by savages, partly by the inevitable conditions of their life and partly unintentionally by their greater care of the individuals most valuable to them, would probably approach closest to the character of a species; and I believe this is the case. Now the characteristic mark of a species, next, if not equal in importance to its sterility when crossed with another species, and indeed almost the only other character (without we beg the question and affirm the essence of a species is its

not having descended from a parent common to any other form)
is the similarity of the individuals composing the species, or
in the language of agriculturalists their "trueness."

Difference Between "Races" and "Species" in Fertility When Crossed

The sterility of species, or of their offspring, when crossed has,
however, received more attention than the uniformity in character of the individuals composing the species. It is exceedingly
natural that such sterility [8] should have been long thought the
certain characteristic of species. For it is obvious that if the
allied different forms which we meet with in the same country
could cross together, instead of finding a number of distinct species, we should have a confused and blending series. The fact
however of a perfect gradation in the degree of sterility between
species, and the circumstance of some species most closely allied (for instance many species of crocus and European heaths)
refusing to breed together, whereas other species, widely different, and even belonging to distinct genera, as the fowl and
the peacock, pheasant and grouse, *Azalea* and *Rhododendron,*
Thuja and *Juniperus,* breeding together ought to have caused
a doubt whether the sterility did not depend on other causes,
distinct from a law, coincident with their creation. I may here
remark that the fact whether one species will or will not breed
with another is far less important than the sterility of the offspring when produced; for even some domestic races differ so
greatly in size (as the great stag-greyhound and lap-dog, or
cart-horse and Burmese ponies) that union is nearly impossible; and what is less generally known is, that in plants Kölreuter has shown by hundreds of experiments that the pollen
of one species will fecundate the germen of another species,
whereas the pollen of this latter will never act on the germen
of the former; so that the simple fact of mutual impregnation
certainly has no relation whatever to the distinctness in crea-

[8] Note in the original: "If domestic animals are descended from several species and *become* fertile *inter se,* then one can see they gain fertility by becoming adapted to new conditions and certainly domestic
animals can withstand changes of climate without loss of fertility in an
astonishing manner."

tion of the two forms. When two species are attempted to be crossed which are so distantly allied that offspring are never produced, it has been observed in some cases that the pollen commences its proper action by exserting its tube, and the germen commences swelling, though soon afterwards it decays. In the next stage in the series, hybrid offspring are produced though only rarely and few in number, and these are absolutely sterile: then we have hybrid offspring more numerous, and occasionally, though very rarely, breeding with either parent, as is the case with the common mule. Again, other hybrids, though infertile *inter se,* will breed *quite* freely with either parent, or with a third species, and will yield offspring generally infertile, but sometimes fertile; and these latter again will breed with either parent, or with a third or fourth species: thus Kölreuter blended together many forms. Lastly it is now admitted by those botanists who have longest contended against the admission that in certain families the hybrid offspring of many of the species are sometimes perfectly fertile in the first generation when bred together: indeed in some few cases Mr. Herbert found that the hybrids were decidedly more fertile than either of their pure parents. There is no way to escape from the admission that the hybrids from some species of plants are fertile, except by declaring that no form shall be considered as a species if it produces with another species fertile offspring: but this is begging the question. It has often been stated that different species of animals have a sexual repugnance towards each other; I can find no evidence of this; it appears as if they merely did not excite each other's passions. I do not believe that in this respect there is any essential distinction between animals and plants; and in the latter there cannot be a feeling of repugnance.

Causes of Sterility in Hybrids

The difference in nature between species which causes the greater or lesser degree of sterility in their offspring appears, according to Herbert and Kölreuter, to be connected much less with external form, size or structure than with constitutional peculiarities; by which is meant their adaptation to different climates, food and situation, etc.: these peculiarities of constitution probably affect the entire frame, and no one part in particular.[9]

From the foregoing facts I think we must admit that there exists a perfect gradation in fertility between species which when crossed are quite fertile (as in *Rhododendron, Calceolaria,* etc.), and indeed in an extraordinary degree fertile (as in *Crinum*), and those species which never produce offspring, but which by certain effects (as the exsertion of the pollen-tube) evidence their alliance. Hence, I conceive, we must give up sterility, although undoubtedly in a lesser or greater degree of very frequent occurrence, as an unfailing mark by which species can be distinguished from races, i.e. from those forms which have descended from a common stock.

Infertility from Causes Distinct from Hybridisation

Let us see whether there are any analogous facts which will throw any light on this subject, and will tend to explain why the offspring of certain species when crossed should be sterile, and not others, without requiring a distinct law connected with their creation to that effect. Great numbers, probably a large majority of animals when caught by man and removed from their natural conditions, although taken very young, rendered quite tame, living to a good old age, and apparently quite healthy, seem incapable under these circumstances of breeding.[10] I do not refer to animals kept in menageries, such as at the Zoological Gardens, many of which, however, appear healthy and live long and unite but do not produce; but to animals caught and left partly at liberty in their native country. Rengger enumerates several caught young and rendered tame, which he kept in Paraguay, and which would not breed:

[9] Note in the original: "Yet this seems introductory to the case of the heaths and crocuses above mentioned."

[10] Note in original: "Animals seem more often made sterile by being taken out of their native condition than plants, and so are more sterile when crossed.

"We have one broad fact that sterility in hybrids is not closely related to external difference, and these are what man alone gets by selection."

the hunting leopard or cheetah and elephant offer other instances; as do bears in Europe, and the twenty-five species of hawks, belonging to different genera, thousands of which have been kept for hawking and have lived for long periods in perfect vigour. When the expense and trouble of procuring a succession of young animals in a wild state be borne in mind, one may feel sure that no trouble has been spared in endeavours to make them breed. So clearly marked is this difference in different kinds of animals, when captured by man, that St. Hilaire makes two great classes of animals useful to man: the *tame,* which will not breed, and the *domestic,* which will breed in domestication. From certain singular facts we might have supposed that the non-breeding of animals was owing to some perversion of instinct. But we meet with exactly the same class of facts in plants: I do not refer to the large number of cases where the climate does not permit the seed or fruit to ripen, but where the flowers do not "set" owing to some imperfection of the ovule or pollen. The latter, which alone can be distinctly examined, is often manifestly imperfect, as anyone with a microscope can observe by comparing the pollen of the Persian and Chinese lilacs with the common lilac; the two former species (I may add) are equally sterile in Italy as in this country. Many of the American bog plants here produce little or no pollen, whilst the Indian species of the same genera freely produce it. Lindley observes that sterility is the bane of the horticulturist: Linnaeus has remarked on the sterility of nearly all alpine flowers when cultivated in a lowland district. Perhaps the immense class of double flowers chiefly owe their structure to an excess of food acting on parts rendered slightly sterile and less capable of performing their true function, and therefore liable to be rendered monstrous, which monstrosity, like any other disease, is inherited and rendered common. So far from domestication being in itself unfavourable to fertility, it is well known that when an organism is once capable of submission to such conditions its fertility is increased beyond the natural limit. According to agriculturists, slight changes of conditions, that is of food or habitation, and likewise crosses with races slightly different, increase the vigour and probably the fertility of their offspring. It would appear also that even a great change of condition, for instance, transportal from temperate countries to

India, in many cases does not in the least affect fertility, although it does health and length of life and the period of maturity. When sterility is induced by domestication it is of the same kind, and varies in degree, exactly as with hybrids: for be it remembered that the most sterile hybrid is no way monstrous; its organs are perfect, but they do not act, and minute microscopical investigations show that they are in the same state as those of pure species in the intervals of the breeding season. The defective pollen in the cases above alluded to precisely resembles that of hybrids. The occasional breeding of hybrids, as of the common mule, may be aptly compared to the most rare but occasional reproduction of elephants in captivity. The cause of many exotic geraniums producing (although in vigorous health) imperfect pollen seems to be connected with the period when water is given them; but in the far greater majority of cases we cannot form any conjecture on what exact cause the sterility of organisms taken from their natural conditions depends. Why, for instance, the cheetah will not breed whilst the common cat and ferret (the latter generally kept shut up in a small box) do; why the elephant will not while the pig will abundantly; why the partridge and grouse in their own country will not, whilst several species of pheasants, the guinea-fowl from the deserts of Africa and the peacock from the jungles of India, will. We must, however, feel convinced that it depends on some constitutional peculiarities in these beings not suited to their new condition; though not necessarily causing an ill state of health. Ought we then to wonder much that those hybrids which have been produced by the crossing of species with different constitutional tendencies (which tendencies we know to be eminently inheritable) should be sterile: it does not seem improbable that the cross from an alpine and lowland plant should have its constitutional powers deranged, in nearly the same manner as when the parent alpine plant is brought into a lowland district. Analogy, however, is a deceitful guide, and it would be rash to affirm, although it may appear probable, that the sterility of hybrids is due to the constitutional peculiarities of one parent being disturbed by being blended with those of the other parent in exactly the same manner as it is caused in some organic beings when placed by man out of their natural conditions. Although this would be rash, it would, I think, be still more rash, seeing that sterility is no more incidental to *all* cross-bred productions than it is to all organic beings when captured by man, to assert that the sterility of certain hybrids proved a distinct creation of their parents.

But it may be objected (however little the sterility of certain hybrids is connected with the distinct creations of species), how comes it, if species are only races produced by natural selection, that when crossed they so frequently produce sterile offspring, whereas in the offspring of those races confessedly produced by the arts of man there is no one instance of sterility. There is not much difficulty in this, for the races produced by the natural means above explained will be slowly but steadily selected; will be adapted to various and diverse conditions, and to these conditions they will be rigidly confined for immense periods of time; hence we may suppose that they would acquire different constitutional peculiarities adapted to the stations they occupy; and on the constitutional differences between species their sterility, according to the best authorities, depends. On the other hand man selects by external appearance [11]; from his ignorance, and from not having any test at least comparable in delicacy to the natural struggle for food, continued at intervals through the life of each individual, he cannot eliminate fine shades of constitution, dependent on invisible differences in the fluids or solids of the body; again, from the value which he attaches to each individual, he asserts his utmost power in contravening the natural tendency of the most vigorous to survive. Man, moreover, especially in the earlier ages, cannot have kept his conditions of life constant, and in later ages his stock pure. Until man selects two varieties from the same stock, adapted to two climates or to other different external conditions, and confines such rigidly for one or several thousand years to such conditions, always selecting the individuals best adapted to them, he cannot be said to have even commenced the experiment. Moreover, the organic beings which man has longest had under domestication have been those which were of the greatest use to him, and one chief element of their use-

[11] It is not clear where these notes in original were meant to go: "Mere difference of structure no guide to what will or will not cross. First step gained by races keeping apart."

fulness, especially in the earlier ages, must have been their capacity to undergo sudden transportals into various climates, and at the same time to retain their fertility, which in itself implies that in such respects their constitutional peculiarities were not closely limited. If the opinion already mentioned be correct, that most of the domestic animals in their present state have descended from the fertile commixture of wild races or species, we have indeed little reason now to expect infertility between any cross of stock thus descended.

It is worthy of remark, that as many organic beings, when taken by man out of their natural conditions, have their reproductive system so affected as to be incapable of propagation, so, we saw in the first chapter, that although organic beings when taken by man do propagate freely, their offspring after some generations vary or sport to a degree which can only be explained by their reproductive system being in some way affected. Again, when species cross, their offspring are generally sterile; but it was found by Kölreuter that when hybrids are capable of breeding with either parent, or with other species, that their offspring are subject after some generations to excessive variation. Agriculturists, also, affirm that the offspring from mongrels, after the first generation, vary much. Hence we see that both sterility and variation in the succeeding generations are consequent both on the removal of individual species from their natural states and on species crossing. The connexion between these facts may be accidental, but they certainly appear to elucidate and support each other—on the principle of the reproductive system of all organic beings being eminently sensitive to any disturbance, whether from removal or commixture, in their constitutional relations to the conditions to which they are exposed.

Points of Resemblance Between "Races" and "Species"

Races and reputed species agree in some respects, although differing from causes which, we have seen, we can in some degree understand, in the fertility and "trueness" of their offspring. In the first place, there is no clear sign by which to distinguish races from species, as is evident from the great difficulty experienced by naturalists in attempting to discriminate them. As far as external characters are concerned, many of the races which are descended from the same stock differ far more than true species of the same genus; look at the willow-wrens, some of which skilful ornithologists can hardly distinguish from each other except by their nests; look at the wild swans, and compare the distinct species of these genera with the races of domestic ducks, poultry, and pigeons; and so again with plants, compare the cabbages, almonds, peaches, and nectarines, etc., with the species of many genera. St. Hilaire has even remarked that there is a greater difference in size between races, as in dogs (for he believes all have descended from one stock), than between the species of any one genus; nor is this surprising, considering that amount of food and consequently of growth is the element of change over which man has most power. I may refer to a former statement, that breeders believe the growth of one part or strong action of one function causes a decrease in other parts; for this seems in some degree analogous to the law of "organic compensation," which many naturalists believe holds good. To give an instance of this law of compensation—those species of carnivora which have the canine teeth greatly developed have certain molar teeth deficient; or again, in that division of the crustaceans in which the tail is much developed, the thorax is little so, and the converse. The points of difference between different races are often strikingly analogous to those between species of the same genus: trifling spots or marks of colour [12] (as the bars on pigeon's wings) are often preserved in races of plants and animals, precisely in the same manner as similar trifling characters often pervade all the species of a genus, and even of a family. Flowers in varying their colours often become veined and spotted and the leaves become divided like true species: it is known that the varieties of the same plant never have red, blue, and yellow flowers, though the hyacinth makes a very

[12] Note in original: "Boitard and Corbié on outer edging red in tail of bird—so bars on wing, white or black or brown, or white edged with black or . . . : analogous to marks running through genera but with different colours. Tail coloured in pigeons."

near approach to an exception; [13] and different species of the same genus seldom, though sometimes they have flowers of these three colours. Dun-coloured horses having a dark stripe down their backs, and certain domestic asses having transverse bars on their legs, afford striking examples of a variation analogous in character to the distinctive marks of other species of the same genus.

External Characters of Hybrids and Mongrels

There is, however, as it appears to me, a more important method of comparison between species and races, namely the character of the offspring when species are crossed and when races are crossed: I believe, in no one respect, except in sterility, is there any difference. It would, I think, be a marvellous fact, if species have been formed by distinct acts of creation, that they should act upon each other in uniting, like races descended from a common stock. In the first place, by repeated crossing one species can absorb and wholly obliterate the characters of another, or of several other species, in the same manner as one race will absorb by crossing another race. Marvellous, that one act of creation should absorb another or even several acts of creation! The offspring of species, that is, hybrids, and the offspring of races, that is, mongrels, resemble each other in being either intermediate in character (as is most frequent in hybrids) or in resembling sometimes closely one and sometimes the other parent; in both the offspring produced by the same act of conception sometimes differ in their degree of resemblance; both hybrids and mongrels sometimes retain a certain part or organ very like that of either parent, both as we have seen, become in succeeding generations variable; and this tendency to vary can be transmitted by both; in both for many generations there is a strong tendency to reversion to their ancestral form. In the case of a hybrid laburnum and of a supposed mongrel vine different parts of the same plants took after each of their two parents. In the hybrids from some species, and in the mongrel of some races, the offspring differ according

---

[13] Note in original: *"Oxalis* and gentian."

as which of the two species, or of the two races, is the father (as in the common mule and hinny) and which the mother. Some races will breed together which differ so greatly in size that the dam often perishes in labour: so it is with some species when crossed; when the dam of one species has borne offspring to the male of another species, her succeeding offspring are sometimes stained (as in Lord Morton's mare by the quagga, wonderful as the fact is) by this first cross; so agriculturists positively affirm is the case when a pig or sheep of one breed has produced offspring by the sire of another breed.

Summary of Second Chapter

Let us sum up this second chapter. If slight variations do occur in organic beings in a state of nature; if changes of condition from geological causes do produce in the course of ages effects analogous to those of domestication on any, however few, organisms; and how can we doubt it, from what is actually known, and from what may be presumed, since thousands of organisms taken by man for sundry uses, and placed in new conditions, have varied; if such variations tend to be hereditary; and how can we doubt it, when we see shades of expression, peculiar manners, monstrosities of the strangest kinds, diseases, and a multitude of other peculiarities, which characterise and form, being inherited, the endless races (there are twelve hundred kinds of cabbages) of our domestic plants and animals; if we admit that every organism maintains its place by an almost periodically recurrent struggle; and how can we doubt it, when we know that all beings tend to increase in a geometrical ratio (as is instantly seen when the conditions become for a time more favourable), whereas on an average the amount of food must remain constant; if so, there will be a natural means of selection, tending to preserve those individuals with any slight deviations of structure more favourable to the then existing conditions, and tending to destroy any with deviations of an opposite nature. If the above propositions be correct, and there be no law of nature limiting the possible amount of variation, new races of beings will—perhaps only rarely, and only in some few districts—be formed.

Limits of Variation

That a limit to variation does exist in nature is assumed by
most authors, though I am unable to discover a single fact on
which this belief is grounded. One of the commonest statements
is that plants do not become acclimatised; and I have even observed that kinds not raised by seed, but propagated by cuttings, etc., are instanced. A good instance has, however, been
advanced in the case of kidney beans, which it is believed are
now as tender as when first introduced. Even if we overlook
the frequent introduction of seed from warmer countries, let
me observe that as long as the seeds are gathered promiscu-
ously from the bed, without continual observation and *careful*
selection of those plants which have stood the climate best
during their whole growth, the experiment of acclimatisation
has hardly been begun. Are not all those plants and animals,
of which we have the greatest number of races, the oldest domesticated? Considering the quite recent progress [14] of systematic agriculture and horticulture, is it not opposed to every fact,
that we have exhausted the capacity of variation in our cattle
and in our corn, even if we have done so in some trivial points,
as their fatness or kind of wool? Will anyone say, that if horticulture continues to flourish during the next few centuries,
we shall not have numerous new kinds of the potato and dahlia?
But take two varieties of each of these plants, and adapt them
to certain fixed conditions and prevent any cross for five thousand years, and then again vary their conditions; try many climates and situations; and who will predict the number and
degrees of difference which might arise from these stocks? I
repeat that we know nothing of any limit to the possible amount
of variation, and therefore to the number and differences of
the races, which might be produced by the natural means of
selection, so infinitely more efficient than the agency of man.
Races thus produced would probably be very "true"; and if
from having been adapted to different conditions of existence,
they possessed different constitutions, if suddenly removed to

[14] Note in original: "History of pigeons shows increase of peculiarities
during last years."

some new station, they would perhaps be sterile and their off-spring would perhaps be infertile. Such races would be indistinguishable from species. But is there any evidence that the species, which surround us on all sides, have been thus produced? This is a question which an examination of the economy of nature we might expect would answer either in the affirmative or negative.[15]

CHAPTER III

On the Variation of Instincts and Other Mental Attributes under Domestication and in State of Nature; on the Difficulties in This Subject; and on Analogous Difficulties with Respect to Corporeal Structures

Variation of Mental Attributes under Domestication

I have as yet only alluded to the mental qualities which differ greatly in different species. Let me here premise that, as will be seen in the second part, there is no evidence and consequently no attempt to show that *all* existing organisms have descended from any one common parent-stock, but that only those have so descended which, in the language of naturalists, are clearly related to each other. Hence the facts and reasoning advanced in this chapter do not apply to the first origin of the senses, or of the chief mental attributes, such as of memory, attention, reasoning, etc., by which most or all of the great related groups are characterised, any more than they apply to the first origin of life, or growth, or the power of reproduction. The application of such facts as I have collected is merely to the differences of the primary mental qualities and of the instincts in the species of the several great groups. In domestic animals every observer has remarked in how great a

[15] Note in original: "Certainly ought to be here introduced, viz., difficulty in forming such organ, as eye, by selection."

degree, in the individuals of the same species, the dispositions, namely courage, pertinacity, suspicion, restlessness, confidence, temper, pugnaciousness, affection, care of their young, sagacity, etc., vary. It would require a most able metaphysician to explain how many primary qualities of the mind must be changed to cause these diversities of complex dispositions. From these dispositions being inherited, of which the testimony is unanimous, families and breeds arise, varying in these respects. I may instance the good and ill temper of different stocks of bees and of horses, the pugnacity and courage of game fowls, the pertinacity of certain dogs, as bull-dogs, and the sagacity of others, for restlessness and suspicion compare a wild rabbit reared with the greatest care from its earliest age with the extreme tameness of the domestic breed of the same animal. The offspring of the domestic dogs which have run wild in Cuba,[1] though caught quite young, are most difficult to tame, probably nearly as much so as the original parent-stock from which the domestic dog descended. The habitual *"periods"* of different families of the same species differ, for instance, in the time of year of reproduction, and the period of life when the capacity is acquired, and the hour of roosting (in Malay fowls), etc. These periodical habits are perhaps essentially corporeal, and may be compared to nearly similar habits in plants, which are known to vary extremely. Consensual movements (as called by Müller) vary and are inherited— such as the cantering and ambling paces in horses, the tumbling of pigeons, and perhaps the hand-writing, which is sometimes so similar between father and sons, may be ranked in this class. *Manners,* and even tricks which perhaps are only *peculiar* manners, according to W. Hunter and my father, are distinctly inherited in cases where children have lost their parent in early infancy. The inheritance of expression, which often reveals the finest shades of character, is familiar to everyone.

Again the tastes and pleasures of different breeds vary; thus the shepherd-dog delights in chasing the sheep, but has no wish to kill them—the terrier (see Knight) delights in killing vermin, and the spaniel in finding game. But it is impossible to separate their mental peculiarities in the way I have done:

[1] In the margin occurs the name of Poeppig.

the tumbling of pigeons, which I have instanced as a consensual movement, might be called a trick and is associated with a taste for flying in a close flock at a great height. Certain breeds of fowls have a taste for roosting in trees. The different actions of pointers and setters might have been adduced in the same class, as might the peculiar *manner* of hunting of the spaniel. Even in the same breed of dogs, namely in fox-hounds, it is the fixed opinion of those best able to judge that the different pups are born with different tendencies; some are best to find their fox in the cover; some are apt to run straggling, some are best to make casts and to recover the lost scent, etc.; and that these peculiarities undoubtedly are transmitted to their progeny. Or again the tendency to point might be adduced as a distinct habit which has become inherited—as might the tendency of a true sheep dog (as I have been assured is the case) to run round the flock instead of directly at them, as is the case with other young dogs when attempted to be taught. The "transandantes" sheep [2] in Spain, which for some centuries have been yearly taken a journey of several hundred miles from one province to another, know when the time comes, and show the greatest restlessness (like migratory birds in confinement), and are prevented with difficulty from starting by themselves, which they sometimes do, and find their own way. There is a case on good evidence [3] of a sheep which, when she lambed, would return across a mountainous country to her own birthplace, although at other times of year not of a rambling disposition. Her lambs inherited this same disposition, and would go to produce their young on the farm whence their parent came; and so troublesome was this habit that the whole family was destroyed.

These facts must lead to the conviction, justly wonderful as it is, that almost infinitely numerous shades of disposition, of tastes, of peculiar movements, and even of individual actions, can be modified or acquired by one individual and transmitted to its offspring. One is forced to admit that mental phenomena (no doubt through their intimate connexion with the brain) can be inherited, like infinitely numerous and fine differences of corporeal structure. In the same manner as peculiarities of

---

[2] Note in original: "Several authors."
[3] In the margin "Hogg" occurs as authority for this fact.

corporeal structure slowly acquired or lost during mature life (especially cognizant in disease), as well as congenital peculiarities, are transmitted: so it appears to be with the mind. The inherited paces in the horse have no doubt been acquired by compulsion during the lives of the parents: and temper and tameness may be modified in a breed by the treatment which the individuals receive. Knowing that a pig has been taught to point, one would suppose that this quality in pointer-dogs was the simple result of habit, but some facts, with respect to the occasional appearance of a similar quality in other dogs, would make one suspect that it originally appeared in a less perfect degree, *"by chance,"* that is, from a congenital tendency in the parent of the breed of pointers. One cannot believe that the tumbling, and high flight in a compact body, of one breed of pigeons has been taught; and in the case of the slight differences in the manner of hunting in young fox-hounds, they are doubtless congenital. The inheritance of the foregoing and similar mental phenomena ought perhaps to create less surprise, from the reflection that in no case do individual acts of reasoning, or movements, or other phenomena connected with consciousness, appear to be transmitted. An action, even a very complicated one, when from long practice it is performed unconsciously without any effort (and indeed in the case of many peculiarities of manners opposed to the will) is said, according to a common expression, to be performed "instinctively." Those cases of languages, and of songs, learnt in early childhood and *quite* forgotten, being *perfectly* repeated during the unconsciousness of illness, appear to me only a few degrees less wonderful than if they had been transmitted to a second generation.

Hereditary Habits Compared with Instincts

The chief characteristics of true instincts appear to be their invariability and non-improvement during the mature age of the individual animal: the absence of knowledge of the end, for which the action is performed, being associated, however, sometimes with a degree of reason; being subject to mistakes and being associated with certain states of the body or times of the year or day. In most of these respects there is a resemblance in the above detailed cases of the mental qualities acquired or modified during domestication. No doubt the instincts of wild animals are more uniform than those habits or qualities modified or recently acquired under domestication, in the same manner and from the same causes that the corporeal structure in this state is less uniform than in beings in their natural conditions. I have seen a young pointer point as fixedly, the first day it was taken out, as any old dog; Magendie says this was the case with a retriever which he himself reared: the tumbling of pigeons is not probably improved by age: we have seen in the case above given that the young sheep inherited the migratory tendency to their particular birth-place the first time they lambed. This last fact offers an instance of a domestic instinct being associated with a state of body; as do the "transandantes" sheep with a time of year. Ordinarily the acquired instincts of domestic animals seem to require a certain degree of education (as generally in pointers and retrievers) to be perfectly developed: perhaps this holds good amongst wild animals in rather a greater degree than is generally supposed; for instance, in the singing of birds, and in the knowledge of proper herbs in ruminants. It seems pretty clear that bees transmit knowledge from generation to generation. Lord Brougham insists strongly on ignorance of the end proposed being eminently characteristic of true instincts; and this appears to me to apply to many acquired hereditary habits; for instance, in the case of the young pointer alluded to before, which pointed so steadfastly the first day that we were obliged several times to carry him away. This puppy not only pointed at sheep, at large white stones, and at every little bird, but likewise "backed" the other pointers: this young dog must have been as unconscious for what end he was pointing, namely, to facilitate his master's killing game to eat, as is a butterfly which lays her eggs on a cabbage, that her caterpillars would eat the leaves. So a horse that ambles instinctively manifestly is ignorant that he performs that peculiar pace for the ease of man; and if man had never existed, he would never have ambled. The young pointer pointing at white stones appears to be as much a mistake of its acquired instinct, as in the case of flesh-flies laying their eggs on certain flowers instead of putrefying meat. However true the ignorance of the end may generally be, one sees that instincts are associated with some degree of reason; for instance, in the case of the tailor-bird, who spins threads with which to make her nest yet will use artificial threads when she can procure them; so it has been known that an old pointer has broken his point and gone round a hedge to drive out a bird towards his master.[4]

There is one other quite distinct method by which the instincts or habits acquired under domestication may be compared with those given by nature, by a test of a fundamental kind: I mean the comparison of the mental powers of mongrels and hybrids. Now the instincts, or habits, tastes, and dispositions of one *breed* of animals, when crossed with another breed, for instance a shepherd-dog with a harrier, are blended and appear in the same curiously mixed degree, both in the first and succeeding generations, exactly as happens when one *species* is crossed with another. This would hardly be the case if there was any fundamental difference between the domestic and natural instinct [5]; if the former were, to use a metaphorical expression, merely superficial.

Variation in the Mental Attributes of Wild Animals

With respect to the variation of the mental powers of animals in a wild state, we know that there is a considerable difference

[4] In the margin is written "Retriever killing one bird." This refers to the cases given in the *Descent of Man,* 2nd ed. (in one vol.), p. 78, of a retriever being puzzled how to deal with a wounded and a dead bird, killed the former and carried both at once. This was the only known instance of her wilfully injuring game.

[5] Note in original: "Give some definition of instinct, or at least give chief attributes. The term instinct is often used in a sense which implies no more than that the animal does the action in question. Faculties and instincts may I think be imperfectly separated. The mole has the faculty of scratching burrows, and the instinct to apply it. The bird of passage has the faculty of finding its way and the instinct to put it in action at certain periods. It can hardly be said to have the faculty of knowing the time, for it can possess no means, without indeed it be some consciousness of passing sensations. Think over all habitual actions and see whether faculties and instincts can be separated. We have faculty of waking in the night, if an instinct prompted us to do something at certain hour of night or day. Savages finding their way. Wrangle's account—probably a faculty inexplicable by the possessor. There are besides faculties '*means,*' as conversion of larvae into neuters and queens. I think all this generally implied, anyhow useful."

in the disposition of different individuals of the same species, as is recognised by all those who have had the charge of animals in a menagerie. With respect to the wildness of animals, that is fear directed particularly against man, which appears to be as true an instinct as the dread of a young mouse of a cat, we have excellent evidence that it is slowly acquired and becomes hereditary. It is also certain that, in a natural state, individuals of the same species lose or do not practise their migratory instincts—as woodcocks in Madeira. With respect to any variation in the more complicated instincts, it is obviously most difficult to detect, even more so than in the case of corporeal structure, of which it has been admitted the variation is exceedingly small, and perhaps scarcely any in the majority of species at any one period. Yet, to take one excellent case of instinct, namely the nests of birds, those who have paid most attention to the subject maintain that not only certain individuals (species?) seem to be able to build very imperfectly, but that a difference in skill may not unfrequently be detected between individuals.[6] Certain birds, moreover, adapt their nests to circumstances; the water-ouzel makes no vault when she builds under cover of a rock—the sparrow builds very differently when its nest is in a tree or in a hole, and the golden-crested wren sometimes suspends its nest below and sometimes places it *on* the branches of trees.

Principles of Selection Applicable to Instincts

As the instincts of a species are fully as important to its preservation and multiplication as its corporeal structure, it is evident that if there be the slightest congenital differences in the instincts and habits, or if certain individuals during their lives are induced or compelled to vary their habits, and if such differences are in the smallest degree more favourable, under slightly modified external conditions, to their preservation, such individuals must in the long run have a better *chance* of being preserved and of multiplying. If this be admitted, a series of small changes may, as in the case of corporeal structure,

[6] This sentence agrees with the MS., but is clearly in need of correction.

work great changes in the mental powers, habits, and instincts of any species.

Difficulties in the Acquirement of Complex Instincts by Selection

Everyone will at first be inclined to explain (as I did for a long time) that many of the more complicated and wonderful instincts could not be acquired in the manner here supposed. The second part of this work is devoted to the general consideration of how far the general economy of nature justifies or opposes the belief that related species and genera are descended from common stocks; but we may here consider whether the instincts of animals offer such a *prima facie* case of impossibility of gradual acquirement, as to justify the rejection of any such theory, however strongly it may be supported by other facts. I beg to repeat that I wish here to consider not the *probability* but the *possibility* of complicated instincts having been acquired by the slow and long-continued selection of very slight (either congenital or produced by habit) modifications of foregoing simpler instincts; each modification being as useful and necessary, to the species practising it, as the most complicated kind.

First, to take the case of birds'-nests; of existing species (almost infinitely few in comparison with the multitude which must have existed, since the period of the New Red Sandstone of North America, of whose habits we must always remain ignorant) a tolerably perfect series could be made from eggs laid on the bare ground, to others with a few sticks just laid round them, to a simple nest like the wood-pigeon's, to others more and more complicated: now if, as is asserted, there occasionally exist slight differences in the building powers of an individual, and if, which is at least probable, such differences would tend to be inherited, then we can see that it is at least *possible* that the nidificatory instincts may have been acquired by the gradual selection, during thousands and thousands of generations, of the eggs and young of those individuals whose nests were in some degree better adapted to the preservation of their young, under the then existing conditions. One of the most surprising instincts on record is that of the Australian bush-turkey, whose eggs are hatched by the heat generated from a huge pile of fermenting materials, which it heaps together: but here the habits of an allied species show how this instinct *might possibly* have been acquired. This second species inhabits a tropical district, where the heat of the sun is sufficient to hatch its eggs; this bird, burying its eggs, apparently for concealment, under a lesser heap of rubbish, but of a dry nature, so as not to ferment. Now suppose this bird to range slowly into a climate which was cooler, and where leaves were more abundant, in that case, those individuals, which chanced to have their collecting instinct strongest developed, would make a somewhat larger pile, and the eggs, aided during some colder season, under the slightly cooler climate by the heat of incipient fermentation, would in the long run be more freely hatched and would probably produce young ones with the same more highly developed collecting tendencies; of these again, those with the best developed powers would again tend to rear most young. Thus this strange instinct might *possibly* be acquired, every individual bird being as ignorant of the laws of fermentation, and the consequent development of heat, as we know they must be.

Secondly, to take the case of animals' feigning death (as it is commonly expressed) to escape danger. In the case of insects, a perfect series can be shown, from some insects, which momentarily stand still, to others which for a second slightly contract their legs, to others which will remain immovably drawn together for a quarter of an hour, and may be torn asunder or roasted at a slow fire, without evincing the smallest sign of sensation. No one will doubt that the length of time, during which each remains immovable, is well adapted to escape the dangers to which it is most exposed, and few will deny the *possibility* of the change from one degree to another, by the means and at the rate already explained. Thinking it, however, wonderful (though not impossible) that the attitude of death should have been acquired by methods which imply no imitation, I compared several species, when feigning, as is said, death, with others of the same species really dead, and their attitudes were in no one case the same.

Thirdly, in considering many instincts it is useful to *endeavour* to separate the faculty by which they perform it, and the mental power which urges to the performance, which is more properly called an instinct. We have an instinct to eat, we have jaws, etc., to give us the faculty to do so. These faculties are often unknown to us: bats, with their eyes destroyed, can avoid strings suspended across a room; we know not at present by what faculty they do this. Thus also, with migratory birds, it is a wonderful instinct which urges them at certain times of the year to direct their course in certain directions, but it is a faculty by which they know the time and find their way. With respect to time,[7] man without seeing the sun can judge to a certain extent of the hour, as must those cattle which come down from the inland mountains to feed on seaweed left bare at the changing hour of low-water.[8] A hawk (d'Orbigny) seems certainly to have acquired a knowledge of a period of every twenty-one days. In the cases already given of the sheep which travelled to their birth-place to cast their lambs, and the sheep in Spain which know their time of march, we may conjecture that the tendency to move is associated, we may then call it instinctively, with some corporeal sensations. With respect to direction we can easily conceive how a tendency to travel in a certain course may possibly have been acquired, although we must remain ignorant how birds are able to preserve any direction whatever in a dark night over the wide ocean. I may observe that the power of some savage races of mankind to find their way, although perhaps wholly different from the faculty of birds, is nearly as unintelligible to us. Bellinghausen, a skilful navigator, describes with the utmost wonder the manner in which some Esquimaux guided him to a certain point, by a course never straight, through newly formed hummocks of ice, on a thick foggy day, when he with a compass found it impossible, from having no landmarks, and from their course being so extremely crooked, to preserve any sort of uniform direction: so it is with Australian savages in thick forests. In North and South America many birds slowly travel northward and southward, urged on by the food they find as the seasons change; let them continue to do this, till, as in

[7] Note in the original in an unknown handwriting: "At the time when corn was pitched in the market instead of sold by sample, the geese in the town fields of Newcastle used to know market day and come in to pick up the corn spilt."
[8] Note in original: "MacCulloch and others."

the case of the sheep in Spain, it has become an urgent instinctive desire, and they will gradually accelerate their journey. They would cross narrow rivers, and if these were converted by subsidence into narrow estuaries, and gradually during centuries to arms of the sea, still we may suppose their restless desire of travelling onwards would impel them to cross such an arm, even if it had become of great width beyond their span of vision. How they are able to preserve a course in any direction, I have said, is a faculty unknown to us. To give another illustration of the means by which I conceive it *possible* that the direction of migrations have been determined: Elk and reindeer in North America annually cross, as if they could marvellously smell or see at the distance of a hundred miles, a wide tract of absolute desert, to arrive at certain islands where there is a scanty supply of food; the changes of temperature, which geology proclaims, render it probable that this desert tract formerly supported some vegetation, and thus these quadrupeds might have been annually led on, till they reached the more fertile spots, and so acquired, like the sheep of Spain, their migratory powers.

Fourthly, with respect to the combs of the hive-bee; here again we must look to some faculty or means by which they make their hexagonal cells, without indeed we view these instincts as mere machines. At present such a faculty is quite unknown: Mr. Waterhouse supposes that several bees are led by their instinct to excavate a mass of wax to a certain thinness, and that the result of this is that hexagons necessarily remain. Whether this or some other theory be true, some such means they must possess. They abound, however, with true instincts, which are the most wonderful that are known. If we examine the little that is known concerning the habits of other species of bees, we find much simpler instincts: the humble bee merely fills rude balls of wax with honey and aggregates them together with little order in a rough nest of grass. If we knew the instinct of all the bees, which ever had existed, it is not improbable that we should have instincts of every degree of complexity, from actions as simple as a bird making a nest, and rearing her young, to the wonderful architecture and government of the hive-bee; at least such is *possible,* which is all that I am here considering.

Finally, I will briefly consider under the same point of view one other class of instincts, which have often been advanced as truly wonderful, namely, parents bringing food to their young which they themselves neither like nor partake of; for instance, the common sparrow, a granivorous bird, feeding its young with caterpillars. We might of course look into the case still earlier, and seek how an instinct in the parent, of feeding its young at all, was first derived; but it is useless to waste time in conjectures on a series of gradations from the young feeding themselves and being slightly and occasionally assisted in their search, to their entire food being brought to them. With respect to the parent bringing a different kind of food from its own kind, we may suppose either that the remote stock, whence the sparrow and other congenerous birds have descended, was insectivorous, and that its own habits and structure have been changed, whilst its ancient instincts with respect to its young have remained unchanged; or we may suppose that the parents have been induced to vary slightly the food of their young, by a slight scarcity of the proper kind (or by the instincts of some individuals not being so truly developed), and in this case those young which were most capable of surviving were necessarily most often preserved, and would themselves in time become parents, and would be similarly compelled to alter their food for their young. In the case of those animals, the young of which feed themselves, changes in their instincts for food, and in their structure, might be selected from slight variations, just as in mature animals. Again, where the food of the young depends on where the mother places her eggs, as in the case of the caterpillars of the cabbage-butterfly, we may suppose that the parent stock of the species deposited her eggs sometimes on one kind and sometimes on another of congenerous plants (as some species now do), and if the cabbage suited the caterpillars better than any other plant, the caterpillars of those butterflies, which had chosen the cabbage, would be most plentifully reared, and would produce butterflies more apt to lay their eggs on the cabbage than on the other congenerous plants.

However vague and unphilosophical these conjectures may appear, they serve, I think, to show that one's first impulse utterly to reject any theory whatever implying a gradual acquirement of these instincts, which for ages have excited man's admiration, may at least be delayed. Once grant that dispositions, tastes, actions, or habits can be slightly modified, either by slight congenital differences (we must suppose in the brain) or by the force of external circumstances, and that such slight modifications can be rendered inheritable—a proposition which no one can reject—and it will be difficult to put any limit to the complexity and wonder of the tastes and habits which may *possibly* be thus acquired.

Difficulties in the Acquirement by Selection of Complex Corporeal Structures

After the past discussion it will perhaps be convenient here to consider whether any particular corporeal organs, or the entire structure of any animals, are so wonderful as to justify the rejection *prima facie* of our theory. In the case of the eye, as with the more complicated instincts, no doubt one's first impulse is to utterly reject every such theory. But if the eye from its most complicated form can be shown to graduate into an exceedingly simple state—if selection can produce the smallest change, and if such a series exists, then it is clear (for in this work we have nothing to do with the first origin of organs in their simplest forms) that it may *possibly* have been acquired by gradual selection of slight, but in each case useful, deviations, and that each eye throughout the animal kingdom is not only most useful, but *perfect* for its possessor. Every naturalist, when he meets with any new and singular organ, always expects to find, and looks for, other and simpler modifications of it in other beings. In the case of the eye, we have a multitude of different forms, more or less simple, not graduating into each other, but separated by sudden gaps or intervals; but we must recollect how incomparably greater would the multitude of visual structures be if we had the eyes of every fossil which ever existed. We shall discuss the probable vast proportion of the extinct to the recent in the succeeding part. Notwithstanding the large series of existing forms, it is most difficult even to conjecture by what intermediate stages very many simple organs could possibly have graduated into complex ones: but it should be here borne in mind, that a part having originally a wholly different function may on the theory of gradual selection be slowly worked into quite another use; the gradations of forms, from which naturalists believe in the hypothetical metamorphosis of part of the ear into the swimming bladder in fishes, and in insects of legs into jaws, show the manner in which this is possible. As under domestication, modifications of structure take place, without any continued selection, which man finds very useful, or valuable for curiosity (as the hooked calyx of the teazle, or the ruff round some pigeons' necks), so in a state of nature some small modifications, apparently beautifully adapted to certain ends, may perhaps be produced from the accidents of the reproductive system, and be at once propagated without long-continued selection of small deviations towards that structure. In conjecturing by what stages any complicated organ in a species may have arrived at its present state, although we may look to the analogous organs in other existing species, we should do this merely to aid and guide our imaginations; for to know the real stages we must look only through one line of species, to one ancient stock, from which the species in question has descended. In considering the eye of a quadruped, for instance, though we may look at the eye of a molluscous animal or of an insect, as a proof how simple an organ will serve some of the ends of vision; and at the eye of a fish as a nearer guide of the manner of simplification; we must remember that it is a mere chance (assuming for a moment the truth of our theory) if any existing organic being has preserved any one organ, in exactly the same condition, as it existed in the ancient species at remote geological periods.

The nature or condition of certain structures has been thought by some naturalists to be of no use to the possessor, but to have been formed wholly for the good of other species; thus certain fruit and seeds have been thought to have been made nutritious for certain animals—numbers of insects, especially in their larval state, to exist for the same end—certain fish to be bright coloured to aid certain birds of prey in catching them, etc. Now could this be proved (which I am far from admitting) the theory of natural selection would be quite overthrown; for it is evident that selection depending on the advantage over others of one individual with some slight deviation would never produce a structure or quality profitable only to another species. No doubt one being takes advantage of qualities in another, and may even cause its extermination; but this is far from proving that this quality was produced for such an end. It may be advantageous to a plant to have its seed attractive to animals, if one out of a hundred or a thousand escapes being digested, and thus aids dissemination: the bright colours of a fish may be of some advantage to it, or more probably may result from exposure to certain conditions in favourable haunts for food, *notwithstanding* it becomes subject to be caught more easily by certain birds.

If instead of looking, as above, at certain individual organs, in order to speculate on the stages by which their parts have been matured and selected, we consider an individual animal, we meet with the same or greater difficulty but which, I believe, as in the case of single organs, rests entirely on our ignorance. It may be asked by what intermediate forms could, for instance, a bat possibly have passed; but the same question might have been asked with respect to the seal, if we had not been familiar with the otter and other semi-aquatic carnivorous quadrupeds. But in the case of the bat, who can say what might have been the habits of some parent form with less developed wings, when we now have insectivorous opossums and herbivorous squirrels fitted for merely gliding through the air.[9] One species of bat is at present partly aquatic in its habits.[10] Woodpeckers and tree-frogs are especially adapted, as their names express, for climbing trees; yet we have species of both inhabiting the open plains of La Plata, where a tree does not exist. I might argue from this circumstance that a structure eminently fitted for climbing trees might descend from forms inhabiting a country where a tree did not exist. Notwithstanding these and a multitude of other well-known facts, it has been maintained by several authors that one species, for instance of the carnivorous order, could not pass into another, for instance into an otter, because in its transitional state its habits would not be adapted to any proper conditions of life; but the jaguar is a thoroughly terrestrial quadruped in its structure, yet it takes freely to the

[9] Note in original: "No one will dispute that the gliding is most useful, probably necessary for the species in question."
[10] Note in original: "Is this the *Galeopithecus?* I forget."

water and catches many fish; will it be said that it is *impossible* that the conditions of its country might become such that the jaguar should be driven to feed more on fish than they now do; and in that case is it impossible, is it not probable, that any the slightest deviation in its instincts, its form of body, in the width of its feet, and in the extension of the skin (which already unites the base of its toes) would give such individuals a better *chance* of surviving and propagating young with similar, barely perceptible (though thoroughly exercised) deviations? [11] Who will say what could thus be effected in the course of ten thousand generations? Who can answer the same question with respect to instincts? If no one can, the *possibility* (for we are not in this chapter considering the *probability*) of simple organs or organic beings being modified by natural selection and the effects of external agencies into complicated ones ought not to be absolutely rejected.

[11] Note in original: "See Richardson a far better case of a polecat animal, which half-year is aquatic."

PART II:

On the Evidence Favourable and
Opposed to the View That Species
Are Naturally Formed Races,
Descended from Common Stocks

CHAPTER IV

On the Number of Intermediate Forms Required on the Theory of Common Descent: and on Their Absence in a Fossil State

I must here premise that, according to the view ordinarily received, the myriads of organisms which have during past and present times peopled this world have been created by so many distinct acts of creation. It is impossible to reason concerning the will of the Creator, and therefore, according to this view, we can see no cause why or why not the individual organism should have been created on any fixed scheme. That all the organisms of this world have been produced on a scheme is certain from their general affinities; and if this scheme can be shown to be the same with that which would result from allied organic beings descending from common stocks, it becomes highly improbable that they have been separately created by individual acts of the will of a Creator. For as well might it be said that, although the planets move in courses conformably to the law of gravity, yet we ought to attribute the course of each planet to the individual act of the will of the Creator. It is in every case more conformable with what we know of the government of this earth that the Creator should have imposed only general laws. As long as no method was known by which races could become exquisitely adapted to various ends, whilst the existence of species was thought to be proved by the sterility of their offspring, it was allowable to attribute each organism to an individual act of creation. But in the two former chapters it has (I think) been shown that the production, under existing conditions, of exquisitely adapted species is at least *possible*. Is there then any direct evidence in favour or against this view? I believe that the geographical distribution of organic beings in past and present times, the kind of affinity linking them together, their so-called "metamorphic" and "abortive" organs, appear in favour of this view. On the other hand, the imperfect evidence of the continuousness of the organic series, which, we shall immediately see, is required on our theory, is against it; and is the most weighty objection. The evidence, however, even on this point, as far as it goes, is favourable; and considering the imperfection of our knowledge, especially with respect to past ages, it would be surprising if evidence drawn from such sources were not also imperfect.

As I suppose that species have been formed in an analogous manner with the varieties of the domesticated animals and plants, so must there have existed intermediate forms between all the species of the same group, not differing more than recognised varieties differ. It must not be supposed necessary that there should have existed forms exactly intermediate in character between any two species of a genus, or even between any two varieties of a species; but it is necessary that there should have existed every intermediate form between the one species or variety of the common parent, and likewise between the second species or variety, and this same common parent. Thus it does not necessarily follow that there ever has existed series of intermediate sub-varieties (differing no more than the occasional seedlings from the same seed-capsule), between broccoli and common red cabbage; but it is certain that there has existed between broccoli and the wild parent cabbage a series of such intermediate seedlings, and again between red cabbage and the wild parent cabbage: so that the broccoli and red cabbage are linked together, but not *necessarily* by directly intermediate forms. It is of course possible that there *may* have been directly intermediate forms, for the broccoli may have long since descended from a common red cabbage, and this from the wild cabbage. So, on my theory, it must have been with species of the same genus. Still more must the supposition be avoided that there has necessarily ever existed (though one *may* have descended from the other) directly intermediate forms between any two genera or families—for instance between the genus *Sus* and the tapir; although it is necessary that intermediate forms (not differing more than the varieties of our domestic animals) should have existed between *Sus* and some unknown parent form, and tapir with this same parent form. The latter may have differed more from *Sus* and tapir than these two genera now differ from each other. In this sense, according to our theory, there has been a gradual passage (the steps not being wider apart than our domestic varieties) between the species of the same genus, between genera of the same family, and between families of the same order, and so on, as far as facts, hereafter to be given, lead us; and the number of forms which must have at former periods existed, thus to make good this passage between different species, genera, and families, must have been almost infinitely great.

What evidence is there of a number of intermediate forms having existed, making a passage in the above sense, between the species of the same groups? Some naturalists have supposed that if every fossil which now lies entombed, together with all existing species, were collected together, a perfect series in every great class would be formed. Considering the enormous number of species requisite to effect this, especially in the above sense of the forms not being *directly* intermediate between the existing species and genera, but only intermediate by being linked through a common but often widely different ancestor, I think this supposition highly improbable. I am however far from underrating the probable number of fossilised species: no one who has attended to the wonderful progress of palaeontology during the last few years will doubt that we as yet have found only an exceedingly small fraction of the species buried in the crust of the earth. Although the almost infinitely numerous intermediate forms in no one class may have been preserved, it does not follow that they have not existed. The fossils which have been discovered, it is important to remark, do tend, the little way they go, to make good the series; for as observed by Buckland they all fall into or between existing groups. Moreover, those that fall between our existing groups fall in, according to the manner required by our theory, for they do not directly connect two existing species of different groups, but they connect the groups themselves: thus the Pachydermata and Ruminantia are now separated by several characters, for instance the Pachydermata [1] have both a tibia and fibula, whilst Ruminantia have only a tibia; now the fossil *Macrauchenia* has a leg bone exactly intermediate in this respect, and likewise has some other intermediate characters. But the *Macrauchenia* does not connect any one species of Pachydermata with some one other of Ruminantia but it shows that these two groups have at one time been less widely divided. So have fish and reptiles been at one time more closely connected in some points than they now are. Generally in those groups in which there has been most change, the more ancient the fossil, if not identical with recent, the more often it falls between existing groups, or into small existing groups which now lie between other large existing groups. Cases like the foregoing, of which there are many, form steps, though few and far between, in a series of the kind required by my theory.

As I have admitted the high improbability that if every fossil were disinterred they would compose in each of the divisions of nature a perfect series of the kind required, consequently I freely admit that if those geologists are in the right who consider the lowest known formation as contemporaneous with the first appearances of life, or the several formations as at all closely consecutive; or any one formation as containing a nearly perfect record of the organisms which existed during the whole period of its deposition in that quarter of the globe; if such propositions are to be accepted, my theory must be abandoned.

If the Palaeozoic system is really contemporaneous with the first appearance of life, my theory must be abandoned, both inasmuch as it limits *from shortness of time* the total

---

[1] The following sentence in the margin appears to refer to pachyderms and ruminants: "There can be no doubt, if we banish all fossils, existing groups stand more separate." The following occurs between the lines: "The earliest forms would be such as others could radiate from."

number of forms which can have existed on this world, and because the organisms, as fish, mollusca,[2] and star-fish found in its lower beds, cannot be considered as the parent forms of all the successive species in these classes. But no one has yet overturned the arguments of Hutton and Lyell that the lowest formations known to us are only those which have escaped being metamorphosed . . . ; if we argued from some considerable districts, we might have supposed that even the Cretaceous system was that in which life first appeared. From the number of distant points, however, in which the Silurian system has been found to be the lowest, and not always metamorphosed, there are some objections to Hutton's and Lyell's view; but we must not forget that the now existing land forms only one-fifth part of the superficies of the globe, and that this fraction is only imperfectly known. With respect to the fewness of the organisms found in the Silurian and other Palaeozoic formations, there is less difficulty, inasmuch as (besides their gradual obliteration) we can expect formations of this vast antiquity to escape entire denudation only when they have been accumulated over a wide area, and have been subsequently protected by vast superimposed deposits: now this could generally only hold good with deposits accumulating in a wide and deep ocean, and therefore unfavourable to the presence of many living things. A mere narrow and not very thick strip of matter, deposited along a coast where organisms most abound, would have no chance of escaping denudation and being preserved to the present time from such immensely distant ages.

If the several known formations are at all nearly consecutive in time, and preserve a fair record of the organisms which have existed, my theory must be abandoned. But when we consider the great changes in mineralogical nature and texture between successive formations, what vast and entire changes in the geography of the surrounding countries must generally have been effected, thus wholly to have changed the nature of the deposits on the same area. What time such changes must have required! Moreover, how often has it not been found

---

[2] Pencil insertion by the author: "The parent-forms of Mollusca would probably differ greatly from all recent—it is not directly that any one division of Mollusca would descend from first time unaltered, whilst others had become metamorphosed from it."

that between two conformable and apparently immediately successive deposits a vast pile of water-worn matter is interpolated in an adjoining district. We have no means of conjecturing in many cases how long a period [3] has elapsed between successive formations, for the species are often wholly different: as remarked by Lyell, in some cases probably as long a period has elapsed between formations as the whole Tertiary system, itself broken by wide gaps.

Consult the writings of anyone who has particularly attended to any one stage in the Tertiary system (and indeed of every system) and see how deeply impressed he is with the time required for its accumulation. Reflect on years elapsed in many cases since the latest beds containing only living species have been formed; see what Jordan Smith says of the twenty thousand years since the last bed, which is above the boulder formation in Scotland, has been upraised; or of the far longer period since the recent beds of Sweden have been upraised four hundred feet, what an enormous period the boulder formation must have required, and yet how insignificant are the records (although there has been plenty of elevation to bring up submarine deposits) of the shells, which we know existed at that time. Think, then, over the entire length of the Tertiary epoch, and think over the probable length of the intervals separating the secondary deposits. Of these deposits, moreover, those consisting of sand and pebbles have seldom been favourable, either to the embedment or to the preservation of fossils.

Nor can it be admitted as probable that any one Secondary formation contains a fair record even of those organisms which are most easily preserved, namely, hard marine bodies. In how many cases have we not certain evidence that between the deposition of apparently closely consecutive beds the lower one existed for an unknown time as land, covered with trees. Some of the Secondary formations which contain most marine remains appear to have been formed in a wide and not deep sea, and therefore only those marine animals which live in such situations would be preserved.[4] In all cases, on indented

---

[3] Note in original: "Reflect on coming in of the chalk, extending from Iceland to the Crimea."

[4] Note in original: "Neither highest or lowest fish (i.e. *Myxine* or *Lepidosiren*) could be preserved in intelligible condition in fossils."

rocky coasts, or any other coast, where sediment is not accumulating, although often highly favourable to marine animals, none can be embedded: where pure sand and pebbles
are accumulating few or none will be preserved. I may here
instance the great western line of the South American coast,
tenanted by many peculiar animals, of which none probably
will be preserved to a distant epoch. From these causes, and
especially from such deposits as are formed along a line of
coast, steep above and below water, being necessarily of little
width, and therefore more likely to be subsequently denuded
and worn away, we can see why it is improbable that our Secondary deposits contain a fair record of the marine fauna of
any one period. The East Indian Archipelago offers an area,
as large as most of our Secondary deposits, in which there are
wide and shallow seas, teeming with marine animals, and in
which sediment is accumulating; now supposing that all the
hard marine animals, or rather those having hard parts to preserve, were preserved to a future age, excepting those which
lived on rocky shores where no sediment or only sand and
gravel were accumulating, and excepting those embedded along
the steeper coasts, where only a narrow fringe of sediment was
accumulating, supposing all this, how poor a notion would a
person at a future age have of the marine fauna of the present
day. Lyell has compared the geological series to a work of
which only the few latter but not consecutive chapters have
been preserved; and out of which, it may be added, very many
leaves have been torn, the remaining ones only illustrating a
scanty portion of the fauna of each period. On this view, the
records of anteceding ages confirm my theory; on any other
they destroy it.

Finally, if we narrow the question into, why do we not find
in some instances every intermediate form between any two
species?, the answer may well be that the average duration of
each specific form (as we have good reason to believe) is immense in years, and that the transition could, according to my
theory, be effected only by numberless small gradations; and
therefore that we should require for this end a most perfect
record, which the foregoing reasoning teaches us not to expect.
It might be thought that in a vertical section of great thickness in the same formation some of the species ought to be

found to vary in the upper and lower parts, but it may be doubted whether any formation has gone on accumulating without any break for a period as long as the duration of a species; and if it had done so, we should require a series of specimens from every part. How rare must be the chance of sediment accumulating for some twenty or thirty thousand years on the same spot, with the bottom subsiding, so that a proper depth might be preserved for any one species to continue living: what an amount of subsidence would be thus required, and this subsidence must not destroy the source whence the sediment continued to be derived. In the case of terrestrial animals, what chance is there when the present time is become a pleistocene formation (at an earlier period than this, sufficient elevation to expose marine beds could not be expected), what chance is there that future geologists will make out the innumerable transitional sub-varieties, through which the short-horned and long-horned cattle (so different in shape of body) have been derived from the same parent stock? Yet this transition has been effected in the *same country,* and in a far *shorter time,* than would be probable in a wild state, both contingencies highly favourable for the future hypothetical geologists being enabled to trace the variation.

CHAPTER V

Gradual Appearance and Disappearance of Species

In the Tertiary system, in the last uplifted beds, we find all the species recent and living in the immediate vicinity; in rather older beds we find only recent species, but some not living in the immediate vicinity; we then find beds with two or three or a few more extinct or very rare species; then considerably more extinct species, but with gaps in the regular increase; and finally we have beds with only two or three or not one living species. Most geologists believe that the gaps in the percentage, that is the sudden increments, in the number of the extinct species in the stages of the Tertiary system are due to the imperfection of the geological record. Hence we are led to believe that the species in the Tertiary system have been gradually introduced; and from analogy to carry on the same view to the Secondary formations. In these latter, however, entire groups of species generally come in abruptly; but this would naturally result, if, as argued in the foregoing chapter, these Secondary deposits are separated by wide epochs. Moreover it is important to observe that, with our increase of knowledge, the gaps between the older formations become fewer and smaller; geologists of a few years' standing remember how beautifully has the Devonian system [1] come in between the Carboniferous and Silurian formations. I need hardly observe that the slow and gradual appearance of new forms follows from our theory, for to form a new species, an old one must not only be plastic in its organisation, becoming so probably from changes in the conditions of its existence, but a place in the natural economy of the district must [be made,] come to exist, for the selection of some new modification of its structure, better fitted to the surrounding conditions than are the other individuals of the same or other species.[2]

In the Tertiary system the same facts, which make us admit as probable that new species have slowly appeared, lead to the admission that old ones have slowly disappeared, not several together, but one after another; and by analogy one is induced to extend this belief to the Secondary and Palaeozoic epochs. In some cases, as the subsidence of a flat country, or the breaking or the joining of an isthmus, and the sudden inroad of many new and destructive species, extinction might be locally sudden. The view entertained by many geologists, that each fauna of each Secondary epoch has been suddenly de-

---

[1] In the margin the author has written "Lonsdale." This refers to W. Lonsdale's paper, "Notes on the age of the Limestone of South Devonshire," *Geol. Soc. Trans.*, p. 721, Series 2, Vol. v (1840). According to Mr. H. B. Woodward (*History of the Geological Society of London* [1907], p. 107), "Lonsdale's 'important and original suggestion of the existence of an intermediary type of Palaeozoic fossils, since called Devonian,' led to a change which was then 'the greatest ever made at one time in the classification of our English formations.'" Mr. Woodward's quotations are from Murchison and Buckland.

[2] Note in original: "Better begin with this. If species really, after catastrophes, created in showers over world, my theory false."

stroyed over the whole world, so that no succession could be left for the production of new forms, is subversive of my theory, but I see no grounds whatever to admit such a view. On the contrary, the law, which has been made out, with reference to distinct epochs, by independent observers, namely, that the wider the geographical range of a species the longer is its duration in time, seems entirely opposed to any universal extermination.[3] The fact of species of mammiferous animals and fish being renewed at a quicker rate than mollusca, though both aquatic; and of these the terrestrial genera being renewed quicker than the marine; and the marine mollusca being again renewed quicker than the Infusorial animalcula, all seem to show that the extinction and renewal of species does not depend on general catastrophes, but on the particular relations of the several classes to the conditions to which they are exposed.[4]

Some authors seem to consider the fact of a few species' having survived [5] amidst a number of extinct forms (as is the case with a tortoise and a crocodile out of the vast number of extinct sub-Himalayan fossils) as strongly opposed to the view of species being mutable. No doubt this would be the case, if it were presupposed with Lamarck that there was some inherent tendency to change and development in all species, for which supposition I see no evidence. As we see some species at present adapted to a wide range of conditions, so we may suppose that such species would survive unchanged and unexterminated for a long time; time generally being from geological causes a correlative of changing conditions. How at present one species becomes adapted to a wide range, and another species to a restricted range of conditions, is of difficult explanation.

Extinction of Species

The extinction of the larger quadrupeds, of which we imagine we better know the conditions of existence, has been thought

[3] Opposite to this passage the author has written "d'Archiac, Forbes, Lyell."
[4] The author gives as authorities the names of Lyell, Forbes, and Ehrenberg.
[5] The author gives Falconer as his authority.

little less wonderful than the appearance of new species; and has, I think, chiefly led to the belief of universal catastrophes. When considering the wonderful disappearance within a late period, whilst recent shells were living, of the numerous great and small mammifers of South America, one is strongly induced to join with the catastrophists. I believe, however, that very erroneous views are held on this subject. As far as is historically known, the disappearance of species from any one country has been slow—the species becoming rarer and rarer, locally extinct, and finally lost. It may be objected that this has been effected by man's direct agency, or by his indirect agency in altering the state of the country; in this latter case, however, it would be difficult to draw any just distinction between his agency and natural agencies. But we now know in the later Tertiary deposits that shells become rarer and rarer in the successive beds, and finally disappear: it has happened, also, that shells common in a fossil state, and thought to have been extinct, have been found to be still living species, but very *rare* ones. If the rule is that organisms become extinct by becoming rarer and rarer, we ought not to view their extinction, even in the case of the larger quadrupeds, as anything wonderful and out of the common course of events. For no naturalist thinks it wonderful that one species of a genus should be rare and another abundant, notwithstanding he be quite incapable of explaining the causes of the comparative rareness. Why is one species of willow-wren or hawk or woodpecker common in England, and another extremely rare: why at the Cape of Good Hope is one species of *Rhinoceros* or antelope far more abundant than other species? Why again is the same species much more abundant in one district of a country than in another district? No doubt there are in each case good causes: but they are unknown and unperceived by us. May we not then safely infer that as certain causes are acting *unperceived* around us, and are making one species to be common and another exceedingly rare, that they might equally well cause the final extinction of some species without being perceived by us? We should always bear in mind that there is a recurrent struggle for life in every organism, and that in every country a destroying agency is always counteracting the geometrical tendency to increase in every species; and yet without our being able to tell

with certainty at what period of life, or at what period of the
year, the destruction falls the heaviest. Ought we then to expect
to trace the steps by which this destroying power, always at
work and scarcely perceived by us, becomes increased, and yet
if it continues to increase ever so slowly (without the fertility
of the species in question be likewise increased) the average
number of the individuals of that species must decrease, and become finally lost. I may give a single instance of a check causing
local extermination which might long have escaped discovery;
the horse, though swarming in a wild state in La Plata, and likewise under apparently the most unfavourable conditions in the
scorched and alternately flooded plains of Caraccas, will not
in a wild state extend beyond a certain degree of latitude into
the intermediate country of Paraguay; this is owing to a certain fly depositing its eggs on the navels of the foals: as, however, man with a *little* care can rear horses in a tame state
abundantly in Paraguay, the problem of its extinction is probably complicated by the greater exposure of the wild horse
to occasional famine from the droughts, to the attacks of the
jaguar and other such evils. In the Falkland Islands the check
to the *increase* of the wild horse is said to be loss of the sucking
foals, from the stallions compelling the mares to travel across
bogs and rocks in search of food: if the pasture on these islands
decreased a little, the horse, perhaps, would cease to exist in a
wild state, not from the absolute want of food, but from the
impatience of the stallions urging the mares to travel whilst the
foals were too young.

From our more intimate acquaintance with domestic animals, we cannot conceive their extinction without some glaring
agency; we forget that they would undoubtedly in a state of
nature (where other animals are ready to fill up their place) be
acted on in some part of their lives by a destroying agency,
keeping their numbers on an average constant. If the common
ox was known only as a wild South African species, we should
feel no surprise at hearing that it was a very rare species; and
this rarity would be a stage towards its extinction. Even in man,
so infinitely better known than any other inhabitant of this
world, how impossible it has been found, without statistical
calculations, to judge of the proportions of births and deaths,
of the duration of life, and of the increase and decrease of

population; and still less of the causes of such changes: and yet, as has so often been repeated, decrease in numbers or rarity seems to be the high-road to extinction. To marvel at the extermination of a species appears to me to be the same thing as to know that illness is the road to death—to look at illness as an ordinary event, nevertheless to conclude, when the sick man dies, that his death has been caused by some unknown and violent agency.

In a future part of this work we shall show that, as a general rule, groups of allied species gradually appear and disappear, one after the other, on the face of the earth, like the individuals of the same species: and we shall then endeavour to show the probable cause of this remarkable fact.

CHAPTER VI

On the Geographical Distribution of Organic Beings in Past and Present Times

For convenience's sake I shall divide this chapter into three sections.[1] In the first place I shall endeavour to state the laws of the distribution of existing beings, as far as our present object is concerned; in the second, that of extinct; and in the third section I shall consider how far these laws accord with the theory of allied species having a common descent.

1. Distribution of the Inhabitants in the Different Continents

In the following discussion I shall chiefly refer to terrestrial mammifers, inasmuch as they are better known; their differences in different countries, strongly marked; and especially

---

[1] In the MS. the author has here written in the margin: "If same species appear at two spots at once, fatal to my theory."

as the necessary means of their transport are more evident, and confusion, from the accidental conveyance by man of a species from one district to another district, is less likely to arise. It is known that all mammifers (as well as all other organisms) are united in one great system; but that the different species, genera, or families of the same order inhabit different quarters of the globe. If we divide the land into two divisions, according to the amount of difference, and disregarding the numbers of the terrestrial mammifers inhabiting them, we shall have first Australia, including New Guinea; and secondly the rest of the world; if we make a three-fold division, we shall have Australia, South America, and the rest of the world; I must observe that North America is in some respects neutral land, from possessing some South American forms, but I believe it is more closely allied (as it certainly is in its birds, plants, and shells) with Europe. If our division had been four-fold, we should have had Australia, South America, Madagascar (though inhabited by few mammifers), and the remaining land; if five-fold, Africa, especially the southern eastern parts, would have to be separated from the remainder of the world. These differences in the mammiferous inhabitants of the several main divisions of the globe cannot, it is well known, be explained by corresponding differences in their conditions; how similar are parts of tropical America and Africa; and accordingly we find some *analogous* resemblances—thus both have monkeys, both large feline animals, both large Lepidoptera, and large dung-feeding beetles; both have palms and epiphytes; and yet the essential difference between their productions is as great as between those of the arid plains of the Cape of Good Hope and the grass-covered savannahs of La Plata.[2] Consider the distribution of the Marsupialia, which are eminently characteristic of Australia, and in a lesser degree of South America; when we reflect that animals of this division, feeding both on animal and vegetable matter, frequent the dry open or wooded plains and mountains of Australia, the humid impenetrable forests of

[2] Opposite this passage is written *"not botanically,"* in Sir J. D. Hooker's hand. The word *palms* is underlined three times and followed by three exclamation marks. An explanatory note is added in the margin, "singular paucity of palms and epiphytes in Trop. Africa compared with Trop. America and Ind. Or." (i.e. East Indies).

New Guinea and Brazil; the dry rocky mountains of Chile, and the grassy plains of Banda Oriental, we must look to some other cause than the nature of the country for their absence in Africa and other quarters of the world.

Furthermore it may be observed that *all* the organisms inhabiting any country are not perfectly adapted to it; I mean by not being perfectly adapted only that some few other organisms can generally be found better adapted to the country than some of the aborigines. We must admit this when we consider the enormous number of horses and cattle which have run wild during the three last centuries in the uninhabited parts of S. Domingo, Cuba, and South America; for these animals must have supplanted some aboriginal ones. I might also adduce the same fact in Australia, but perhaps it will be objected that thirty or forty years has not been a sufficient period to test this power of struggling with and overcoming the aborigines. We know the European mouse is driving before it that of New Zealand, as the Norway rat has driven before it the old English species in England. Scarcely an island can be named where casually introduced plants have not supplanted some of the native species: in La Plata the cardoon covers square leagues of country on which some South American plants must once have grown: the commonest weed over the whole of India is an introduced Mexican poppy. The geologist who knows that slow changes are in progress, replacing land and water, will easily perceive that even if all the organisms of any country had originally been the best adapted to it, this could hardly continue so during succeeding ages without either extermination, or changes, first in the relative proportional numbers of the inhabitants of the country, and finally in their constitutions and structure.

Inspection of a map of the world at once shows that the five divisions, separated according to the greatest amount of difference in the mammifers inhabiting them, are likewise those most widely separated from each other by barriers which mammifers cannot pass: thus Australia is separated from New Guinea and some small adjoining islets only by a narrow and shallow strait; whereas New Guinea and its adjoining islets are cut off from the other East Indian islands by deep water. These latter islands I may remark, which fall into the great Asiatic group, are separated from each other and the continent only by shallow water; and where this is the case we may suppose, from geological oscillations of level, that generally there has been recent union. South America, including the southern part of Mexico, is cut off from North America by the West Indies, and the great tableland of Mexico, except by a mere fringe of tropical forests along the coast: it is owing, perhaps, to this fringe that North America possesses some South American forms. Madagascar is entirely isolated. Africa is also to a great extent isolated, although it approaches, by many promontories and by lines of shallower sea, to Europe and Asia: southern Africa, which is the most distinct in its mammiferous inhabitants, is separated from the northern portion by the Great Sahara Desert and the tableland of Abyssinia. That the distribution of organisms is related to barriers, stopping their progress, we clearly see by comparing the distribution of marine and terrestrial productions. The marine animals being different on the two sides of land tenanted by the same terrestrial animals, thus the shells are wholly different on the opposite sides of the temperate parts of South America, as they are in the Red Sea and the Mediterranean. We can at once perceive that the destruction of a barrier would permit two geographical groups of organisms to fuse and blend into one. But the original cause of groups being different on opposite sides of a barrier can only be understood on the hypothesis of each organism having been created or produced on one spot or area, and afterwards migrating as widely as its means of transport and subsistence permitted it.

Relation of Range in Genera and Species

It is generally [3] found that where a genus or group ranges over nearly the entire world many of the species composing the group have wide ranges: on the other hand, where a group is restricted to any one country, the species composing it generally have restricted ranges in that country. Thus among mammifers the feline and canine genera are widely distributed, and many of the

[3] Note in original: "The same laws seem to govern distribution of species and genera, and individuals in time and space."

individual species have enormous ranges (the genus *Mus* I believe, however, is a strong exception to the rule). Mr. Gould informs me that the rule holds with birds, as in the owl genus, which is mundane, and many of the species range widely. The rule holds also with land and fresh-water mollusca, with butterflies, and very generally with plants. As instances of the converse rule, I may give that division of the monkeys which is confined to South America, and amongst plants, the cacti, confined to the same continent, the species of both of which have generally narrow ranges. On the ordinary theory of the separate creation of each species, the cause of these relations is not obvious; we can see no reason, because many allied species have been created in the several main divisions of the world, that several of these species should have wide ranges; and on the other hand, that species of the same group should have narrow ranges if all have been created in one main division of the world. As the result of such and probably many other unknown relations, it is found that, even in the same great classes of beings, the different divisions of the world are characterised by either merely different species, or genera, or even families: thus in cats, mice, foxes, South America differs from Asia and Africa only in species; in her pigs, camels, and monkeys the difference is generic or greater. Again, whilst southern Africa and Australia differ more widely in their mammalia than do Africa and South America, they are more closely (though indeed very distantly) allied in their plants.

Distribution of the Inhabitants in the Same Continent

If we now look at the distribution of the organisms in any one of the above main divisions of the world, we shall find it split up into many regions, with all or nearly all their species distinct, but yet partaking of one common character. This similarity of type in the subdivisions of a great region is equally well known with the dissimilarity of the inhabitants of the several great regions; but it has been less often insisted on, though more worthy of remark. Thus for instance, if in Africa or South America we go from south to north, or from lowland to upland, or from a humid to a drier part, we find wholly different species of those genera or groups which characterise the continent over which we are passing. In these subdivisions we may clearly observe, as in the main divisions of the world, that sub-barriers divide different groups of species, although the opposite sides of such sub-barriers may possess nearly the same climate, and may be in other respects nearly similar: thus it is on the opposite sides of the Cordillera of Chile, and in a lesser degree on the opposite sides of the Rocky Mountains. Deserts, arms of the sea, and even rivers form the barriers; mere preoccupied space seems sufficient in several cases: thus eastern and western Australia, in the same latitude, with very similar climate and soils, have scarcely a plant, and few animals or birds, in common, although all belong to the peculiar genera characterizing Australia. It is in short impossible to explain the differences in the inhabitants, either of the main divisions of the world, or of these sub-divisions, by the differences in their physical conditions, and by the adaptation of their inhabitants. Some other cause must intervene.

We can see that the destruction of sub-barriers would cause (as before remarked in the case of the main divisions) two sub-divisions to blend into one; and we can only suppose that the original difference in the species, on the opposite sides of sub-barriers, is due to the creation or production of species in distinct areas, from which they have wandered till arrested by such sub-barriers. Although thus far is pretty clear, it may be asked, why, when species in the same main division of the world were produced on opposite sides of a sub-barrier, both when exposed to similar conditions and when exposed to widely different influences (as on alpine and lowland tracts, as on arid and humid soils, as in cold and hot climates), have they invariably been formed on a similar type, and that type confined to this one division of the world? Why, when an ostrich was produced in the southern parts of America, was it formed on the American type, instead of on the African or on Australian types? Why, when hare-like and rabbit-like animals were formed to live on the savannahs of La Plata, were they produced on the peculiar rodent type of South America, instead of on the true [4] hare type of North America, Asia, and Africa?

[4] Note in original: "There is a hare in South America—so bad example."

Why, when burrowing rodents and camel-like animals were formed to tenant the Cordillera, were they formed on the same type with their representatives on the plains? Why were the mice, and many birds of different species on the opposite sides of the Cordillera, but exposed to a very similar climate and soil, created on the same peculiar South American type? Why were the plants in eastern and western Australia, though wholly different as species, formed on the same peculiar Australian types? The generality of the rule, in so many places and under such different circumstances, makes it highly remarkable and seems to demand some explanation.

Insular Faunas

If we now look to the character of the inhabitants of small islands, we shall find that those situated close to other land have a similar fauna with that land,[5] whilst those at a considerable distance from other land often possess an almost entirely peculiar fauna. The Galapagos Archipelago is a remarkable instance of this latter fact; here almost every bird, its one mammifer, its reptiles, land and sea shells, and even fish are almost all peculiar and distinct species, not found in any other quarter of the world: so are the majority of its plants. But although situated at the distance of between five hundred and six hundred miles from the South American coast, it is impossible to even glance at a large part of its fauna, especially at the birds, without at once seeing that they belong to the American type. Hence, in fact, groups of islands thus circumstanced form merely small but well-defined sub-divisions of the larger geographical divisions. But the fact is in such cases far more striking: for taking the Galapagos Archipelago as an instance; in the first place we must feel convinced, seeing that every island is wholly volcanic and bristles with craters, that in a geological sense the whole is of recent origin comparatively with a continent; and as the species are nearly all peculiar, we must conclude that they have in the same sense recently been produced on this very spot; and although in the nature

[5] Between the lines, above the words "with that land," the author wrote "Cause, formerly joined, no one doubts after Lyell."

of the soil, and in a lesser degree in the climate, there is a wide difference with the nearer part of the South American coast, we see that the inhabitants have been formed on the same closely allied type. On the other hand, these islands, as far as their physical conditions are concerned, resemble closely the Cape Verde volcanic group, and yet how wholly unlike are the productions of these two archipelagos. The Cape Verde group, to which may be added the Canary Islands, are allied in their inhabitants (of which many are peculiar species) to the coast of Africa and southern Europe, in precisely the same manner as the Galapagos Archipelago is allied to America. We here clearly see that mere geographical proximity affects, more than any relation of adaptation, the character of species. How many islands in the Pacific exist far more like in their physical conditions to Juan Fernandez than this island is to the coast of Chile, distant three hundred miles; why then, except from mere proximity, should this island alone be tenanted by two very peculiar species of humming-birds—that form of birds which is so exclusively American? Innumerable other similar cases might be adduced.

The Galapagos Archipelago offers another, even more remarkable, example of the class of facts we are here considering. Most of its genera are, as we have said, American, many of them are mundane, or found everywhere, and some are quite or nearly confined to this archipelago. The islands are of absolutely similar composition, and exposed to the same climate; most of them are in sight of each other; and yet several of the islands are inhabited, each by peculiar species (or in some cases perhaps only varieties) of some of the genera characterizing the archipelago. So that the small group of the Galapagos Islands typifies, and follows exactly, the same laws in the distribution of its inhabitants as a great continent. How wonderful it is that two or three closely similar but distinct species of a mocking thrush should have been produced on three neighbouring and absolutely similar islands; and that these three species of mocking thrush should be closely related to the other species inhabiting wholly different climates and different districts of America, and only in America. No similar case so striking as this of the Galapagos has hitherto been observed;

and this difference of the productions in the different islands may perhaps be partly explained by the depth of the sea between them (showing that they could not have been united within recent geological periods), and by the currents of the sea sweeping *straight* between them—and by storms of wind being rare, through which means seeds and birds could be blown, or drifted, from one island to another. There are however some similar facts: it is said that the different, though neighbouring, islands of the East Indian Archipelago are inhabited by some different species of the same genera; and at the Sandwich group some of the islands have each their peculiar species of the same genera of plants.

Islands standing quite isolated within the intratropical oceans have generally very peculiar floras, related, though feebly (as in the case of St. Helena, where almost every species is distinct), with the nearest continent: Tristan d'Acunha is feebly related, I believe, in its plants both to Africa and South America, not by having species in common, but by the genera to which they belong.[6] The floras of the numerous scattered islands of the Pacific are related to each other and to all the surrounding continents; but it has been said that they have more of an Indo-Asiatic than American character. This is somewhat remarkable, as America is nearer to all the Eastern islands, and lies in the direction of the trade-wind and prevailing currents; on the other hand, all the heaviest gales come from the Asiatic side. But even with the aid of these gales, it is not obvious on the ordinary theory of creation how the possibility of migration (without we suppose, with extreme improbability, that each species with an Indo-Asiatic character has actually travelled from the Asiatic shores, where such species do not now exist) explains this Asiatic character in the plants of the Pacific. This is no more obvious than that (as before remarked) there should exist a relation between the creation of closely allied species in several regions of the world, and the fact of many such species having wide ranges; and on the other hand, of allied species confined to one region of the world having in that region narrow ranges.

[6] It is impossible to make out the precise form which the author intended to give to this sentence, but the meaning is clear.

Alpine Floras

We will now turn to the floras of mountain summits which are well known to differ from the floras of the neighbouring lowlands. In certain characters, such as dwarfness of stature, hariness, etc., the species from the most distant mountains frequently resemble each other—a kind of analogy like that for instance of the succulency of most desert plants. Besides this analogy, alpine plants present some eminently curious facts in their distribution. In some cases the summits of mountains, although immensely distant from each other, are clothed by the same identical species which are likewise the same with those growing on the likewise very distant arctic shores. In other cases, although few or none of the species may be actually identical, they are closely related; whilst the plants of the lowland districts surrounding the two mountains in question will be wholly dissimilar. As mountain summits, as far as their plants are concerned, are islands rising out of an ocean of land in which the alpine species cannot live, nor across which is there any known means of transport, this fact appears directly opposed to the conclusion which we have come to form considering the general distribution of organisms both on continents and on islands—namely, that the degree of relationship between the inhabitants of two points depends on the completeness and nature of the barriers between those points. I believe, however, this anomalous case admits, as we shall presently see, of some explanation. We might have expected that the flora of a mountain summit would have presented the same relation to the flora of the surrounding lowland country, which any isolated part of a continent does to the whole, or an island does to the mainland, from which it is separated by a rather wide space of sea. This in fact is the case with the plants clothing the summits of *some* mountains, which mountains it may be observed are particularly isolated; for instance, all the species are peculiar, but they belong to the forms characteristic of the surrounding continent, on the mountains of Caraccas, of Van Diemen's Land, and of the Cape of Good Hope. On some other mountains, for instance Tierra del Fuego and in Brazil, some of the plants though distinct species are South American forms; whilst others are allied to or are identical with the alpine species of Europe. In islands of which the lowland flora is distinct but allied to that of the nearest continent, the alpine plants are sometimes (or perhaps mostly) eminently peculiar and distinct, this is the case on Teneriffe, and in a lesser degree even on some of the Mediterranean islands.

If all alpine floras had been characterized like that of the mountain of Caraccas, or of Van Diemen's Land, etc., whatever explanation is possible of the general laws of geographical distribution would have applied to them. But the apparently anomalous case just given, namely, of the mountains of Europe, of some mountains in the United States (Dr. Boott), and of the summits of the Himalaya (Royle) having many identical species in common conjointly with the arctic regions, and many species, though not identical, closely allied, require a separate explanation. The fact likewise of several of the species on the mountains of Tierra del Fuego (and in a lesser degree on the mountains of Brazil) not belonging to American forms, but to those of Europe, though so immensely remote, requires also a separate explanation.

Cause of the Similarity in the Floras of Some Distant Mountains

Now we may with confidence affirm, from the number of the then floating icebergs and low descent of the glaciers, that within a period so near that species of shells have remained the same the whole of Central Europe and of North America (and perhaps of Eastern Asia) possessed a very cold climate; and therefore it is probable that the floras of these districts were the same as the present arctic one—as is known to have been to some degree the case with then existing sea-shells, and those now living on the arctic shores. At this period the mountains must have been covered with ice of which we have evidence in the surfaces polished and scored by glaciers. What then would be the natural and almost inevitable effects of the gradual change into the present more temperate climate? [7]

[7] In the margin the author has written "(Forbes)."

The ice and snow would disappear from the mountains, and as new plants from the more temperate regions of the south migrated northward, replacing the arctic plants, these latter would crawl up the now uncovered mountains, and likewise be driven northward to the present arctic shores. If the arctic flora of that period was a nearly uniform one, as the present one is, then we should have the same plants on these mountain summits and on the present arctic shores. On this view the arctic flora of that period must have been a widely extended one, more so than even the present one; but considering how similar the physical conditions must always be of land bordering on perpetual frost, this does not appear a great difficulty; and may we not venture to suppose that the almost infinitely numerous icebergs, charged with great masses of rocks, soil, and *brushwood* [8] and often driven high up on distant beaches, might have been the means of widely distributing the seeds of the same species?

I will only hazard one other observation, namely, that during the change from an extremely cold climate to a more temperate one the conditions, both on lowland and mountain, would be singularly favourable for the diffusion of any existing plants, which could live on land, just freed from the rigour of eternal winter; for it would possess no inhabitants; and we cannot doubt that *preoccupation* [9] is the chief bar to the diffusion of plants. For amongst many other facts, how otherwise can we explain the circumstance that the plants on the opposite, though similarly constituted, sides of a wide river in eastern Europe (as I was informed by Humboldt) should be widely different; across which river birds, swimming quadrupeds, and the wind must often transport seeds; we can only suppose that plants already occupying the soil and freely seeding check the germination of occasionally transported seeds.

At about the same period when icebergs were transporting boulders in North America as far as 36° south, where the cotton tree now grows in South America, in latitude 42° (where

---

[8] Note in original: "Perhaps vitality checked by cold and so prevented germinating."

[9] A note by the author gives "many authors" apparently as authority for this statement.

the land is now clothed with forests having an almost tropical aspect with the trees bearing epiphytes and intertwined with canes), the same ice action was going on; is it not then in some degree probable that at this period the whole tropical parts of the two Americas possessed [10] (as Falconer asserts that India did) a more temperate climate? In this case the alpine plants of the long chain of the Cordillera would have descended much lower and there would have been a broad high-road connecting those parts of North and South America which were then frigid. As the present climate supervened, the plants occupying the districts which now are become in both hemispheres temperate and even semi-tropical must have been driven to the arctic and antarctic regions; and only a few of the loftiest points of the Cordillera can have retained their former connecting flora. The transverse chain of Chiquitos might perhaps in a similar manner during the ice-action period have served as a connecting road (though a broken one) for alpine plants to become dispersed from the Cordillera to the highlands of Brazil. It may be observed that some (though not strong) reasons can be assigned for believing that at about this same period the two Americas were not so thoroughly divided as they now are by the West Indies and tableland of Mexico. I will only further remark that the present most singularly close similarity in the vegetation of the lowlands of Kerguelen's Land and of Tierra del Fuego (Hooker), though so far apart, may perhaps be explained by the dissemination of seeds during this same cold period, by means of icebergs, as before alluded to.[11]

Finally, I think we may safely grant from the foregoing facts and reasoning that the anomalous similarity in the vegetation of certain very distant mountain summits is not in truth opposed to the conclusion of the intimate relation subsisting between proximity in space (in accordance with the means of transport in each class) and the degree of affinity of the inhabitants of any two countries. In the case of several quite isolated mountains, we have seen that the general law holds good.

---

[10] Opposite to this passage, in the margin, the author has written: "too hypothetical."

[11] Note by the author: "Similarity of flora of coral islands easily explained."

Whether the Same Species Has Been Created More Than Once

As the fact of the same species of plants' having been found on mountain summits immensely remote has been one chief cause of the belief of some species' having been contemporaneously produced or created at two different points, I will here briefly discuss this subject. On the ordinary theory of creation, we can see no reason why on two similar mountain summits two similar species may not have been created; but the opposite view, independently of its simplicity, has been generally received from the analogy of the general distribution of all organisms, in which (as shown in this chapter) we almost always find that great and continuous barriers separate distinct series; and we are naturally led to suppose that the two series have been separately created. When taking a more limited view we see a river, with a quite similar country on both sides, with one side well stocked with a certain animal and on the other side not one (as is the case with the bizcacha on the opposite sides of the Plata), we are at once led to conclude that the bizcacha was produced on some one point or area on the western side of the river. Considering our ignorance of the many strange chances of diffusion by birds (which occasionally wander to immense distances) and quadrupeds swallowing seeds and ova (as in the case of the flying water-beetle which disgorged the eggs of a fish), and of whirlwinds carrying seeds and animals into strong upper currents (as in the case of volcanic ashes and showers of hay, grain, and fish), and of the possibility of species having survived for short periods at intermediate spots and afterwards becoming extinct there; and considering our knowledge of the great changes which *have* taken place from subsidence and elevation in the surface of the earth, and of our ignorance of the greater changes which *may have* taken place, we ought to be very slow in admitting the probability of double creations. In the case of plants on mountain summits, I think I have shown how almost necessarily they would, under the past conditions of the northern hemisphere, be as similar as are the plants on the present arctic shores; and this ought to teach us a lesson of caution.

But the strongest argument against double creations may be drawn from considering the case of mammifers in which, from their nature and from the size of their offspring, the means of distribution are more in view. There are no cases where the same species is found in *very remote* localities, except where there is a continuous belt of land: the arctic region perhaps offers the strongest exception, and here we know that animals are transported on icebergs.[12] The cases of lesser difficulty may all receive a more or less simple explanation; I will give only one instance; the nutria, I believe, on the eastern coast of South America live exclusively in fresh-water rivers, and I was much surprised how they could have got into rivulets, widely apart, on the coast of Patagonia; but on the opposite coast I found these quadrupeds living exclusively in the sea, and hence their migration along the Patagonian coast is not surprising. There is no case of the same mammifer being found on an island far from the coast and on the mainland, as happens with plants. On the idea of double creations it would be strange if the same species of several plants should have been created in Australia and Europe; and no one instance of the same species of mammifer having been created, or aboriginally existing, in two as nearly remote and equally isolated points. It is more philosophical, in such cases, as that of some plants being found in Australia and Europe, to admit that we are ignorant of the means of transport. I will allude only to one other case, namely that of the *Mydas,* an alpine animal, found only on the distant peaks of the mountains of Java: who will pretend to deny that during the ice period of the northern and southern hemispheres, and when India is believed to have been colder, the climate might not have permitted this animal to haunt a lower country, and thus to have passed along the ridges from summit to summit? Mr. Lyell has further observed that, *as in space, so in time,* there is no reason to believe that after the extinction of a species, the self-same form has ever reappeared. I think, then, we may notwithstanding the many cases of difficulty, conclude with some confidence that every species has been created or produced on a single point or area.

---

[12] Note by the author: "Many authors."

On the Number of Species, and of the Classes
to Which They Belong in Different Regions

The last fact in geographical distribution, which, as far as I
can see, in any way concerns the origin of species, relates to
the absolute number and nature of the organic beings inhabiting different tracts of land. Although every species is admirably
adapted (but not necessarily better adapted than every other
species, as we have seen in the great increase of introduced
species) to the country and station it frequents; yet it has been
shown that the entire difference between the species in distant
countries cannot possibly be explained by the difference of the
physical conditions of these countries. In the same manner, I
believe, neither the number of the species, nor the nature of
the great classes to which they belong, can possibly in all cases
be explained by the conditions of their country. New Zealand,
a linear island stretching over about seven hundred miles of
latitude, with forests, marshes, plains, and mountains reaching
to the limits of eternal snow, has far more diversified habitats
than an equal area at the Cape of Good Hope; and yet, I believe, at the Cape of Good Hope there are, of phanerogamic
plants, from five to ten times the number of species as in all
New Zealand. Why on the theory of absolute creations should
this large and diversified island only have from four hundred
to five hundred (Dieffenbach?) phanerogamic plants? and why
should the Cape of Good Hope, characterised by the uniformity of its scenery, swarm with more species of plants than
probably any other quarter of the world? Why on the ordinary
theory should the Galapagos Islands abound with terrestrial
reptiles? And why should many equal-sized islands in the Pacific be without a single one or with only one or two species?
Why should the great island of New Zealand be without one
mammiferous quadruped except the mouse, and that was probably introduced with the aborigines? Why should not one island (it can be shown, I think, that the mammifers of Mauritius and St. Iago have all been introduced) in the open ocean
possess a mammiferous quadruped? Let it not be said that
quadrupeds cannot live in islands, for we know that cattle,

horses, and pigs during a long period have run wild in the West Indian and Falkland Islands; pigs at St. Helena; goats at Tahiti; asses in the Canary Islands; dogs in Cuba; cats at Ascension; rabbits at Madeira and the Falklands; monkeys at St. Iago and the Mauritius; even elephants during a long time in one of the very small Sooloo Islands; and European mice on very many of the smallest islands far from the habitations of man. Nor let it be assumed that quadrupeds are more slowly created and hence that the oceanic islands, which generally are of volcanic formation, are of too recent origin to possess them; for we know (Lyell) that new forms of quadrupeds succeed each other quicker than Mollusca or Reptilia. Nor let it be assumed (though such an assumption would be no explanation) that quadrupeds cannot be created on small islands; for islands not lying in mid-ocean do possess their peculiar quadrupeds; thus many of the smaller islands of the East Indian Archipelago possess quadrupeds; as does Fernando Po on the west coast of Africa; as the Falkland Islands possess a peculiar wolf-like fox; so do the Galapagos Islands a peculiar mouse of the South American type. These two last are most remarkable cases with which I am acquainted; inasmuch as the islands lie further from other land. It is possible that the Galapagos mouse may have been introduced in some ship from the South American coast (though the species is at present unknown there), for the aboriginal species soon haunts the goods of man, as I noticed in the roof of a newly erected shed in a desert country south of the Plata. The Falkland Islands, though between two hundred and three hundred miles from the South American coast, may in one sense be considered as intimately connected with it; for it is certain that formerly many icebergs loaded with boulders were stranded on its southern coast, and the old canoes which are occasionally now stranded show that the currents still set from Tierra del Fuego. This fact, however, does not explain the presence of the *Canis antarcticus* on the Falkland Islands, unless we suppose that it formerly lived on the mainland and became extinct there, whilst it survived on these islands, to which it was borne (as happens with its northern congener, the common wolf) on an iceberg, but this fact removes the anomaly of an island, in appearance effectually separated from other land, having its own species of quadruped, and makes the case like that of Java and Sumatra, each having their own rhinoceros.

Before summing up all the facts given in this section on the present condition of organic beings, and endeavouring to see how far they admit of explanation, it will be convenient to state all such facts in the past geographical distribution of extinct beings as seem anyway to concern the theory of descent.

2. Geographical Distribution of Extinct Organisms

I have stated that if the land of the entire world be divided into (we will say) three sections, according to the amount of difference of the terrestrial mammifers inhabiting them, we shall have three unequal divisions of first Australia and its dependent islands, second South America, third Europe, Asia, and Africa. If we now look to the mammifers which inhabited these three divisions during the later Tertiary periods, we shall find them almost as distinct as at the present day, and intimately related in each division to the existing forms in that division. This is wonderfully the case with the several fossil marsupial genera in the caverns of New South Wales and even more wonderfully so in South America, where we have the same peculiar group of monkeys, of a guanaco-like animal, of many rodents, of the marsupial *Didelphys,* of armadillos and other Edentata. This last family is at present very characteristic of South America, and in a late Tertiary epoch it was even more so, as is shown by the numerous enormous animals of the megatheroid family, some of which were protected by an osseous armour like that, but on a gigantic scale, of the recent armadillo. Lastly, over Europe the remains of the several deer, oxen, bears, foxes, beavers, field-mice, show a relation to the present inhabitants of this region; and the contemporaneous remains of the elephant, rhinoceros, hippopotamus, hyaena, show a relation with the grand Africo-Asiatic division of the world. In Asia the fossil mammifers of the Himalaya (though mingled with forms long extinct in Europe) are equally related to the existing forms of the Africo-Asiatic division; but especially to those of India itself. As the gigantic and now extinct quadrupeds of Europe have naturally excited more attention than the other and smaller remains, the relation between the past and present mammiferous inhabitants of Europe has not been sufficiently attended to. But in fact the mammifers of Europe are at present nearly as much Africo-Asiatic as they were formerly when Europe had its elephants and rhinoceroses, etc.: Europe neither now nor then possessed peculiar groups as does Australia and South America. The extinction of certain peculiar forms in one quarter does not make the remaining mammifers of that quarter less related to its own great division of the world: though Tierra del Fuego possesses only a fox, three rodents, and the guanaco, no one (as these all belong to South American types, but not to the most characteristic forms) would doubt for one minute as to classifying this district with South America; and if fossil Edentata, marsupials, and monkeys were to be found in Tierra del Fuego, it would not make this district more truly South American than it now is. So it is with Europe, and so far as is known with Asia, for the lately past and present mammifers all belong to the Africo-Asiatic division of the world. In every case, I may add, the forms which a country has are of more importance in geographical arrangement than what it has not.

We find some evidence of the same general fact in a relation between the recent and the Tertiary sea-shells, in the different main divisions of the marine world.

This general and most remarkable relation between the lately past and present mammiferous inhabitants of the three main divisions of the world is precisely the same kind of fact as the relation between the different species of the several sub-regions of any one of the main divisions. As we usually associate great physical changes with the total extinction of one series of beings, and its succession by another series, this identity of relation between the past and the present races of beings in the same quarters of the globe is more striking than the same relation between existing beings in different sub-regions: but in truth we have no reason for supposing that a change in the conditions has in any of these cases supervened, greater than that now existing between the temperate and tropical, or between the highlands and lowlands of the same main divisions, now tenanted by related beings. Finally, then, we clearly see that in each main division of the world the same relation holds good between its inhabitants in time as over space.

Changes in Geographical Distribution

If, however, we look closer, we shall find that even Australia, in possessing a terrestrial pachyderm, was so far less distinct from the rest of the world than it now is; so was South America in possessing the *Mastodon,* horse, [hyaena],[13] and antelope. North America, as I have remarked, is now, in its mammifers, in some respects neutral ground between South America and the great Africo-Asiatic division; formerly, in possessing the horse, *Mastodon,* and three megatheroid animals, it was more nearly related to South America; but in the horse and *Mastodon,* and likewise in having the elephant, oxen, sheep, and pigs, it was as much if not more related to the Africo-Asiatic division. Again, northern India was more closely related (in having the giraffe, hippopotamus, and certain musk-deer) to southern Africa than it now is; for southern and eastern Africa deserve, if we divide the world into five parts, to make one division by itself. Turning to the dawn of the Tertiary period, we must, from our ignorance of other portions of the world, confine ourselves to Europe; and at that period, in the presence of marsupials and Edentata, we behold an *entire* blending of those mammiferous forms which now eminently characterise Australia and South America.[14]

If we now look at the distribution of sea-shells, we find the same changes in distribution. The Red Sea and the Mediterranean were more nearly related in these shells than they now are. In different parts of Europe, on the other hand, during the Miocene period, the sea-shells seem to have been more different than at present. In [15] the Tertiary period, according to Lyell, the shells of North America and Europe were less related than at present, and during the Cretaceous still less like;

[13] The word *hyaena* is erased. There appear to be no fossil Hyaenidae in South America.

[14] Note by the author: "And see Eocene European mammals in North America."

[15] Note by the author: "All this requires much verification."

whereas, during this same Cretaceous period, the shells of India and Europe were more like than at present. But going further back to the Carbonaceous period, in North America and Europe, the productions were much more like than they now are. These facts harmonise with the conclusions drawn from the present distribution of organic beings, for we have seen that from species being created in different points or areas the formation of a barrier would cause or make two distinct geographical areas; and the destruction of a barrier would permit their diffusion. And as long-continued geological changes must both destroy and make barriers, we might expect, the further we looked backwards, the more changed should we find the present distribution. This conclusion is worthy of attention, because, finding in widely different parts of the same main division of the world, and in volcanic islands near them, groups of distinct, but related, species; and finding that a singularly analogous relation holds good with respect to the beings of past times, when none of the present species were living, a person might be tempted to believe in some mystical relation between certain areas of the world, and the production of certain organic forms; but we now see that such an assumption would have to be complicated by the admission that such a relation, though holding good for long revolutions of years, is not truly persistent.

I will only add one more observation to this section. Geologists finding in the most remote period with which we are acquainted, namely in the Silurian period, that the shells and other marine productions [16] in North and South America, in Europe, southern Africa, and western Asia, are much more similar than they now are at these distant points, appear to have imagined that in these ancient times the laws of geographical distribution were quite different than what they now are: but we have only to suppose that great continents were extended east and west, and thus did not divide the inhabitants of the temperate and tropical seas, as the continents now do; and it would then become probable that the inhabitants of the seas would be much more similar than they now are. In the immense space of ocean extending from the east coast of Africa to the eastern islands of the Pacific, which

[16] Note by the author: "D'Orbigny shows that this is not so."

space is connected either by lines of tropical coast or by islands
not very distant from each other, we know (Cuming) that
many shells, perhaps even as many as two hundred, are common to the Zanzibar coast, the Philippines, and the eastern islands of the Low or Dangerous Archipelago in the Pacific. This
space equals that from the arctic to the antarctic pole! Pass
over the space of quite open ocean, from the Dangerous Archipelago to the west coast of South America, and every shell is
different: pass over the narrow space of South America, to its
eastern shores, and again every shell is different! Many fish,
I may add, are also common to the Pacific and Indian Oceans.

Summary on the Distribution of Living and Extinct Organic Beings

Let us sum up the several facts now given with respect to the
past and present geographical distribution of organic beings.
In a previous chapter it was shown that species are not exterminated by universal catastrophes, and that they are slowly
produced: we have also seen that each species is probably only
once produced, on one point or area once in time; and that
each diffuses itself, as far as barriers and its conditions of life
permit. If we look at any one main division of land, we find
in the different parts, whether exposed to different conditions
or to the same conditions, many groups of species wholly or
nearly distinct as species, nevertheless intimately related. We
find the inhabitants of islands, though distinct as species, similarly related to the inhabitants of the nearest continent; we find
in some cases that even the different islands of one such group
are inhabited by species distinct, though intimately related to
one another and to those of the nearest continent: thus typify-
ing the distribution of organic beings over the whole world.
We find the floras of distant mountain summits either very similar (which seems to admit, as shown, of a simple explanation)
or very distinct but related to the floras of the surrounding
region; and hence, in this latter case, the floras of two mountain summits, although exposed to closely similar conditions,
will be very different. On the mountain summits of islands,
characterised by peculiar faunas and floras, the plants are often

eminently peculiar. The dissimilarity of the organic beings inhabiting nearly similar countries is best seen by comparing the main divisions of the world; in each of which some districts may be found very similarly exposed, yet the inhabitants are wholly unlike; far more unlike than those in very dissimilar districts in the same main division. We see this strikingly in comparing two volcanic archipelagos, with nearly the same climate, but situated not very far from two different continents; in which case their inhabitants are totally unlike. In the different main divisions of the world, the amount of difference between the organisms, even in the same class, is widely different, each main division having only the species distinct in some families, in other families having the genera distinct. The distribution of aquatic organisms is very different from that of the terrestrial organisms; and necessarily so, from the barriers to their progress being quite unlike. The nature of the conditions in an isolated district will not explain the number of species inhabiting it; nor the absence of one class or the presence of another class. We find that terrestrial mammifers are not present on islands far removed from other land. We see in two regions, that the species though distinct are more or less related, according to the greater or less *possibility* of the transportal in past and present times of species from one to the other region; although we can hardly admit that all the species in such cases have been transported from the first to the second region, and since have become extinct in the first: we see this law in the presence of the fox on the Falkland Islands; in the European character of some of the plants of Tierra del Fuego; in the Indo-Asiatic character of the plants of the Pacific; and in the circumstance of those genera which range widest having many species with wide ranges; and those genera with restricted ranges having species with restricted ranges. Finally, we find in each of the main divisions of the land, and probably of the sea, that the existing organisms are related to those lately extinct.

Looking further backwards we see that the past geographical distribution of organic beings was different from the present; and indeed, considering that geology shows that all our land was once under water, and that where water now extends land is forming, the reverse could hardly have been possible.

Now these several facts, though evidently all more or less connected together, must by the creationist (though the geologist may explain some of the anomalies) be considered as so many ultimate facts. He can only say that it so pleased the Creator that the organic beings of the plains, deserts, mountains, tropical and temperate forests, of South America, should all have some affinity together; that the inhabitants of the Galapagos Archipelago should be related to those of Chile; and that some of the species on the similarly constituted islands of this archipelago, though most closely related, should be distinct; that all its inhabitants should be totally unlike those of the similarly volcanic and arid Cape Verde and Canary Islands; that the plants on the summit of Teneriffe should be eminently peculiar; that the diversified island of New Zealand should have not many plants, and not one, or only one, mammifer; that the mammifers of South America, Australia and Europe should be clearly related to their ancient and exterminated prototypes; and so on with other facts. But it is absolutely opposed to every analogy, drawn from the laws imposed by the Creator on inorganic matter, that facts, when connected, should be considered as ultimate and not the direct consequences of more general laws.

3. An Attempt to Explain the Foregoing Laws of Geographical Distribution, on the Theory of Allied Species Having a Common Descent

First let us recall the circumstances most favourable for variation under domestication, as given in the first chapter, viz., first, a change, or repeated changes, in the conditions to which the organism has been exposed, continued through several seminal (i.e. not by buds or divisions) generations: secondly, steady selection of the slight varieties thus generated with a fixed end in view: thirdly, isolation as perfect as possible of such selected varieties; that is, the preventing their crossing with other forms; this latter condition applies to all terrestrial animals, to most if not all plants and perhaps even to most (or all) aquatic organisms. It will be convenient here to show the advantage of isolation in the formation of a new breed, by comparing the progress of two persons (to neither of whom let time be of any consequence) endeavouring to select and form some very peculiar new breed. Let one of these persons work on the vast herds of cattle in the plains of La Plata, and the other on a small stock of twenty or thirty animals in an island. The latter might have to wait centuries (by the hypothesis, of no importance) before he obtained a "sport" approaching to what he wanted; but when he did and saved the greater number of its offspring and their offspring again, he might hope that his whole little stock would be in some degree affected, so that by continued selection he might gain his end. But on the Pampas, though the man might get his first approach to his desired form sooner, how hopeless would it be to attempt, by saving its offspring amongst so many of the common kind, to affect the whole herd: the effect of this one peculiar "sport" would be quite lost before he could obtain a second original sport of the same kind. If, however, he could separate a small number of cattle, including the offspring of the desirable "sport," he might hope, like the man on the island, to effect his end. If there be organic beings of which two individuals *never* unite, then simple selection whether on a continent or island would be equally serviceable to make a new and desirable breed; and this new breed might be made in surprisingly few years from the great and geometrical powers of propagation to beat out the old breed; as has happened (notwithstanding crossing) where good breeds of dogs and pigs have been introduced into a limited country, for instance, into the islands of the Pacific.

Let us now take the simplest natural case of an islet upheaved by the volcanic or subterranean forces in a deep sea, at such a distance from other lands that only a few organic beings at rare intervals were transported to it, whether borne by the sea (like the seeds of plants to coral-reefs), or by hurricanes, or by floods, or on rafts, or in roots of large trees, or the germs of one plant or animal attached to or in the stomach of some other animal, or by the intervention (in most cases the most probable means) of other islands since sunk or destroyed. It may be remarked that when one part of the earth's crust is raised it is probably the general rule than another part sinks. Let this island go on slowly, century after century, rising foot by foot; and in the course of time we shall have instead of a

small mass of rock, lowland and highland, moist woods and
dry sandy spots, various soils, marshes, streams, and pools:
under water on the sea-shore, instead of a rocky steeply shelv-
ing coast, we shall have in some parts bays with mud, sandy
beaches, and rocky shoals. The formation of the island by itself
must often slightly affect the surrounding climate. It is impossible that the first few transported organisms could be perfectly
adapted to all these stations; and it will be a chance if those
successively transported will be so adapted. The greater number would probably come from the lowlands of the nearest
country; and not even all these would be perfectly adapted to
the new islet whilst it continued low and exposed to coast influences. Moreover, as it is certain that all organisms are nearly
as much adapted in their structure to the other inhabitants
of their country as they are to its physical conditions, so the
mere fact that a *few* beings (and these taken in great degree by
chance) were in the first case transported to the islet would in
itself greatly modify their conditions. As the island continued
rising we might also expect an occasional new visitant; and I
repeat that even one new being must often affect beyond our
calculation by occupying the room and taking part of the subsistence of another (and this again from another and so on)
several or many other organisms. Now as the first transported
and any occasional successive visitants spread or tended to
spread over the growing island, they would undoubtedly be
exposed through several generations to new and varying conditions: it might also easily happen that some of the species
on an average might obtain an increase of food, or food of a
more nourishing quality.[17] According then to every analogy
with what we have seen takes place in every country, with
nearly every organic being under domestication, we might expect that some of the inhabitants of the island would "sport,"
or have their organisation rendered in some degree plastic.
As the number of the inhabitants are supposed to be few and
as all these cannot be so well adapted to their new and varying
conditions as they were in their native country and habitat, we
cannot believe that every place or office in the economy of

[17] In the MS. "some of the species . . . nourishing quality" is doubtfully erased. It seems clear that he doubted whether such a problemati-
cal supply of food would be likely to cause variation.

the island would be as well filled as on a continent where the
number of aboriginal species is far greater and where they
consequently hold a more strictly limited place. We might therefore expect on our island that although very many slight variations were of no use to the plastic individuals, yet that occasionally in the course of a century an individual might be
born of which the structure or constitution in some slight degree would allow it better to fill up some office in the insular
economy and to struggle against other species. If such were
the case the individual and its offspring would have a better
chance of surviving and of beating out its parent form; and if
(as is probable) it and its offspring crossed with the unvaried
parent form, yet the number of the individuals being not very
great, there would be a chance of the new and more serviceable
form being nevertheless in some slight degree preserved. The
struggle for existence would go on annually selecting such individuals until a new race or species was formed. Either few
or all the first visitants to the island might become modified,
according as the physical conditions of the island and those
resulting from the kind and number of other transported species
were different from those of the parent country—according to
the difficulties offered to fresh immigration—and according to
the length of time since the first inhabitants were introduced.
It is obvious that whatever was the country, generally the
nearest from which the first tenants were transported, they
would show an affinity, even if all had become modified, to
the natives of that country and even if the inhabitants of the
same source had been modified. On this view we can at once
understand the cause and meaning of the affinity of the fauna
and flora of the Galapagos Islands with that of the coast of
South America; and consequently why the inhabitants of these
islands show not the smallest affinity with those inhabiting other
volcanic islands, with a very similar climate and soil, near the
coast of Africa.

To return once again to our island, if by the continued action of the subterranean forces other neighbouring islands were
formed, these would generally be stocked by the inhabitants
of the first island, or by a few immigrants from the neighbouring mainland; but if considerable obstacles were interposed to
any communication between the terrestrial productions of these

islands, and their conditions were different (perhaps only by the number of different species on each island), a form transported from one island to another might become altered in the same manner as one from the continent; and we should have several of the islands tenanted by representative races or species, as is so wonderfully the case with the different islands of the Galapagos Archipelago. As the islands become mountainous, if mountain-species were not introduced, as could rarely happen, a greater amount of variation and selection would be requisite to adapt the species, which originally came from the lowlands of the nearest continent, to the mountain summits than to the lower districts of our islands. For the lowland species from the continent would have first to struggle against other species and other conditions on the coast-land of the island, and so probably become modified by the selection of its best fitted varieties, then to undergo the same process when the land had attained a moderate elevation; and then lastly when it had become alpine. Hence we can understand why the faunas of insular mountain summits are, as in the case of Teneriffe, eminently peculiar. Putting on one side the case of a widely extended flora being driven up the mountain summits, during a change of climate from cold to temperate, we can see why in other cases the floras of mountain summits (or as I have called them islands in a sea of land) should be tenanted by peculiar species, but related to those of the surrounding lowlands, as are the inhabitants of a real island in the sea to those of the nearest continent.[18]

Let us now consider the effect of a change of climate or of other conditions on the inhabitants of a continent and of an isolated island without any great change of level. On a continent the chief effects would be changes in the numerical proportion of the individuals of the different species; for whether the climate became warmer or colder, drier or damper, more uniform or extreme, some species are at present adapted to its diversified districts; if for instance it became cooler, species would migrate from its more temperate parts and from its higher land; if damper, from its damper regions, etc. On a

[18] In the MS. the author has added between the lines, "As world has been getting hotter, there has been radiation from high-lands—old view? —curious; I presume Diluvian in origin."

small and isolated island, however, with few species, and these not adapted to much diversified conditions, such changes instead of merely increasing the number of certain species already adapted to such conditions, and decreasing the number of other species, would be apt to affect the constitutions of some of the insular species: thus if the island became damper it might well happen that there were no species living in any part of it adapted to the consequences resulting from more moisture. In this case therefore, and still more (as we have seen) during the production of new stations from the elevation of the land, an island would be a far more fertile source, as far as we can judge, of new specific forms than a continent. The new forms thus generated on an island, we might expect, would occasionally be transported by accident or through long-continued geographical changes be enabled to emigrate and thus become slowly diffused.

But if we look to the origin of a continent almost every geologist will admit that in most cases it will have first existed as separate islands which gradually increased in size; and therefore all that which has been said concerning the probable changes of the forms tenanting a small archipelago is applicable to a continent in its early state. Furthermore, a geologist who reflects on the geological history of Europe (the only region well known) will admit that it has been many times depressed, raised, and left stationary. During the sinking of a continent and the probable generally accompanying changes of climate the effect would be little, *except* on the numerical proportions and in the extinction (from the lessening of rivers, the drying of marshes, and the conversion of high-lands into low, etc.) of some or of many of the species. As soon however as the continent became divided into many isolated portions or islands, preventing free immigration from one part to another, the effect of climatic and other changes on the species would be greater. But let the now broken continent, forming isolated islands, begin to rise and new stations thus to be formed, exactly as in the first case of the upheaved volcanic islet, and we shall have equally favourable conditions for the modification of old forms, that is the formation of new races or species. Let the islands become reunited into a continent; and then the new and old forms would all spread, as far as barriers, the means of transportal, and the preoccupation of the land by other species, would permit. Some of the new species or races would probably become extinct, and some perhaps would cross and blend together. We should thus have a multitude of forms, adapted to all kinds of slightly different stations, and to diverse groups of either antagonist or food-serving species. The oftener these oscillations of level had taken place (and therefore generally the older the land) the greater the number of species which would tend to be formed. The inhabitants of a continent being thus derived in the first stage from the same original parents, and subsequently from the inhabitants of one wide area, since often broken up and reunited, all would be obviously related together and the inhabitants of the most *dissimilar* stations on the same continent would be more closely allied than the inhabitants of two very *similar* stations on two of the main divisions of the world.

I need hardly point out that we now can obviously see why the number of species in two districts, independently of the number of stations in such districts, should be in some cases as widely different as in New Zealand and the Cape of Good Hope. We can see, knowing the difficulty in the transport of terrestrial mammals, why islands far from mainlands do not possess them; we see the general reason, namely, accidental transport (though not the precise reason), why certain islands should, and others should not, possess members of the class of reptiles. We can see why an ancient channel of communication between two distant points, as the Cordillera probably was between southern Chile and the United States during the former cold periods; and icebergs between the Falkland Islands and Tierra del Fuego; and gales, at a former or present time, between the Asiatic shores of the Pacific and eastern islands in this ocean; is connected with (or we may now say causes) an affinity between the species, though distinct, in two such districts. We can see how the better chance of diffusion, from several of the species of any genus having wide ranges in their own countries, explains the presence of other species of the same genus in other countries; and on the other hand, of species of restricted powers of ranging, forming genera with restricted ranges.

As everyone would be surprised if two exactly similar but

peculiar varieties of any species were raised by man by long continued selection in two different countries, or at two very different periods, so we ought not to expect that an exactly similar form would be produced from the modification of an old one in two distinct countries or at two distinct periods. For in such places and times they would probably be exposed to somewhat different climates and almost certainly to different associates. Hence we can see why each species appears to have been produced singly, in space and in time. I need hardly remark that, according to this theory of descent, there is no necessity of modification in a species when it reaches a new and isolated country. If it be able to survive and if slight variations better adapted to the new conditions are not selected, it might retain (as far as we can see) its old form for an indefinite time. As we see that some sub-varieties produced under domestication are more variable than others, so in nature, perhaps, some species and genera are more variable than others. The same precise form, however, would probably be seldom preserved through successive geological periods, or in widely and differently conditioned countries.

Finally, during the long periods of time and probably of oscillations of level necessary for the formation of a continent, we may conclude (as above explained) that many forms would become extinct. These extinct forms, and those surviving (whether or not modified and changed in structure), will all be related in each continent in the same manner and degree, as are the inhabitants of any two different sub-regions in that same continent. I do not mean to say that, for instance, the present marsupials of Australia or Edentata and rodents of South America have descended from any one of the few fossils of the same orders which have been discovered in these countries. It is possible that, in a very few instances, this may be the case; but generally they must be considered as merely co-descendants of common stocks. I believe in this, from the improbability, considering the vast number of species, which (as explained in the last chapter) must by our theory have existed, that the *comparatively* few fossils which have been found should chance to be the immediate and linear progenitors of those now existing. Recent as the yet discovered fossil mammifers of South America are, who will pretend to say that very many intermediate forms may not have existed? Moreover, we shall see in the ensuing chapter that the very existence of genera and species can be explained only by a few species of each epoch leaving modified successors or new species to a future period; and the more distant that future period, the fewer will be the *linear* heirs of the former epoch. As by our theory, all mammifers must have descended from the same parent stock, so is it necessary that each land now possessing terrestrial mammifers shall at some time have been so far united to other land as to permit the passage of mammifers; and it accords with this necessity that in looking far back into the earth's history we find, first changes in the geographical distribution, and secondly a period when the mammiferous forms most distinctive of two of the present main divisions of the world were living together.

I think then I am justified in asserting that most of the above enumerated and often trivial points in the geographical distribution of past and present organisms (which points must be viewed by the creationists as so many ultimate facts) follow as a simple consequence of specific forms being mutable and of their being adapted by natural selection to diverse ends, conjoined with their powers of dispersal, and the geologico-geographical changes now in slow progress and which undoubtedly have taken place. This large class of facts being thus explained, far more than counterbalances many separate difficulties and apparent objections in convincing my mind of the truth of this theory of common descent.

Improbability of Finding Fossil Forms Intermediate Between Existing Species

There is one observation of considerable importance that may be here introduced, with regard to the improbability of the chief transitional forms between any two species being found fossil. With respect to the finer shades of transition, I have before remarked that no one has any cause to expect to trace them in a fossil state, without he be bold enough to imagine that geologists at a future epoch will be able to trace from fossil bones the gradations between the short-horns, Herefordshire, and Alderney breeds of cattle. I have attempted to show that rising islands, in process of formation, must be the best nurseries of new specific forms, and these points are the least favourable for the embedment of fossils: I appeal, as evidence, to the state of the *numerous* scattered islands in the several great oceans: how rarely do any sedimentary deposits occur on them; and when present they are mere narrow fringes of no great antiquity, which the sea is generally wearing away and destroying. The cause of this lies in isolated islands being generally volcanic and rising points; and the effects of subterranean elevation is to bring up the surrounding newly deposited strata within the destroying action of the coast-waves: the strata, deposited at greater distances, and therefore in the depths of the ocean, will be almost barren of organic remains. These remarks may be generalised: periods of subsidence will always be most favourable to an accumulation of great thicknesses of strata, and consequently to their long preservation; for without one formation be protected by successive strata, it will seldom be preserved to a distant age, owing to the enormous amount of denudation, which seems to be a general contingent of time. I may refer, as evidence of this remark, to the vast amount of subsidence evident in the great pile of the European formations, from the Silurian epoch to the end of the Secondary, and perhaps to even a later period. Periods of elevation on the other hand cannot be favourable to the accumulation of strata and their preservation to distant ages, from the circumstance just alluded to, viz., of elevation tending to bring to the surface the circum-littoral strata (always abounding most in fossils) and destroying them. The bottom of tracts of deep water (little favourable, however, to life) must be excepted from this unfavourable influence of elevation. In the quite open ocean, probably no sediment is accumulating, or at a rate so slow as not to preserve fossil remains, which will always be subject to disintegration. Caverns, no doubt, will be equally likely to preserve terrestrial fossils in periods of elevation and of subsidence; but whether it be owing to the enormous amount of denudation, which all land seems to have undergone, no cavern with fossil bones has been found belonging to the Secondary period.

Hence many more remains will be preserved to a distant

age, in any region of the world, during periods of its subsidence than of its elevation.

But during the subsidence of a tract of land, its inhabitants
(as before shown) will from the decrease of space and of the
diversity of its stations, and from the land being fully preoccupied by species fitted to diversified means of subsistence,
be little liable to modification from selection, although many
may, or rather must, become extinct. With respect to its cir-
cum-marine inhabitants, although during a change from a
continent to a *great* archipelago, the number of stations fitted
for marine beings will be increased, their means of diffusion
(an important check to change of form) will be greatly improved; for a continent stretching north and south, or a quite
open space of ocean, seems to be to them the only barrier.
On the other hand, during the elevation of a small archipelago
and its conversion into a continent, we have, whilst the number
of stations are increasing, both for aquatic and terrestrial
productions, and whilst these stations are not fully preoccupied
by perfectly adapted species, the most favourable conditions
for the selection of new specific forms; but few of them in their
early transitional states will be preserved to a distant epoch.
We must wait during an enormous lapse of time, until long-
continued subsidence shall have taken the place in this quarter
of the world of the elevatory process, for the best conditions of
the embedment and the preservation of its inhabitants. Generally the great mass of the strata in every country, from having
been chiefly accumulated during subsidence, will be the tomb
not of transitional forms but of those either becoming extinct
or remaining unmodified.

The state of our knowledge, and the slowness of the changes
of level, do not permit us to test the truth of these remarks,
by observing whether there are more transitional or "fine" (as
naturalists would term them) species, on a rising and enlarging
tract of land, than on an area of subsidence. Nor do I know
whether there are more "fine" species on isolated volcanic
islands in process of formation, than on a continent; but I may
remark that at the Galapagos Archipelago the number of
forms which according to some naturalists are true species,
and according to others are mere races, is considerable: this
particularly applies to the different species or races of the same

genera inhabiting the different islands of this archipelago. Furthermore it may be added (as bearing on the great facts discussed in this chapter) that when naturalists confine their attention to any one country, they have comparatively little difficulty in determining what forms to call species and what to call varieties; that is, those which can or cannot be traced or shown to be probably descendants of some other form: but the difficulty increases as species are brought from many stations, countries, and islands. It was this increasing (but I believe in few cases insuperable) difficulty which seems chiefly to have urged Lamarck to the conclusion that species are mutable.

CHAPTER VII

On the Nature of the Affinities and Classification of Organic Beings [1]

Gradual Appearance and Disappearance of Groups

It has been observed from the earliest times that organic beings fall into groups, and these groups into others of several values, such as species into genera, and then into sub-families, into families, orders, etc. The same fact holds with those beings which no longer exist. Groups of species seem to follow the same laws in their appearance and extinction as do the individuals of any one species: we have reason to believe that, first, a few species appear, that their numbers increase; and that, when tending to extinction, the numbers of the species decrease, till finally the group becomes extinct, in the same way as a species becomes extinct, by the individuals becoming rarer and rarer.

[1] In the present *Essay* the author adds a note: "The obviousness of the fact [i.e. the natural grouping of organisms] alone prevents it being remarkable. It is scarcely explicable by creationist: groups of aquatic, of vegetable feeders and carnivorous, etc., might resemble each other; but why as it is. So with plants, analogical resemblance thus accounted for. Must not here enter into details."

Moreover, groups, like the individuals of a species, appear to become extinct at different times in different countries. The *Palaeotherium* was extinct much sooner in Europe than in India: the *Trigonia* was extinct in early ages in Europe, but now lives in the seas of Australia. As it happens that one species of a family will endure for a much longer period than another species, so we find that some whole groups, such as Mollusca, tend to retain their forms, or to remain persistent, for longer periods than other groups, for instance than the Mammalia. Groups therefore, in their appearance, extinction, and rate of change or succession, seem to follow nearly the same laws with the individuals of a species.

What Is the Natural System?

The proper arrangement of species into groups, according to the natural system, is the object of all naturalists; but scarcely two naturalists will give the same answer to the question, What is the natural system and how are we to recognize it? The most important characters it might be thought (as it was by the earliest classifiers) ought to be drawn from those parts of the structure which determine its habits and place in the economy of nature, which we may call the final end of its existence. But nothing is further from the truth than this; how much external resemblance there is between the little otter (*Chironectes*) of Guiana and the common otter; or again between the common swallow and the swift; and who can doubt that the means and ends of their existence are closely similar, yet how grossly wrong would be the classification which put close to each other a marsupial and placental animal, and two birds with widely different skeletons. Relations, such as in the two latter cases, or as that between the whale and fishes, are denominated "analogical," or are sometimes described as "relations of adaption." They are infinitely numerous and often very singular, but are of no use in the classification of the higher groups. How it comes that certain parts of the structure, by which the habits and functions of the species are settled, are of no use in classification, whilst other parts, formed at the same time, are of the greatest, it would be difficult to say, on the theory of separate creations.

Some authors as Lamarck, Whewell, etc., believe that the
degree of affinity on the natural system depends on the degrees
of resemblance in organs more or less physiologically important
for the preservation of life. This scale of importance in the organs
is admitted to be of difficult discovery. But quite independent of
this, the proposition, as a general rule, must be rejected as false;
though it may be partially true. For it is universally admitted
that the same part or organ, which is of the highest service in
classification in one group, is of very little use in another group,
though in both groups, as far as we can see, the part or organ is
of equal physiological importance: moreover, characters quite
unimportant physiologically, such as whether the covering of the
body consists of hair or feathers, whether the nostrils communicated with the mouth, etc., are of the highest generality in classification; even colour, which is so inconstant in many species,
will sometimes well characterise even a whole group of species.
Lastly, the fact, that no one character is of so much importance
in determining to what great group an organism belongs, as the
forms through which the embryo passes from the germ upwards
to maturity, cannot be reconciled with the idea that natural
classification follows according to the degrees of resemblance in
the parts of most physiological importance. The affinity of the
common rock-barnacle with the crustaceans can hardly be perceived in more than a single character in its mature state, but
whilst young, locomotive, and furnished with eyes, its affinity
cannot be mistaken. The cause of the greater value of characters,
drawn from the early stages of life, can, as we shall in a succeeding chapter see, be in a considerable degree explained, on the
theory of descent, although inexplicable on the views of the
creationist.

Practically, naturalists seem to classify according to the resemblance of those parts or organs which in related groups are
most uniform, or vary least: thus the aestivation, or manner in
which the petals, etc., are folded over each other, is found to
afford an unvarying character in most families of plants, and
accordingly any difference in this respect would be sufficient to
cause the rejection of a species from many families; but in the
Rubiaceae the aestivation is a varying character, and a botanist
would not lay much stress on it, in deciding whether or not to
class a new species in this family. But this rule is obviously so

arbitrary a formula that most naturalists seem to be convinced that something ulterior is represented by the natural system; they appear to think that we only discover by such similarities what the arrangement of the system is, not that such similarities make the system. We can only thus understand Linnaeus's well-known saying, that the characters do not make the genus; but that the genus gives the characters: for a classification, independent of characters, is here presupposed. Hence many naturalists have said that the natural system reveals the plan of the Creator: but without it be specified whether order in time or place, or what else is meant by the plan of the Creator, such expressions appear to me to leave the question exactly where it was.

Some naturalists consider that the geographical position of a species may enter into the consideration of the group into which it should be placed; and most naturalists (either tacitly or openly) give value to the different groups, not solely by their relative differences in structure, but by the number of forms included in them. Thus a genus containing a few species might be, and has often been, raised into a family on the discovery of several other species. Many natural families are retained, although most closely related to other families, from including a great number of closely similar species. The more logical naturalist would perhaps, if he could, reject these two contingents in classification. From these circumstances, and especially from the undefined objects and criterions of the natural system, the number of divisions, such as genera, sub-families, families, etc., has been quite arbitrary; without the clearest definition, how can it be possible to decide whether two groups of species are of equal value, and of what value, whether they should both be called genera or families; or whether one should be a genus, and the other a family? [2]

On the Kind of Relation between Distinct Groups

I have only one other remark on the affinities of organic beings; that is, when two quite distinct groups approach each other, the

[2] Footnote by the author: "I discuss this because if Quinarism true, I false." The Quinary system is set forth in W. S. Macleay's *Horae Entomologicae* (1821).

approach is *generally* generic and not special; I can explain this most easily by an example: of all rodents the bizcacha, by certain peculiarities in its reproductive system, approaches nearest to the marsupials; of all marsupials the *Phascolomys,* on the other hand, appears to approach in the form of its teeth and intestines nearest to the rodents; but there is no special relation between these two genera; the bizcacha is no nearer related to the *Phascolomys* than to any other marsupial in the points in which it approaches this division; nor again is the *Phascolomys,* in the points of structure in which it approaches the rodents, any nearer related to the bizcacha than to any other rodent. Other examples might have been chosen, but I have given (from Waterhouse) this example as it illustrates another point, namely, the difficulty of determining what are analogical or adaptive and what real affinities; it seems that the teeth of the *Phascolomys* though *appearing closely* to resemble those of a rodent are found to be built on the marsupial type; and it is thought that these teeth and consequently the intestines may have been adapted to the peculiar life of this animal and therefore may not show any real relation. The structure in the bizcacha that connects it with the marsupials does not seem a peculiarity related to its manner of life, and I imagine that no one would doubt that this shows a real affinity, though not more with any one marsupial species than with another. The difficulty of determining what relations are real and what analogical is far from surprising when no one pretends to define the meaning of the term relation or the ulterior object of all classification. We shall immediately see on the theory of descent how it comes that there should be "real" and "analogical" affinities; and why the former alone should be of value in classification—difficulties which it would be I believe impossible to explain on the ordinary theory of separate creations.

Classification of Races or Varieties

Let us now for a few moments turn to the classification of the generally acknowledged varieties and sub-divisions of our domestic beings; we shall find them systematically arranged in groups of higher and higher value. De Candolle has treated the varieties of the cabbage exactly as he would have done a natural family with various divisions and sub-divisions. In dogs again we have one main division which may be called the *family* of hounds; of these, there are several (we will call them) *genera,* such as blood-hounds, fox-hounds, and harriers; and of each of these we have different *species,* as the blood-hound of Cuba and that of England; and of the latter again we have breeds truly producing their own kind, which may be called races or varieties. Here we see a classification practically used which typifies on a lesser scale that which holds good in nature. But amongst true species in the natural system and amongst domestic races the number of divisions or groups, instituted between those most alike and those most unlike, seems to be quite arbitrary. The number of the forms in both cases seems practically, whether or not it ought theoretically, to influence the denomination of groups including them. In both, geographical distribution has sometimes been used as an aid to classification; amongst varieties, I may instance the cattle of India or the sheep of Siberia, which from possessing some characters in common permit a classification of Indian and European cattle, or Siberian and European sheep. Amongst domestic varieties we have even something very like the relations of "analogy" or "adaptation"; thus the common and Swedish turnip are both artificial varieties which strikingly resemble each other, and they fill nearly the same end in the economy of the farm-yard; but although the Swede so much more resembles a turnip than its presumed parent the field cabbage, no one thinks of putting it out of the cabbage into the turnips. Thus the greyhound and racehorse, having been selected and trained for extreme fleetness for short distances, present an analogical resemblance of the same kind as, but less striking than, that between the little otter (marsupial) of Guiana and the common otter; though these two otters are really less related than are the horse and dog. We are even cautioned by authors treating of varieties to follow the *natural* in contradistinction of an artificial system and not, for instance, to class two varieties of the pineapple near each other, because their fruits accidentally resemble each other closely (though the fruit may be called the *final end* of this plant in the economy of its world, the hothouse), but to judge from the general resemblance of the entire plants. Lastly, varieties often become extinct; sometimes from unexplained causes, sometimes from accident, but more often from the production of more useful varieties, and the less useful ones being destroyed or bred out.

I think it cannot be doubted that the main cause of all the varieties which have descended from the aboriginal dog or dogs, or from the aboriginal wild cabbage, not being equally like or unlike—but on the contrary, obviously falling into groups and sub-groups—must in chief part be attributed to different degrees of true relationship; for instance, that the different kinds of blood-hound have descended from one stock, whilst the harriers have descended from another stock, and that both these have descended from a different stock from that which has been the parent of the several kinds of greyhound. We often hear of a florist having some choice variety and breeding from it a whole group of sub-varieties more or less characterised by the peculiarities of the parent. The case of the peach and nectarine, each with their many varieties, might have been introduced. No doubt the relationship of our different domestic breeds has been obscured in an extreme degree by their crossing; and likewise from the slight difference between many breeds it has probably often happened that a "sport" from one breed has less closely resembled its parent breed than some other breed, and has therefore been classed with the latter. Moreover the effects of a similar climate may in some cases have more than counterbalanced the similarity, consequent on a common descent, though I should think the similarity of the breeds of cattle of India or sheep of Siberia was far more probably due to the community of their descent than to the effects of climate on animals descended from different stocks.

Notwithstanding these great sources of difficulty, I apprehend everyone would admit, that if it were possible, a genealogical classification of our domestic varieties would be the most satisfactory one; and as far as varieties were concerned would be the natural system: in some cases it has been followed. In attempting to follow out this object a person would have to class a variety, whose parentage he did not know, by its external characters; but he would have a distinct ulterior object in view, namely, its descent in the same manner as a regular systematist seems also to have an ulterior but undefined end in all his classifications. Like the regular systematist he would not care whether his characters were drawn from more or less important organs as long as

he found in the tribe which he was examining that the characters
from such parts were persistent; thus amongst cattle he does
value a character drawn from the form of the horns more than
from the proportions of the limbs and whole body, for he finds
that the shape of the horns is to a considerable degree persistent
amongst cattle,[3] whilst the bones of the limbs and body vary. No
doubt as a frequent rule the more important the organ, as being
less related to external influences, the less liable it is to variation;
but he would expect that according to the object for which the
races had been selected, parts more or less important might
differ; so that characters drawn from parts generally most liable
to vary, as colour, might in some instances be highly serviceable
—as is the case. He would admit that general resemblances
scarcely definable by language might sometimes serve to allocate
a species by its nearest relation. He would be able to assign a
clear reason why the close similarity of the fruit in two varieties
of pineapple, and of the so-called root in the common and
Swedish turnips, and why the similar gracefulness of form in
the greyhound and racehorse, are characters of little value in
classification; namely, because they are the result not of community of descent but either of selection for a common end, or
of the effects of similar external conditions.

Classification of "Races" and Species Similar

Thus seeing that both the classifiers of species and of varieties
work by the same means, make similar distinctions in the value
of the characters, and meet with similar difficulties, and that both
seem to have in their classification an ulterior object in view; I
cannot avoid strongly suspecting that the same cause which has
made amongst our domestic varieties groups and sub-groups has
made similar groups (but of higher values) amongst species; and
that this cause is the greater or less propinquity of actual descent.
The simple fact of species, both those long since extinct and those
now living, being divisible into genera, families, orders, etc.—
divisions analogous to those into which varieties are divisible—
is otherwise an inexplicable fact, and only not remarkable from
its familiarity.

[3] In the margin Marshall is given as the authority.

Origin of Genera and Families

Let us suppose for example that a species spreads and arrives at six or more different regions, or being already diffused over one wide area, let this area be divided into six distinct regions, exposed to different conditions, and with stations slightly different, not fully occupied with other species, so that six different races or species were formed by selection, each best fitted to its new habits and station. I must remark that in every case, if a species becomes modified in any one sub-region, it is probable that it will become modified in some other of the sub-regions over which it is diffused, for its organization is shown to be capable of being rendered plastic; its diffusion proves that it is able to struggle with the other inhabitants of the several sub-regions; and as the organic beings of every great region are in some degree allied, and as even the physical conditions are often in some respects alike, we might expect that a modification in structure, which gave our species some advantage over antagonist species in one sub-region, would be followed by other modifications in other of the sub-regions. The races or new species supposed to be formed would be closely related to each other; and would either form a new genus or sub-genus, or would rank (probably forming a slightly different section) in the genus to which the parent species belonged. In the course of ages, and during the contingent physical changes, it is probable that some of the six new species would be destroyed; but the same advantage, whatever it may have been (whether mere tendency to vary, or some peculiarity of organisation, power of mind, or means of distribution), which in the parent-species and in its six selected and changed species-offspring caused them to prevail over other antagonist species would generally tend to preserve some or many of them for a long period. If then, two or three of the six species were preserved, they in their turn would, during continued changes, give rise to as many small groups of species: if the parents of these small groups were closely similar, the new species would form one great genus, barely perhaps divisible into two or three sections: but if the parents were considerably unlike, their species-offspring would, from inheriting most of the peculiarities of their parent-stocks, form either two or more sub-

genera or (if the course of selection tended in different ways)
genera. And lastly species descending from different species of
the newly formed genera would form new genera, and such
genera collectively would form a family.

The extermination of species follows from changes in the external conditions, and from the increase or immigration of more
favoured species: and as those species which are undergoing
modification in any one great region (or indeed over the world)
will very often be allied ones from (as just explained) partaking
of many characters, and therefore advantages in common, so the
species whose place the new or more favoured ones are seizing,
from partaking of a common inferiority (whether in any particular point of structure, or of general powers of mind, of means
of distribution, of capacity for variation, etc.), will be apt to be
allied. Consequently species of the same genus will slowly, one
after the other, *tend* to become rarer and rarer in numbers, and
finally extinct; and as each last species of several allied genera
fails, even the family will become extinct. There may of course
be occasional exceptions to the entire destruction of any genus
or family. From what has gone before, we have seen that the
slow and successive formation of several new species from the
same stock will make a new genus, and the slow and successive
formation of several other new species from another stock will
make another genus; and if these two stocks were allied, such
genera will make a new family. Now, as far as our knowledge
serves, it is in this slow and gradual manner that groups of
species appear on, and disappear from, the face of the earth.

The manner in which, according to our theory, the arrangement of species in groups is due to partial extinction will perhaps
be rendered clearer in the following way. Let us suppose in any
one great class, for instance in the Mammalia, that every species
and every variety, during each successive age, had sent down one
unaltered descendant (either fossil or living) to the present time;
we should then have had one enormous series, including by small
gradations every known mammiferous form; and consequently
the existence of groups,[4] or chasms in the series, which in some
parts are in greater width, and in some of less, is solely due to
former species, and whole groups of species, not having thus sent
down descendants to the present time.

---

[4] The author probably intended to write "groups separated by chasms."

With respect to the "analogical" or "adaptive" resemblances between organic beings which are not really related, I will only add that probably the isolation of different groups of species is an important element in the production of such characters: thus we can easily see, in a large increasing island, or even a continent like Australia, stocked with only certain orders of the main classes, that the conditions would be highly favourable for species from these orders to become adapted to play parts in the economy of nature, which in other countries were performed by tribes especially adapted to such parts. We can understand how it might happen that an otter-like animal might have been formed in Australia by slow selection from the more carnivorous marsupial types; thus we can understand that curious case in the southern hemisphere, where there are no auks (but many petrels), of a petrel having been modified into the external general form so as to play the same office in nature with the auks of the northern hemisphere; although the habits and form of the petrels and auks are normally so wholly different. It follows, from our theory, that two orders must have descended from one common stock at an immensely remote epoch; and we can perceive when a species in either order, or in both, shows some affinity to the other order, why the affinity is usually generic and not particular —that is why the bizcacha amongst rodents, in the points in which it is related to the marsupial, is related to the whole group, and not particularly to the *Phascolomys,* which of all Marsupialia is related most to the rodents. For the bizcacha is related to the present Marsupialia only from being related to their common parent-stock; and not to any one species in particular. And generally, it may be observed in the writings of most naturalists, that when an organism is described as intermediate between two *great* groups, its relations are not to particular species of either group, but to both groups, as wholes. A little reflection will show how exceptions (as that of the *Lepidosiren,* a fish closely related to *particular* reptiles) might occur, namely, from a few descendants of those species, which at a very early period branched out from a common parent-stock and so formed the two orders or groups, having survived, in nearly their original state, to the present time.

Finally, then, we see that all the leading facts in the affinities and classification of organic beings can be explained on the theory of the natural system being simply a genealogical one. The similarity of the principles in classifying domestic varieties and true species, both those living and extinct, is at once explained; the rules followed and difficulties met with being the same. The existence of genera, families, orders, etc., and their mutual relations naturally ensues from extinction going on at all periods amongst the diverging descendants of a common stock. These terms of affinity, relations, families, adaptive characters, etc., which naturalists cannot avoid using, though metaphorically, cease being so, and are full of plain signification.

CHAPTER VIII

Unity of Type in the Great Classes and Morphological Structures

Unity of Type

Scarcely anything is more wonderful or has been oftener insisted on than that the organic beings in each great class, though living in the most distant climes and at periods immensely remote, though fitted to widely different ends in the economy of nature, yet all in their internal structure evince an obvious uniformity. What, for instance, is more wonderful than that the hand to clasp, the foot or hoof to walk, the bat's wing to fly, the porpoise's fin to swim, should all be built on the same plan? And that the bones in their position and number should be so similar that they can all be classed and called by the same names? Occasionally some of the bones are merely represented by an apparently useless smooth style, or are soldered closely to other bones, but the unity of type is not by this destroyed, and hardly rendered less clear. We see in this fact some deep bond of union between the organic beings of the same great classes—to illustrate which is the object and foundation of the natural system. The perception of this bond, I may add, is the evident cause that naturalists make an ill-defined distinction between true and adaptive affinities.

Morphology

There is another allied or rather almost identical class of facts admitted by the least visionary naturalists and included under the name of morphology. These facts show that in an individual organic being, several of its organs consist of some other organ metamorphosed: thus the sepals, petals, stamens, pistils, etc., of every plant can be shown to be metamorphosed leaves; and thus not only can the number, position, and transitional states of these several organs, but likewise their monstrous changes, be most lucidly explained. It is believed that the same laws hold good with the gemmiferous vesicles of zoophytes. In the same manner the number and position of the extraordinarily complicated jaws and palpi of Crustacea and of insects, and likewise their differences in the different groups, all become simple, on the view of these parts, or rather legs and all metamorphosed appendages, being metamorphosed legs. The skulls, again, of the Vertebrata are composed of three metamorphosed vertebrae, and thus we can see a meaning in the number and strange complication of the bony case of the brain. In this latter instance, and in that of the jaws of the Crustacea, it is only necessary to see a series taken from the different groups of each class to admit the truth of these views. It is evident that when in each species of a group its organs consist of some other part metamorphosed that there must also be a "unity of type" in such a group. And in the cases as that above given in which the foot, hand, wing, and paddle are said to be constructed on a uniform type, if we could perceive in such parts or organs traces of an apparent change from some other use or function, we should strictly include such parts or organs in the department of morphology: thus if we could trace in the limbs of the Vertebrata, as we can in their ribs, traces of an apparent change from being processes of the vertebrae, it would be said that in each species of the Vertebrata the limbs were "metamorphosed spinal processes," and that in all the species throughout the class the limbs displayed a "unity of type."

These wonderful parts of the hoof, foot, hand, wing, paddle, both in living and extinct animals, being all constructed on the same framework, and again of the petals, stamina, germens, etc., being metamorphosed leaves, can by the creationist be viewed only as ultimate facts and incapable of explanation; whilst on our theory of descent these facts all necessarily follow: for by this theory all the beings of any one class, say of the mammalia, are supposed to be descended from one parent-stock, and to have been altered by such slight steps as man effects by the selection of chance domestic variations. Now we can see according to this view that a foot might be selected with longer and longer bones, and wider connecting membranes, till it became a swimming organ, and so on till it became an organ by which to flap along the surface or to glide over it, and lastly to fly through the air: but in such changes there would be no tendency to alter the framework of the internal inherited structure. Parts might become lost (as the tail in dogs, or horns in cattle, or the pistils in plants), others might become united together (as in the feet of the Lincolnshire breed of pigs, and in the stamens of many garden flowers); parts of a similar nature might become increased in number (as the vertebrae in the tails of pigs, etc., and the fingers and toes in six-fingered races of men and in the Dorking fowls), but analogous differences are observed in nature and are not considered by naturalists to destroy the uniformity of the types. We can, however, conceive such changes to be carried to such length that the unity of type might be obscured and finally be undistinguishable, and the paddle of the *Plesiosaurus* has been advanced as an instance in which the uniformity of type can hardly be recognized.[1] If, after long and gradual changes in the structure of the co-descendants from any parent stock, evidence (either from monstrosities or from a graduated series) could be still detected of the function which certain parts or organs played in the parent stock, these parts or organs might be strictly determined by their former function with the term "metamorphosed" appended. Naturalists have used this term in the same metaphorical manner as they have been obliged to use the terms of affinity and relation; and when they affirm, for instance, that the jaws of a crab are metamorphosed legs, so that one crab has more legs and fewer jaws than another, they are far from meaning that the jaws, either during the life of the individual crab

---

[1] In the margin C. Bell is given as authority, apparently for the statement about *Plesiosaurus*.

or of its progenitors, were really legs. By our theory this term assumes its literal meaning; and this wonderful fact of the complex jaws of an animal retaining numerous characters, which they would probably have retained if they had really been metamorphosed during many successive generations from true legs, is simply explained.

Embryology

The unity of type in the great classes is shown in another and very striking manner, namely, in the stages through which the embryo passes in coming to maturity. Thus, for instance, at one period of the embryo, the wings of the bat, the hand, hoof, or foot of the quadruped, and the fin of the porpoise do not differ, but consist of a simple undivided bone. At a still earlier period the embryo of the fish, bird, reptile, and mammal all strikingly resemble each other. Let it not be supposed this resemblance is only external; for on dissection, the arteries are found to branch out and run in a peculiar course, wholly unlike that in the full-grown mammal and bird, but much less unlike that in the full-grown fish, for they run as if to aerate blood by branchiae on the neck, of which even the slit-like orifices can be discerned. How wonderful it is that this structure should be present in the embryos of animals about to be developed into such different forms, and of which two great classes respire only in the air. Moreover, as the embryo of the mammal is matured in the parent's body, and that of the bird in an egg in the air, and that of the fish in an egg in the water, we cannot believe that this course of the arteries is related to any external conditions. In all shell-fish (gastropods) the embryo passes through a state analogous to that of the pteropodous Mollusca: amongst insects again, even the most different ones, as the moth, fly, and beetle, the crawling larvae are all closely analogous: amongst the Radiata, the jelly-fish in its embryonic state resembles a polyp, and in a still earlier state an infusorial animalcule—as does likewise the embryo of the polyp. From the part of the embryo of a mammal at one period resembling a fish more than its parent form; from the larvae of all orders of insects more resembling the simpler articulate animals than their parent insects; and from such other cases as the embryo of the jelly-fish resembling a polyp much nearer than the perfect jelly-fish; it has often been asserted that the higher animal in each class passes through the state of a lower animal; for instance, that the mammal amongst the vertebrata passes through the state of a fish: but Müller denies this, and affirms that the young mammal is at no time a fish, as does Owen assert that the embryonic jelly-fish is at no time a polyp, but that mammal and fish, jelly-fish and polyp pass through the same state; the mammal and jelly-fish being only further developed or changed.

As the embryo, in most cases, possesses a less complicated structure than that into which it is to be developed, it might have been thought that the resemblance of the embryo to less complicated forms in the same great class was in some manner a necessary preparation for its higher development; but in fact the embryo, during its growth, may become less, as well as more, complicated. Thus certain female epizoic crustaceans in their mature state have neither eyes nor any organs of locomotion: they consist of a mere sack, with a simple apparatus for digestion and procreation; and when once attached to the body of the fish on which they prey, they never move again during their whole lives: in their embryonic condition, on the other hand, they are furnished with eyes, and with well articulated limbs, actively swim about, and seek their proper object to become attached to. The larvae, also, of some moths are as complicated and are more active than the wingless and limbless females, which never leave their pupa-case, never feed, and never see the daylight.

Attempt to Explain the Facts of Embryology

I think considerable light can be thrown by the theory of descent on these wonderful embryological facts which are common in a greater or less degree to the whole animal kingdom, and in some manner to the vegetable kingdom: on the fact, for instance, of the arteries in the embryonic mammal, bird, reptile, and fish running and branching in the same courses and nearly in the same manner with the arteries in the full-grown fish; on the fact I may add of the high importance to systematic naturalists of the characters and resemblances in the embryonic state, in ascertaining the true position in the natural system of mature organic beings. The following are the considerations which throw light on these curious points.

In the economy, we will say, of a feline animal, the feline structure of the embryo or of the sucking kitten is of quite secondary importance to it; hence, if a feline animal varied (assuming for the time the possibility of this) and if some place in the economy of nature favoured the selection of a longer-limbed variety, it would be quite unimportant to the production by natural selection of a long-limbed breed, whether the limbs of the embryo and kitten were elongated if they *became* so *as soon* as the animal had to provide food for itself. And if it were found after continued selection and the production of several new breeds from one parent-stock that the successive variations had supervened not very early in the youth or embryonic life of each breed (and we have just seen that it is quite unimportant whether it does so or not), then it obviously follows that the young or embryos of the several breeds will continue resembling each other more closely than their adult parents. And again, if two of these breeds became each the parent-stock of several other breeds, forming two genera, the young and embryos of these would still retain a greater resemblance to the one original stock than when in an adult state. Therefore if it could be shown that the period of the slight successive variations does not always supervene at a very early period of life, the greater resemblance or closer unity in type of animals in the young than in the full-grown state would be explained. Before practically [2] endeavouring to discover in our domestic races whether the structure or form of the young has or has not changed in an exactly corresponding degree with the changes of full-grown animals, it will be well to show that it is at least quite *possible* for the primary germinal vesicle to be impressed with a tendency to produce some change on the growing tissues which will not be fully effected till the animal is advanced in life.

From the following peculiarities of structure being inheritable and appearing only when the animal is full-grown—namely, general size, tallness (not consequent on the tallness of the infant), fatness either over the whole body, or local; change of

[2] In the margin is written, "Get young pigeons"; this was afterwards done.

colour in hair and its loss; deposition of bony matter on the legs of horses; blindness and deafness, that is changes of structure in the eye and ear; gout and consequent deposition of chalkstones; and many other diseases, as of the heart and brain, etc.; from all such tendencies being, I repeat, inheritable, we clearly see that the germinal vesicle is impressed with some power which is wonderfully preserved during the production of infinitely numerous cells in the ever changing tissues, till the part ultimately to be affected is formed and the time of life arrived at. We see this clearly when we select cattle with any peculiarity of their horns, or poultry with any peculiarity of their second plumage, for such peculiarities cannot of course reappear till the animal is mature. Hence, it is certainly *possible* that the germinal vesicle may be impressed with a tendency to produce a longlimbed animal, the full proportional length of whose limbs shall appear only when the animal is mature.[3]

In several of the cases just enumerated we know that the first cause of the peculiarity, when *not* inherited, lies in the conditions to which the animal is exposed during mature life; thus to a certain extent general size and fatness, lameness in horses and in a lesser degree blindness, gout, and some other diseases are certainly in some degree caused and accelerated by the habits of life, and these peculiarities when transmitted to the offspring of the affected person reappear at a nearly corresponding time of life. In medical works it is asserted generally that at whatever period an hereditary disease appears in the parent, it tends to reappear in the offspring at the same period. Again, we find that early maturity, the season of reproduction and longevity are transmitted to corresponding periods of life. Dr. Holland has insisted much on children of the same family exhibiting certain diseases in similar and peculiar manners; my father has known three brothers die in very old age in a *singular* comatose state; now to make these latter cases strictly bear, the children of such families ought similarly to suffer at corresponding times of life; this is probably not the case, but such facts show that a tendency in a disease to appear at particular stages of life can be transmitted through the germinal vesicle to different individuals of the same family. It is then certainly possible that diseases affect-

---

[3] In the margin is written: "Aborted organs show, perhaps, something about period at which changes supervene in embryo."

ing widely different periods of life can be transmitted. So little attention is paid to very young domestic animals that I do not know whether any case is on record of selected peculiarities in young animals, for instance, in the first plumage of birds, being transmitted to their young. If, however, we turn to silk-worms, we find that the caterpillars and cocoons (which must correspond to a *very early* period of the embryonic life of mammalia) vary, and that these varieties reappear in the offspring caterpillars and cocoons.

I think these facts are sufficient to render it probable that at whatever period of life any peculiarity (capable of being inherited) appears, whether caused by the action of external influences during mature life, or from an affection of the primary germinal vesicle, it *tends* to reappear in the offspring at the corresponding period of life. Hence (I may add) whatever effect training, that is, the full employment or action of every newly selected slight variation, has in fully developing and increasing such variation, would only show itself in mature age, corresponding to the period of training; in the second chapter I showed that there was in this respect a marked difference in natural and artificial selection, man not regularly exercising or adapting his varieties to new ends, whereas selection by nature presupposes such exercise and adaptation in each selected and changed part. The foregoing facts show and presuppose that slight variations occur at various periods of life *after birth;* the facts of monstrosity, on the other hand, show that many changes take place before birth, for instance, all such cases as extra fingers, harelip, and all sudden and great alterations in structure; and these when inherited reappear during the embryonic period in the offspring. I will only add that at a period even anterior to embryonic life, namely, during the egg state, varieties appear in size and colour (as with the Hertfordshire duck with blackish eggs) which reappear in the eggs; in plants also the capsule and membranes of the seed are very variable and inheritable.

If then the two following propositions are admitted (and I think the first can hardly be doubted), viz., that variation of structure takes place at all times of life, though no doubt far less in amount and seldomer in quite mature life (and then generally taking the form of disease); and secondly, that these variations tend to reappear at a corresponding period of life, which seems at least probable, then we might _a priori_ have expected that in any selected breed the _young_ animal would not partake in a corresponding degree the peculiarities characterising the _full-grown_ parent; though it would in a lesser degree. For during the thousand or ten thousand selections of slight increments in the length of the limbs of individuals necessary to produce a long-limbed breed, we might expect that such increments would take place in different individuals (as we do not certainly know at what period they do take place), some earlier and some later in the embryonic state, and some during early youth; and these increments would reappear in their offspring only at corresponding periods. Hence, the entire length of limb in the new long-limbed breed would only be acquired at the latest period of life, when that one which was latest of the thousand primary increments of length supervened. Consequently, the foetus of the new breed during the earlier part of its existence would remain much less changed in the proportions of its limbs; and the earlier the period the less would the change be.

Whatever may be thought of the facts on which this reasoning is grounded, it shows how the embryos and young of different species might come to remain less changed than their mature parents; and practically we find that the young of our domestic animals, though differing, differ less than their full-grown parents. Thus if we look at the young puppies of the greyhound and bull-dog (the two most obviously modified of the breeds of dog) we find their puppies at the age of six days with legs and noses (the latter measured from the eyes to the tip) of the same length; though in the proportional thicknesses and general appearance of the these parts there is a great difference. So it is with cattle, though the young calves of different breeds are easily recognisable, yet they do not differ so much in their proportions as the full-grown animals. We see this clearly in the fact that it shows the highest skill to select the best forms early in life, either in horses, cattle, or poultry; no one would attempt it only a few hours after birth; and it requires great discrimination to judge with accuracy even during their full youth, and the best judges are sometimes deceived. This shows that the ultimate proportions of the body are not acquired till near mature age. If I had collected sufficient facts to firmly establish the proposition that in artificially selected breeds the embryonic and young animals are not changed in a corresponding degree with their mature parents, I might have omitted all the foregoing reasoning and the attempts to explain how this happens; for we might safely have transferred the proposition to the breeds or species naturally selected; and the ultimate effect would necessarily have been that in a number of races or species descended from a common stock and forming several genera and families the embryos would have resembled each other more closely than full-grown animals. Whatever may have been the form of habits of the parent-stock of the Vertebrata, in whatever course the arteries ran and branched, the selection of variations, supervening after the first formation of the arteries in the embryo, would not tend from variations supervening at corresponding periods to alter their course at that period: hence, the similar course of the arteries in the mammal, bird, reptile, and fish must be looked at as a most ancient record of the embryonic structure of the common parent stock of these four great classes.

A long course of selection might cause a form to become more simple, as well as more complicated; thus the adaptation of a crustaceous animal to live attached during its whole life to the body of a fish might permit with advantage great simplification of structure, and on this view the singular fact of an embryo being more complex than its parent is at once explained.

On the Graduated Complexity in Each Great Class

I may take this opportunity of remarking that naturalists have observed that in most of the great classes a series exists from very complicated to very simple beings; thus in fish, what a range there is between the sand-eel and shark; in the Articulata, between the common crab and the *Daphnia*—between the *Aphis* and butterfly, and between a mite and a spider.[4] Now the observation just made, namely, that selection might tend to simplify, as well as to complicate, explains this; for we can see that during the endless geologico-geographical changes, and consequent isolation of species, a station occupied in other districts by less complicated animals might be left unfilled, and be occupied by a

[4] Note in original: "Scarcely possible to distinguish between non-development and retrograde development."

degraded form of a higher or more complicated class; and it would by no means follow that, when the two regions became united, the degraded organism would give way to the aboriginally lower organism. According to our theory, there is obviously no power tending constantly to exalt species, except the mutual struggle between the different individuals and classes; but from the strong and general hereditary tendency we might expect to find some tendency to progressive complication in the successive production of new organic forms.

Modification by Selection of the Forms of Immature Animals

I have above remarked that the feline form is quite of secondary importance to the embryo and to the kitten. Of course, during any great and prolonged change of structure in the mature animal, it might, and often would be, indispensable that the form of the embryo should be changed; and this could be effected, owing to the hereditary tendency at corresponding ages, by selection, equally well as in mature age: thus if the embryo tended to become, or to remain, either over its whole body or in certain parts, too bulky, the female parent would die or suffer more during parturition; and as in the case of the calves with large hinder quarters, the peculiarity must be either eliminated or the species become extinct. Where an embryonic form has to seek its own food, its structure and adaptation is just as important to the species as that of the full-grown animal; and as we have seen that a peculiarity appearing in a caterpillar (or in a child, as shown by the hereditariness in the milk-teeth) reappears in its offspring, so we can at once see that our common principle of the selection of slight accidental variations would modify and adapt a caterpillar to a new or changing condition, precisely as in the full-grown butterfly. Hence probably it is that caterpillars of different species of the Lepidoptera differ more than those embryos, at a corresponding early period of life, do which remain inactive in the womb of their parents. The parent during successive ages continuing to be adapted by selection for some one object, and the larva for quite another one, we need not wonder at the difference becoming wonderfully great between them; even as great as that between the fixed rock-barnacle and its free, crab-like offspring, which is furnished with eyes and well-articulated, locomotive limbs.

Importance of Embryology in Classification

We are now prepared to perceive why the study of embryonic forms is of such acknowledged importance in classification. For we have seen that a variation, supervening at any time, may aid in the modification and adaptation of the full-grown being; but for the modification of the embryo, only the variations which supervene at a very early period can be seized on and perpetuated by selection: hence there will be less power and less tendency (for the structure of the embryo is mostly unimportant) to modify the young: and hence we might expect to find at this period similarities preserved between different groups of species which had been obscured and quite lost in the full-grown animals. I conceive on the view of separate creations it would be impossible to offer any explanation of the affinities of organic beings thus being plainest and of the greatest importance at that period of life when their structure is not adapted to the final part they have to play in the economy of nature.

Order in Time in Which the Great Classes Have First Appeared

It follows strictly from the above reasoning only that the embryos of (for instance) existing Vertebrata resemble more closely the embryo of the parent-stock of this great class than do full-grown existing Vertebrata resemble their full-grown parent-stock. But it may be argued with much probability that in the earliest and simplest condition of things the parent and embryo must have resembled each other, and that the passage of any animal through embryonic states in its growth is entirely due to subsequent variations affecting *only* the more mature periods of life. If so, the embryos of the existing Vertebrata will shadow forth the full-grown structure of some of those forms of this great class which existed at the earlier periods of the earth's history: and

accordingly, animals with a fish-like structure ought to have preceded birds and mammals; and of fish, that higher organized
division with the vertebrae extending into one division of the
tail ought to have preceded the equal-tailed, because the embryos
of the latter have an unequal tail; and of Crustacea, Entomostraca
ought to have preceded the ordinary crabs and barnacles—
polyps ought to have preceded jelly-fish, and infusorial animal-
cules to have existed before both. This order of precedence in
time in some of these cases is believed to hold good; but I think
our evidence is so exceedingly incomplete regarding the number
and kinds of organisms which have existed during all, especially
the earlier, periods of the earth's history that I should put no
stress on this accordance, even if it held truer than it probably
does in our present state of knowledge.

CHAPTER IX

Abortive or Rudimentary Organs

The Abortive Organs of Naturalists

Parts of structure are said to be "abortive," or when in a still
lower state of development "rudimentary," when the same reasoning power which convinces us that in some cases similar parts
are beautifully adapted to certain ends declares that in others
they are absolutely useless. Thus the rhinoceros, the whale, etc.,
have, when young, small but properly formed teeth, which never
protrude from the jaws; certain bones, and even the entire
extremities, are represented by mere little cylinders or points of
bone, often soldered to other bones: many beetles have exceedingly minute but regularly formed wings lying under their wing-
cases, which latter are united never to be opened: many plants
have, instead of stamens, mere filaments or little knobs; petals
are reduced to scales, and whole flowers to buds, which (as in
the feather hyacinth) never expand. Similar instances are almost
innumerable, and are justly considered wonderful: probably not

one organic being exists in which some part does not bear the stamp of inutility; for what can be clearer, as far as our reasoning powers can reach, than that teeth are for eating, extremities for locomotion, wings for flight, stamens and the entire flowers for reproduction; yet for these clear ends the parts in question are manifestly unfit. Abortive organs are often said to be mere representatives (a metaphorical expression) of similar parts in other organic beings; but in some cases they are more than representatives, for they seem to be the actual organ not fully grown or developed; thus the existence of mammae in the male Vertebrata is one of the oftenest adduced cases of abortion; but we know that these organs in man (and in the bull) have performed their proper function and secreted milk: the cow has normally four mammae and two abortive ones, but these latter in some instances are largely developed and even (?) give milk. Again in flowers, the representatives of stamens and pistils can be traced to be really these parts not developed; Kölreuter has shown, by crossing a dioecious plant (a *Cucubalus*) having a rudimentary pistil with another species having this organ perfect, that in the hybrid offspring the rudimentary part is more developed, though still remaining abortive; now this shows how intimately related in nature the mere rudiment and the fully developed pistil must be.

Abortive organs, which must be considered as useless as far as their ordinary and normal purpose is concerned, are sometimes adapted to other ends: thus the marsupial bones, which properly serve to support the young in the mother's pouch, are present in the male and serve as the fulcrum for muscles connected only with male functions: in the male of the marigold flower the pistil is abortive for its proper end of being impregnated, but serves to sweep the pollen out of the anthers [1] ready to be borne by insects to the perfect pistils in the other florets. It is likely in many cases, yet unknown to us, that abortive organs perform some useful function; but in other cases, for instance in that of teeth embedded in the solid jaw-bone, or of mere knobs, the rudiments of stamens and pistils, the boldest imagination will hardly venture to ascribe to them any function. Abortive parts, even when wholly useless to the individual species, are of great signification in the system of nature; for they are often found to be

[1] This is here stated on the authority of Sprengel.

of very high importance in a natural classification; [2] thus the presence and position of entire abortive flowers, in the grasses, cannot be overlooked in attempting to arrange them according to their true affinities. This corroborates a statement in a previous chapter, viz., that the physiological importance of a part is no index of its importance in classification. Finally, abortive organs often are only developed, proportionally with other parts, in the embryonic or young state of each species; this again, especially considering the classificatory importance of abortive organs, is evidently part of the law (stated in the last chapter) that the higher affinities of organisms are often best seen in the stages towards maturity through which the embryo passes. On the ordinary view of individual creations, I think that scarcely any class of facts in natural history are more wonderful or less capable of receiving explanation.

The Abortive Organs of Physiologists

Physiologists and medical men apply the term "abortive" in a somewhat different sense from naturalists; and their application is probably the primary one; namely, to parts which from accident or disease before birth are not developed or do not grow: thus, when a young animal is born with a little stump in the place of a finger or of the whole extremity, or with a little button instead of a head, or with a mere bead of bony matter instead of a tooth, or with a stump instead of a tail, these parts are said to be aborted. Naturalists on the other hand, as we have seen, apply this term to parts not stunted during the growth of the embryo, but which are as regularly produced in successive generations as any other most essential parts of the structure of the individual: naturalists, therefore, use this term in a metaphorical sense. These two classes of facts, however, blend into each other; by parts accidentally aborted, during the embryonic life of one individual, becoming hereditary in the succeeding generations: thus a cat or dog, born with a stump instead of a tail, tends to transmit stumps to their offspring; and so it is with stumps representing the extremities; and so again with flowers, with de-

[2] In the margin R. Brown is given apparently as the authority for the fact.

fective and rudimentary parts, which are annually produced in new flower-buds and even in successive seedlings. The strong hereditary tendency to reproduce every either congenital or slowly acquired structure, whether useful or injurious to the individual, has been shown in the first part; so that we need feel no surprise at these truly abortive parts becoming hereditary. A curious instance of the force of hereditariness is sometimes seen in two little loose hanging horns, quite useless as far as the function of a horn is concerned, which are produced in hornless races of our domestic cattle. Now I believe no real distinction can be drawn between a stump representing a tail or a horn or the extremities; or a short shrivelled stamen without any pollen; or a dimple in a petal representing a nectary, when such rudiments are regularly reproduced in a race or family, and the true abortive organs of naturalists. And if we had reason to believe (which I think we have not) that all abortive organs had been at some period *suddenly* produced during the embryonic life of an individual, and afterwards become inherited, we should at once have a simple explanation of the origin of abortive and rudimentary organs. In the same manner as during changes of pronunciation certain letters in a word may become useless in pronouncing it, but yet may aid us in searching for its derivation, so we can see that rudimentary organs no longer useful to the individual may be of high importance in ascertaining its descent, that is, its true classification in the natural system.

Abortion from Gradual Disuse

There seems to be some probability that continued disuse of any part or organ, and the selection of individuals with such parts slightly less developed, would in the course of ages produce in organic beings under domesticity races with such parts abortive. We have every reason to believe that every part and organ in an individual becomes fully developed only with exercise of its functions; that it becomes developed in a somewhat lesser degree with less exercise; and if forcibly precluded from all action, such part will often become atrophied. Every peculiarity, let it be remembered, tends, especially where both parents have it, to be inherited. The less power of flight in the common duck compared with the wild must be partly attributed to disuse during successive generations, and as the wing is properly adapted to flight, we must consider our domestic duck in the first stage towards the state of the *Apteryx,* in which the wings are so curiously abortive. Some naturalists have attributed (and possibly with truth) the falling ears so characteristic of most domestic dogs, some rabbits, oxen, cats, goats, horses, etc., as the effects of the lesser use of the muscles of these flexible parts during successive generations of inactive life; and muscles which cannot perform their functions must be considered verging towards abortion. In flowers, again, we see the gradual abortion during successive seedlings (though this is more properly a conversion) of stamens into imperfect petals, and finally into perfect petals. When the eye is blinded in early life the optic nerve sometimes becomes atrophied; may we not believe that where this organ, as is the case with the subterranean mole-like tuco-tuco (*Ctenomys*), is frequently impaired and lost, that in the course of generations the whole organ might become abortive, as it normally is in some burrowing quadrupeds having nearly similar habits with the tuco-tuco?

In as far then as it is admitted as probable that the effects of disuse (together with occasional true and sudden abortions during the embryonic period) would cause a part to be less developed, and finally to become abortive and useless; then during the infinitely numerous changes of habits in the many descendants from a common stock, we might fairly have expected that cases of organs becoming abortive would have been numerous. The preservation of the stump of the tail, as usually happens when an animal is born tailless, we can only explain by the strength of the hereditary principle and by the period in embryo when affected [3]: but on the theory of disuse gradually obliterating a part, we can see, according to the principles explained in the last chapter (viz., of hereditariness at corresponding periods of life, together with the use and disuse of the part in question not being brought into play in early or embryonic life), that organs or parts would tend not to be utterly obliterated, but to be reduced to that state in which they existed in early embryonic life. Owen often speaks of a part in a full grown animal being in an "embryonic condition." Moreover we can thus see why abortive

[3] These words seem to have been inserted as an afterthought.

organs are most developed at an early period of life. Again, by gradual selection, we can see how an organ rendered abortive in its primary use might be converted to other purposes; a duck's wing might come to serve for a fin, as does that of the penguin; an abortive bone might come to serve, by the slow increment and change of place in the muscular fibres, as a fulcrum for a new series of muscles; the pistil of the marigold might become abortive as a reproductive part, but be continued in its function of sweeping the pollen out of the anthers; for if in this latter respect the abortion had not been checked by selection, the species must have become extinct from the pollen remaining enclosed in the capsules of the anthers.

Finally then I must repeat that these wonderful facts of organs formed with traces of exquisite care, but now either absolutely useless or adapted to ends wholly different from their ordinary end, being present and forming part of the structure of almost every inhabitant of this world, both in long past and present times; being best developed and often only discoverable at a very early embryonic period, and being full of signification in arranging the long series of organic beings in a natural system— these wonderful facts not only receive a simple explanation on the theory of long-continued selection of many species from a few common parent-stocks, but necessarily follow from this theory. If this theory be rejected, these facts remain quite inexplicable; without indeed we rank as an explanation such loose metaphors as that of de Candolle's, in which the kingdom of nature is compared to a well-covered table, and the abortive organs are considered as put in for the sake of symmetry!

CHAPTER X

Recapitulation and Conclusion

Recapitulation

I will now recapitulate the course of this work, more fully with respect to the former parts, and briefly as to the latter. In the first chapter we have seen that most, if not all, organic beings, when taken by man out of their natural condition, and bred during several generations, vary; that is, variation is partly due to the direct effect of the new external influences, and partly to the indirect effect on the reproductive system rendering the organisation of the offspring in some degree plastic. Of the variations thus produced, man when uncivilised naturally preserves the life, and therefore unintentionally breeds from those individuals most useful to him in his different states: when even semi-civilised, he intentionally separates and breeds from such individuals. Every part of the structure seems occasionally to vary in a very slight degree, and the extent to which all kinds of peculiarities in mind and body, when congenital and when slowly acquired either from external influences, from exercise, or from disuse are inherited, is truly wonderful. When several breeds are once formed, then crossing is the most fertile source of new breeds. Variation must be ruled, of course, by the health of the new race, by the tendency to return to the ancestral forms, and by unknown laws determining the proportional increase and symmetry of the body. The amount of variation, which has been effected under domestication, is quite unknown in the majority of domestic beings.

In the second chapter it was shown that wild organisms undoubtedly vary in some slight degree: and that the kind of variation, though much less in degree, is similar to that of domestic organisms. It is highly probable that every organic being, if subjected during several generations to new and varying conditions, would vary. It is certain that organisms, living in an *isolated* country which is undergoing geological changes, must in the course of time be so subjected to new conditions; moreover an organism, when by chance transported into a new station, for instance into an island, will often be exposed to new conditions, and be surrounded by a new series of organic beings. If there were no power at work selecting every slight variation, which opened new sources of subsistence to a being thus situated, the effects of crossing, the chance of death and the constant tendency to reversion to the old parent-form, would prevent the production of new races. If there were any selective agency at work it seems impossible to assign any limit to the complexity and beauty of the adaptive structures, which *might* thus be produced: for certainly the limit of possible variation of organic beings, either in a wild or domestic state, is not known.

It was then shown, from the geometrically increasing tendency of each species to multiply (as evidenced from what we know of mankind and of other animals when favoured by circumstances), and from the means of subsistence of each species on an *average* remaining constant, that during some part of the life of each, or during every few generations, there must be a severe struggle for existence; and that less than a grain in the balance will determine which individuals shall live and which perish. In a country, therefore, undergoing changes, and cut off from the free immigration of species better adapted to the new station and conditions, it cannot be doubted that there is a most powerful means of selection, *tending* to preserve even the slightest variation, which aided the subsistence or defence of those organic beings, during any part of their whole existence, whose organisation had been rendered plastic. Moreover, in animals in which the sexes are distinct, there is a sexual struggle, by which the most vigorous, and consequently the best adapted, will oftener procreate their kind.

A new race thus formed by natural selection would be undistinguishable from a species. For comparing, on the one hand, the several species of a genus, and on the other hand several domestic races from a common stock, we cannot discriminate them by the amount of external difference, but only, first, by domestic races not remaining so constant or being so "true" as species are; and secondly by races always producing fertile offspring when crossed. And it was then shown that a race naturally selected—from the variation being slower—from the selection steadily leading towards the same ends, and from every new slight change in structure being adapted (as is implied by its selection) to the new conditions and being fully exercised, and lastly from the freedom from occasional crosses with other species, would almost necessarily be "truer" than a race selected by ignorant or capricious and short-lived man. With respect to the sterility of species when crossed, it was shown not to be a universal character, and when present to vary in degree: sterility also was shown probably to depend less on external than on constitutional differences. And it was shown that when individual animals and plants are placed under new conditions, they become, without losing their healths, as sterile, in the same manner and to the same degree, as hybrids; and it is therefore conceivable that the cross-bred offspring between two species, having different constitutions, might have its constitution affected in the same peculiar manner as when an individual animal or plant is placed under new conditions. Man in selecting domestic races has little wish and still less power to adapt the whole frame to new conditions; in nature, however, where each species survives by a struggle against other species and external nature, the result must be very different.

Races descending from the same stock were then compared with species of the same genus, and they were found to present some striking analogies. The offspring also of races when crossed, that is mongrels, were compared with the cross-bred offspring of species, that is hybrids, and they were found to resemble each other in all their characters, with the one exception of sterility, and even this, when present, often becomes after some generations variable in degree. The chapter was summed up, and it was shown that no ascertained limit to the amount of variation is known; or could be predicted with due time and changes of condition granted. It was then admitted that although the production of new races, undistinguishable from true species, is probable, we must look to the relations in the past and present geographical distribution of the infinitely numerous beings, by which we are surrounded—to their affinities and to their structure—for any direct evidence.

In the third chapter the inheritable variations in the mental phenomena of domestic and of wild organic beings were considered. It was shown that we are not concerned in this work with the first origin of the leading mental qualities; but that tastes, passions, dispositions, consensual movements, and habits all became, either congenitally or during mature life, modified, and were inherited. Several of these modified habits were found to correspond in every essential character with true instincts, and they were found to follow the same laws. Instincts and dispositions, etc., are fully as important to the preservation and increase of a species as its corporeal structure and therefore the natural means of selection would act on and modify them equally with corporeal structures. This being granted, as well as the proposition that mental phenomena are variable, and that the modifications are inheritable, the possibility of the several most complicated instincts being slowly acquired was considered, and it was shown from the very imperfect series in the instincts of the animals now existing, that we are not justified in *prima facie* rejecting a theory of the common descent of allied organisms from the difficulty of imagining the transitional stages in the various now most complicated and wonderful instincts. We were thus led on to consider the same question with respect both to highly complicated organs, and to the aggregate of several such organs, that is individual organic beings; and it was shown, by the same method of taking the existing most imperfect series, that we ought not at once to reject the theory, because we cannot trace the transitional stages in such organs, or conjecture the transitional habits of such individual species.

In Part II the direct evidence of allied forms having descended from the same stock was discussed. It was shown that this theory requires a long series of intermediate forms between the species and groups in the same classes—forms not directly intermediate between existing species, but intermediate with a common parent. It was admitted that if even all the preserved fossils and existing species were collected, such a series would be far from being formed; but it was shown that we have not *good* evidence that the oldest known deposits are contemporaneous with the first appearance of living beings; or that the several subsequent formations are nearly consecutive; or that any one formation preserves a nearly perfect fauna of even the hard marine organisms, which lived in that quarter of the world. Consequently, we have no reason to suppose that more than a small fraction of the organisms which have lived at any one period have ever been preserved; and hence that we ought not to expect to discover the fossilised sub-varieties between any two species. On the other hand, the evidence, though extremely imperfect, drawn from fossil remains, as far as it does go, is in favour of such a series of organisms having existed as that required. This want of evidence of the past existence of almost infinitely numerous intermediate forms is, I conceive, much the weightiest difficulty on the theory of common descent; but I must think that this is due to ignorance necessarily resulting from the imperfection of all geological records.

In the fifth chapter it was shown that new species gradually [1]

[1] The following words were inserted by the author, apparently to replace a doubtful erasure: "The fauna changes singly."

appear, and that the old ones gradually disappear, from the earth; and this strictly accords with our theory. The extinction of species seems to be preceded by their rarity; and if this be so, no one ought to feel more surprise at a species being exterminated than at its being rare. Every species which is not increasing in number must have its geometrical tendency to increase checked by some agency seldom accurately perceived by us. Each slight increase in the power of this unseen checking agency would cause a corresponding decrease in the average numbers of that species, and the species would become rare: we feel not the least surprise at one species of a genus being rare and another abundant; why then should we be surprised at its extinction, when we have good reason to believe that this very rarity is its regular precursor and cause.

In the sixth chapter the leading facts in the geographical distribution of organic beings were considered—namely, the dissimilarity in areas widely and effectively separated, of the organic beings being exposed to very similar conditions (as for instance, within the tropical forests of Africa and America, or on the volcanic islands adjoining them). Also the striking similarity and general relations of the inhabitants of the same great continents, conjoined with a lesser degree of dissimilarity in the inhabitants living on opposite sides of the barriers intersecting it —whether or not these opposite sides are exposed to similar conditions. Also the dissimilarity, though in a still lesser degree, in the inhabitants of different islands in the same archipelago, together with their similarity taken as a whole with the inhabitants of the nearest continent, whatever its character may be. Again, the peculiar relations of alpine floras; the absence of mammifers on the smaller isolated islands; and the comparative fewness of the plants and other organisms on islands with diversified stations: the connexion between the possibility of occasional transportal from one country to another, with an affinity, though not identity, of the organic beings inhabiting them. And lastly, the clear and striking relations between the living and the extinct in the same great divisions of the world; which relation, if we look very far backward, seems to die away. These facts, if we bear in mind the geological changes in progress, all simply follow from the proposition of allied organic beings having lineally descended from common parent-stocks. On the theory of independent creations they must remain, though evidently connected together, inexplicable and disconnected.

In the seventh chapter, the relationship or grouping of extinct and recent species; the appearance and disappearance of groups; the ill-defined objects of the natural classification, not depending on the similarity of organs physiologically important, not being influenced by adaptive or analogical characters, though these often govern the whole economy of the individual, but depending on any character which varies least, and especially on the forms through which the embryo passes, and, as was afterwards shown, on the presence of rudimentary and useless organs. The alliance between the nearest species in *distinct* groups being general and not especial; the close similarity in the rules and objects in classifying domestic races and true species. All these facts were shown to follow on the natural system being a genealogical system.

In the eighth chapter, the unity of structure throughout large groups, in species adapted to the most different lives, and the wonderful metamorphosis (used metaphorically by naturalists) of one part or organ into another, were shown to follow simply on new species being produced by the selection and inheritance of successive *small* changes of structure. The unity of type is wonderfully manifested by the similarity of structure, during the embryonic period, in the species of entire classes. To explain this it was shown that the different races of our domestic animals differ less, during their young state, than when full grown; and consequently, if species are produced like races, the same fact, on a greater scale, might have been expected to hold good with them. This remarkable law of nature was attempted to be explained through establishing, by sundry facts, that slight variations originally appear during all periods of life, and that when inherited they tend to appear at the corresponding period of life; according to these principles, in several species descended from the same parent-stock, their embryos would almost necessarily much more closely resemble each other than they would in their adult state. The importance of these embryonic resemblances, in making out a natural or genealogical classification, thus becomes at once obvious. The occasional greater simplicity of structure in the mature animal than in the embryo; the gradation in complexity of the species in the great classes; the adaptation of the larvae of animals to independent powers of existence; the immense difference in certain animals in their larval and mature states, were all shown on the above principles to present no difficulty.

In the ninth chapter, the frequent and almost general presence of organs and parts, called by naturalists abortive or rudimentary, which, though formed with exquisite care, are generally absolutely useless, though sometimes applied to uses not normal —which cannot be considered as mere representative parts, for they are sometimes capable of performing their proper function, which are always best developed, and sometimes only developed, during a very early period of life, and which are of admitted high importance in classification—were shown to be simply explicable on our theory of common descent.

Why Do We Wish to Reject the Theory of Common Descent?

Thus have many general facts, or laws, been included under one explanation; and the difficulties encountered are those which would naturally result from our acknowledged ignorance. And why should we not admit this theory of descent? Can it be shown that organic beings in a natural state are *all absolutely invariable?* Can it be said that the *limit of variation* or the number of varieties capable of being formed under domestication are known? Can any distinct line be drawn *between a race and a species?* To these three questions we may certainly answer in the negative. As long as species were thought to be divided and defined by an impassable barrier of *sterility,* whilst we were ignorant of geology, and imagined that the *world was of short duration,* and the number of its past inhabitants few, we were justified in assuming individual creations, or in saying with Whewell that the beginnings of all things are hidden from man. Why then do we feel so strong an inclination to reject this theory—especially when the actual case of any two species, or even of any two races, is adduced—and one is asked, have these two originally descended from the same parent womb? I believe it is because we are always slow in admitting any great change of which we do not see the intermediate steps. The mind cannot grasp the full meaning of the term of a million or hundred million years, and cannot consequently add up and perceive the full effects of small successive variations accumulated during almost infinitely many generations. The difficulty is the same with that which, with most geologists, it has taken long years to remove, as when Lyell propounded that great valleys were hollowed out [and long lines of inland cliffs had been formed] by the slow action of the waves of the sea. A man may long view a grand precipice without actually believing, though he may not deny it, that thousands of feet in thickness of solid rock once extended over many square miles where the open sea now rolls; without fully believing that the same sea which he sees beating the rock at his feet has been the sole removing power.

Shall we then allow that the three distinct species of *Rhinoceros* which separately inhabit Java and Sumatra and the neighbouring mainland of Malacca were created, male and female, out of the inorganic materials of these countries? Without any adequate cause, as far as our reason serves, shall we say that they were merely, from living near each other, created very like each other, so as to form a section of the genus dissimilar from the African section, some of the species of which sections inhabit very similar and some very dissimilar stations? Shall we say that without any apparent cause they were created on the same generic type with the ancient woolly rhinoceros of Siberia and of the other species which formerly inhabited the same main division of the world: that they were created, less and less closely related, but still with interbranching affinities, with all the other living and extinct mammalia? That without any apparent adequate cause their short necks should contain the same number of vertebrae with the giraffe; that their thick legs should be built on the same plan with those of the antelope, of the mouse, of the hand of the monkey, of the wing of the bat, and of the fin of the porpoise? That in each of these species the second bone of their leg should show clear traces of two bones having been soldered and united into one; that the complicated bones of their head should become intelligible on the supposition of their having been formed of three expanded vertebrae; that in the jaws of each when dissected young there should exist small teeth which never come to the surface? That in possessing these useless abortive teeth, and in other characters, these three rhinoceroses in their embryonic state should much more closely resemble other mammalia than they do when mature? And lastly, that in a still earlier period of life, their arteries should run and branch as in a fish, to carry the blood to gills which do not exist? Now these three species of rhinoceros closely resemble each other, more closely than many generally acknowledged races of our domestic animals; these three species if domesticated would almost certainly vary, and races adapted to different ends might be selected out of such variations. In this state they would probably breed together, and their offspring would possibly be quite, and probably in some degree, fertile; and in either case, by continued crossing, one of these specific forms might be absorbed and lost in another. I repeat, shall we then say that a pair, or a gravid female, of each of these three species of rhinoceros, were separately created with deceptive appearances of true relationship, with the stamp of inutility on some parts, and of conversion in other parts, out of the inorganic elements of Java, Sumatra, and Malacca? Or have they descended, like our domestic races, from the same parent-stock? For my own part I could no more admit the former proposition than I could admit that the planets move in their courses, and that a stone falls to the ground, not through the intervention of the secondary and appointed law of gravity, but from the direct volition of the Creator.

Before concluding it will be well to show, although this has incidentally appeared, how far the theory of common descent can legitimately be extended. If we once admit that two true species of the same genus can have descended from the same parent, it will not be possible to deny that two species of two genera may also have descended from a common stock. For in some families the genera approach almost as closely as species of the same genus; and in some orders, for instance in the monocotyledonous plants, the families run closely into each other. We do not hesitate to assign a common origin to dogs or cabbages, because they are divided into groups analogous to the groups in nature. Many naturalists indeed admit that all groups are artificial; and that they depend entirely on the extinction of intermediate species. Some naturalists, however, affirm that though driven from considering sterility as the characteristic of species, an entire incapacity to propagate together is the best evidence of the existence of natural genera. Even if we put on one side the undoubted fact that some species of the same genus will not breed together, we cannot possibly admit the above rule, seeing that the grouse and pheasant (considered by some good ornithologists as forming two families), the bull-finch and canary-bird have bred together.

No doubt the more remote two species are from each other, the weaker the arguments become in favour of their common descent. In species of two distinct families the analogy, from the variation of domestic organisms and from the manner of their intermarrying, fails; and the arguments from their geographical distribution quite or almost quite fails. But if we once admit the general principles of this work, as far as a clear unity of type can be made out in groups of species, adapted to play diversified parts in the economy of nature, whether shown in the structure of the embryonic or mature being, and especially if shown by a community of abortive parts, we are legitimately led to admit their community of descent. Naturalists dispute how widely this unity of type extends: most, however, admit that the vertebrata are built on one type; the articulata on another; the mollusca on a third; and the radiata on probably more than one. Plants also appear to fall under three or four great types. On this theory, therefore, all the organisms *yet discovered* are descendants of probably less than ten parent-forms.

Conclusion

My reasons have now been assigned for believing that specific forms are not immutable creations. The terms used by naturalists of affinity, unity of type, adaptive characters, the metamorphosis and abortion of organs cease to be metaphorical expressions and become intelligible facts. We no longer look at an organic being as a savage does at a ship or other great work of art, as at a thing wholly beyond his comprehension, but as a production that has a history which we may search into. How interesting do all instincts become when we speculate on their origin as hereditary habits, or as slight congenital modifications of former instincts perpetuated by the individuals so characterised having been preserved. When we look at every complex instinct and mechanism as the summing up of a long history of contrivances, each most useful to its possessor, nearly in the same way as when we look at a great mechanical invention as the summing up of the labour, the experience, the reason, and even the blunders of numerous workmen. How interesting does the geographical distribution of all organic beings, past and present, become as throwing light on the ancient geography of the world. Geology loses glory from the imperfection of its archives, but it gains in the immensity of its subject. There is much grandeur in looking at every existing organic being either as the lineal successor of some form now buried under thousands of feet of solid rock, or as being the co-descendant of that buried form of some more ancient and utterly lost inhabitant of this world. It accords with what we know of the laws impressed by the Creator on matter that the production and extinction of forms should, like the birth and death of individuals, be the result of secondary means. It is derogatory that the Creator of countless Universes should have made by individual acts of His will the myriads of creeping parasites and worms, which since the earliest dawn of life have swarmed over the land and in the depths of the ocean. We cease to be astonished that a group of animals should have been formed to lay their eggs in the bowels and flesh of other sensitive beings; that some animals should live by and even delight in cruelty; that animals should be led away by false instincts; that annually there should be an incalculable waste of the pollen, eggs, and immature beings; for we see in all this the inevitable consequences of one great law, of the multiplication of organic beings not created immutable. From death, famine, and the struggle for existence, we see that the most exalted end which we are capable of conceiving, namely, the creation of the higher animals, has directly proceeded. Doubtless, our first impression is to disbelieve that any secondary law could produce infinitely numerous organic beings, each characterised by the most exquisite workmanship and widely extended adaptations: it at first accords better with our faculties to suppose that each required the fiat of a Creator. There is a [simple] grandeur in this view of life with its several powers of growth, reproduction, and of sensation having been originally breathed into matter under a few forms, perhaps into only one,[2] and that whilst this planet has gone

[2] These four words are added in pencil between the lines.

cycling onwards according to the fixed laws of gravity and whilst land and water have gone on replacing each other—that from so simple an origin, through the selection of infinitesimal varieties, endless forms most beautiful and most wonderful have been evolved.

The Origin of Species

The Conclusion

. . . It may be asked how far I extend the doctrine of the modification of species. The question is difficult to answer, because the more distinct the forms are which we consider, by so much the arguments in favour of community of descent become fewer in number and less in force. But some arguments of the greatest weight extend very far. All the members of whole classes are connected together by a chain of affinities, and all can be classed on the same principle, in groups subordinate to groups. Fossil remains sometimes tend to fill up very wide intervals between existing orders.

Organs in a rudimentary condition plainly show that an early progenitor had the organ in a fully developed condition; and this in some cases implies an enormous amount of modification in the descendants. Throughout whole classes various structures are formed on the same pattern, and at a very early age the embryos closely resemble each other. Therefore I cannot doubt that the theory of descent with modification embraces all the members of the same great class or kingdom. I believe that animals are descended from at most only four or five progenitors, and plants from an equal or lesser number.

Analogy would lead me one step farther, namely, to the belief that all animals and plants are descended from some one prototype. But analogy may be a deceitful guide. Nevertheless all living things have much in common, in their chemical composition, their cellular structure, their laws of growth, and their liability to injurious influences. We see this even in so trifling a fact as that the same poison often similarly affects plants and animals; or that the poison secreted by the gall-fly produces monstrous growths on the wild rose or oak-tree. With all organic beings, excepting perhaps some of the very lowest, sexual reproduction seems to be essentially similar. With all, as far as is at present known, the germinal vesicle is the same; so that all organisms start from a common origin. If we look even to the two main divisions—namely, to the animal and vegetable kingdoms—certain low forms are so far intermediate in character that naturalists have disputed to which kingdom they should be referred. As Prof. Asa Gray has remarked, "The spores and other reproductive bodies of many of the lower algae may claim to have first a characteristically animal, and then an unequivocally vegetable existence." Therefore, on the principle of natural selection with divergence of character, it does not seem incredible that, from some such low and intermediate form, both animals and plants may have been developed; and, if we admit this, we must likewise admit that all the organic beings which have ever lived on this earth may be descended from some one primordial form. But this inference is chiefly grounded on analogy, and it is immaterial whether or not it be accepted. No doubt it is possible, as Mr. G. H. Lewes has urged, that at the first commencement of life many different forms were evolved; but if so, we may conclude that only a very few have left modified descendants. For, as I have recently remarked in regard to the members of each great kingdom, such as the Vertebrata, Articulata, &c., we have distinct evidence in their embryological, homologous, and rudimentary structures that within each kingdom all the members are descended from a single progenitor.

When the views advanced by me in this volume, and by Mr. Wallace, or when analogous views on the origin of species are generally admitted, we can dimly foresee that there will be a considerable revolution in natural history. Systematists will be able to pursue their labours as at present; but they will not be incessantly haunted by the shadowy doubt whether this or that form be a true species. This, I feel sure and I speak after experience, will be no slight relief. The endless disputes whether or not some fifty species of British brambles are good species will cease. Systematists will have only to decide (not that this will be easy) whether any form be sufficiently constant and distinct from other forms, to be capable of definition; and if definable, whether the differences be sufficiently important to deserve a specific name. This latter point will become a far more essential consideration than it is at present; for differences, however slight, between any two forms, if not blended by intermediate gradations, are looked at by most naturalists as sufficient to raise both forms to the rank of species.

Hereafter we shall be compelled to acknowledge that the only distinction between species and well-marked varieties is that the latter are known, or believed, to be connected at the present day by intermediate gradations, whereas species were formerly thus connected. Hence, without rejecting the consideration of the present existence of intermediate gradations between any two forms, we shall be led to weigh more carefully and to value higher the actual amount of difference between them. It is quite possible that forms now generally acknowledged to be merely varieties may hereafter be thought worthy of specific names; and in this case scientific and common language will come into accordance. In short, we shall have to treat species in the same manner as those naturalists treat genera who admit that genera are merely artificial combinations made for convenience. This may not be a cheering prospect; but we shall at least be freed from the vain search for the undiscovered and undiscoverable essence of the term species.

The other and more general departments of natural history will rise greatly in interest. The terms used by naturalists, of affinity, relationship, community of type, paternity, morphology, adaptive characters, rudimentary and aborted organs, &c., will cease to be metaphorical, and will have a plain signification. When we no longer look at an organic being as a savage looks at a ship, as something wholly beyond his comprehension; when we regard every production of nature as one which has had a long history; when we contemplate every complex structure and instinct as the summing up of many contrivances, each useful to the possessor, in the same way as any great mechanical invention is the summing up of the labour, the experience, the reason, and even the blunders of numerous workmen; when we thus view each organic being, how far more interesting—I speak from experience—does the study of natural history become!

A grand and almost untrodden field of inquiry will be opened on the causes and laws of variation, on correlation, on the effects of use and disuse, on the direct action of external conditions, and so forth. The study of domestic productions will rise immensely in value. A new variety raised by man will be a more important and interesting subject for study than one more species added to the infinitude of already recorded species. Our classifications will come to be, as far as they can be so made, genealogies; and will then truly give what may be called the plan of creation. The rules for classifying will no doubt become simpler when we have a definite object in view. We possess no pedigrees or armorial bearings; and we have to discover and trace the many diverging lines of descent in our natural genealogies by characters of any kind which have long been inherited. Rudimentary organs will speak infallibly with respect to the nature of long-lost structures. Species and groups of species which are called aberrant, and which may fancifully be called living fossils, will aid us in forming a picture of the ancient forms of life. Embryology will often reveal to us the structure, in some degree obscured, of the prototypes of each great class.

When we can feel assured that all the individuals of the same species, and all the closely allied species of most genera, have within a not very remote period descended from one parent, and have migrated from some one birth-place; and when we better know the many means of migration, then, by the light which geology now throws, and will continue to throw, on former changes of climate and of the level of the land, we shall surely be enabled to trace in an admirable manner the former migrations of the inhabitants of the whole world. Even at present, by comparing the differences between the inhabitants of the sea on the opposite sides of a continent, and the nature of the various inhabitants on that continent in relation to their apparent means of immigration, some light can be thrown on ancient geography.

The noble science of geology loses glory from the extreme imperfection of the record. The crust of the earth with its imbedded remains must not be looked at as a well-filled museum, but as a poor collection made at hazard and at rare intervals. The accumulation of each great fossiliferous formation will be recognised as having depended on an unusual concurrence of favourable circumstances, and the blank intervals between the successive stages as having been of vast duration. But we shall be able to gauge with some security the duration of these intervals by a comparison of the preceding and succeeding organic forms. We must be cautious in attempting to correlate as strictly contemporaneous two formations, which do not include many identical species, by the general succession of the forms of life. As species are produced and exterminated by slowly acting and still existing causes, and not by miraculous acts of creation; and as the most important of all causes of organic change is one which is almost independent of altered and perhaps suddenly altered physical conditions, namely, the mutual relation of organism to organism—the improvement of one organism entailing the improvement or the extermination of others; it follows, that the amount of organic change in the fossils of consecutive formations probably serves as a fair measure of the relative, though not actual lapse of time. A number of species, however, keeping in a body might remain for a long period unchanged, whilst within the same period several of these species by migrating into new countries and coming into competition with foreign associates might become modified; so that we must not overrate the accuracy of organic change as a measure of time.

In the future I see open fields for far more important researches. Psychology will be securely based on the foundation already well laid by Mr. Herbert Spencer, that of the necessary acquirement of each mental power and capacity by gradation. Much light will be thrown on the origin of man and his history.

Authors of the highest eminence seem to be fully satisfied with the view that each species has been independently created. To my mind it accords better with what we know of the laws impressed on matter by the Creator that the production and extinction of the past and present inhabitants of the world should have been due to secondary causes, like those determining the birth and death of the individual. When I view all beings not as special creations, but as the lineal descendants of some few beings which lived long before the first bed of the Cambrian system was deposited, they seem to me to become ennobled. Judging from the past, we may safely infer that not one living species will transmit its unaltered likeness to a distant futurity. And of the species now living very few will transmit progeny of any kind to a far distant futurity; for the manner in which all organic beings are grouped shows that the greater number of species in each genus, and all the species in many genera, have left no descendants, but have become utterly extinct. We can so far take a prophetic glance into futurity as to foretell that it will be the common and widely spread species, belonging to the larger and dominant groups within each class, which will ultimately prevail and procreate new and dominant species. As all the living forms of life are the lineal descendants of those which lived long before the Cambrian epoch, we may feel certain that the ordinary succession by generation has never once been broken, and that no cataclysm has desolated the whole world. Hence we may look with some confidence to a secure future of great length. And as natural selection works solely by and for the good of each being, all corporeal and mental endowments will tend to progress towards perfection.

It is interesting to contemplate a tangled bank, clothed with many plants of many kinds, with birds singing on the bushes, with various insects flitting about, and with worms crawling through the damp earth, and to reflect that these elaborately constructed forms, so different from each other, and dependent upon each other in so complex a manner, have all been produced by laws acting around us. These laws, taken in the largest sense, being Growth with Reproduction; Inheritance, which is almost implied by Reproduction; Variability from the indirect and direct action of the conditions of life, and from use and disuse: a Ratio of Increase so high as to lead to a Struggle for Life, and as a consequence to Natural Selection, entailing Divergence of Character and the Extinction of less-improved forms. Thus, from the war of nature, from famine and death, the most exalted object which we are capable of conceiving, namely, the production of the higher animals, directly follows. There is grandeur in this view of life, with its several powers, having been originally breathed by the Creator into a few forms or into one; and that, whilst this planet has gone cycling on according to the fixed law of gravity, from so simple a beginning endless forms most beautiful and most wonderful have been, and are being, evolved.

The Expression of the Emotions
in Man and Animals

Special Expressions

The various species and genera of monkeys express their feelings in many different ways; and this fact is interesting as in some degree bearing on the question whether the so-called races of man should be ranked as distinct species or varieties; for, as we shall see in the following chapters, the different races of man express their emotions and sensations with remarkable uniformity throughout the world. Some of the expressive actions of monkeys are interesting in another way, namely, from being closely analogous to those of man. As I have had no opportunity of observing any one species of the group under all circumstances, my miscellaneous remarks will be best arranged under different states of the mind.

Pleasure, joy, affection. It is not possible to distinguish in monkeys, at least without more experience than I have had, the expression of pleasure or joy from that of affection. Young chimpanzees make a kind of barking noise when pleased by the return of anyone to whom they are attached. When this noise, which the keepers call a laugh, is uttered, the lips are protruded; but so they are under various other emotions. Nevertheless I could perceive that when they were pleased the form of the lips differed a little from that assumed when they were angered. If a young chimpanzee be tickled—and the armpits are particularly sensitive to tickling, as in the case of our children—a more decided chuckling or laughing sound is uttered; though the laughter is sometimes noiseless. The corners of the mouth are then drawn backwards; and this sometimes causes the lower eyelids to be slightly wrinkled. But this wrinkling, which is so characteristic of our own laughter, is more plainly seen in some other monkeys. The teeth in the upper jaw in the chimpanzee are not exposed when they utter their laughing noise, in which respect they differ from us. But their eyes sparkle and grow brighter, as Mr. W. L. Martin,[1] who has particularly attended to their expression, states.

[1] *Natural History of Mammalia* (1841), Vol. I, pp. 383, 410.

Young orangs, when tickled, likewise grin and make a chuckling sound; and Mr. Martin says that their eyes grow brighter. As
soon as their laughter ceases, an expression may be detected
passing over their faces, which, as Mr. Wallace remarked to me,
may be called a smile. I have also noticed something of the same
kind with the chimpanzee. Dr. Duchenne—and I cannot quote
a better authority—informs me that he kept a very tame monkey
in his house for a year; and when he gave it during meal-times
some choice delicacy, he observed that the corners of its mouth
were slightly raised; thus an expression of satisfaction, partaking of the nature of an incipient smile, and resembling that often
seen on the face of man, could be plainly perceived in this animal.

The *Cebus azarae*,[2] when rejoiced at again seeing a beloved
person, utters a peculiar tittering (*kichernden*) sound. It also
expresses agreeable sensations by drawing back the corners of its
mouth, without producing any sound. Rengger calls this movement laughter, but it would be more appropriately called a smile.
The form of the mouth is different when either pain or terror is
expressed, and high shrieks are uttered. Another species of
Cebus in the Zoological Gardens (*C. hypoleucus*) when pleased
makes a reiterated shrill note, and likewise draws back the
corners of its mouth, apparently through the contraction of the
same muscles as with us. So does the Barbary ape (*Inuus ecaudatus*) to an extraordinary degree; and I observed in this monkey that the skin of the lower eyelids then became much
wrinkled. At the same time it rapidly moved its lower jaw or
lips in a spasmodic manner, the teeth being exposed; but the
noise produced was hardly more distinct than that which we
sometimes call silent laughter. Two of the keepers affirmed that
this slight sound was the animal's laughter, and when I expressed some doubt on this head (being at the time quite in-
experienced), they made it attack or rather threaten a hated
Entellus monkey living in the same compartment. Instantly the
whole expression of the face of the Inuus changed; the mouth
was opened much more widely, the canine teeth were more fully
exposed, and a hoarse barking noise was uttered.

The Anubis baboon (*Cynocephalus anubis*) was first insulted
and put into a furious rage, as was easily done, by his keeper,

[2] Rengger (*Säugetheire von Paraquay* [1830], s. 46) kept these monkeys in confinement for seven years in their native country of Paraguay.

who then made friends with him and shook hands. As the reconciliation was effected the baboon rapidly moved up and down his jaws and lips, and looked pleased. When we laugh heartily, a similar movement, or quiver, may be observed more or less distinctly in our jaws; but with man the muscles of the chest are more particularly acted on, whilst with this baboon, and with some other monkeys, it is the muscles of the jaws and lips which are spasmodically affected.

I have already had occasion to remark on the curious manner in which two or three species of Macacus and the *Cynopithecus niger* draw back their ears and utter a slight jabbering noise when they are pleased by being caressed. With the Cynopithecus (fig. 17),* the corners of the mouth are at the same time drawn backwards and upwards, so that the teeth are exposed. Hence this expression would never be recognized by a stranger as one of pleasure. The crest of long hairs on the forehead is depressed, and apparently the whole skin of the head drawn backwards. The eyebrows are thus raised a little, and the eyes assume a staring appearance. The lower eyelids also become slightly wrinkled; but this wrinkling is not conspicuous, owing to the permanent transverse furrows on the face.

Painful emotions and sensations. With monkeys the expression of slight pain, or of any painful emotion, such as grief, vexation, jealousy, &c., is not easily distinguished from that of moderate anger; and these states of mind readily and quickly pass into each other. Grief, however, with some species is certainly exhibited by weeping. A woman who sold a monkey to the Zoological Society believed to have come from Borneo (*Macacus maurus* or *M. inornatus* of Gray) said that it often cried; and Mr. Bartlett, as well as the keeper Mr. Sutton, have repeatedly seen it, when grieved, or even when much pitied, weeping so copiously that the tears rolled down its cheeks. There is, however, something strange about this case, for two specimens subsequently kept in the Gardens, and believed to be the same species, have never been seen to weep, though they were carefully observed by the keeper and myself when much distressed and loudly screaming. Rengger states [3] that the eyes of the *Cebus*

* In the original edition.—S.E.H.

[3] Rengger, ibid. s. 46. Humboldt, *Personal Narrative*, Eng. translat., Vol. IV, p. 527.

azarae fill with tears, but not sufficiently to overflow, when it is prevented getting some much desired object, or is much frightened. Humboldt also asserts that the eyes of the *Callithrix sciureus* "instantly fill with tears when it is seized with fear"; but when this pretty little monkey in the Zoological Gardens was teased so as to cry out loudly, this did not occur. I do not, however, wish to throw the least doubt on the accuracy of Humboldt's statement.

The appearance of dejection in young orangs and chimpanzees when out of health is as plain and almost as pathetic as in the case of our children. This state of mind and body is shown by their listless movements, fallen countenances, dull eyes, and changed complexion.

Anger. This emotion is often exhibited by many kinds of monkeys, and is expressed, as Mr. Martin remarks,[4] in many different ways. "Some species, when irritated, pout the lips, gaze with a fixed and savage glare on their foe, and make repeated short starts as if about to spring forward, uttering at the same time inward guttural sounds. Many display their anger by suddenly advancing, making abrupt starts, at the same time opening the mouth and pursing up the lips, so as to conceal the teeth, while the eyes are daringly fixed on the enemy, as if in savage defiance. Some again, and principally the long-tailed monkeys, or Guenons, display their teeth, and accompany their malicious grins with a sharp, abrupt, reiterated cry." Mr. Sutton confirms the statement that some species uncover their teeth when enraged, whilst others conceal them by the protrusion of their lips; and some kinds draw back their ears. The *Cynopithecus niger,* lately referred to, acts in this manner, at the same time depressing the crest of hair on its forehead, and showing its teeth; so that the movements of the features from anger are nearly the same as those from pleasure; and the two expressions can be distinguished only by those familiar with the animal.

Baboons often show their passion and threaten their enemies in a very odd manner, namely, by opening their mouths widely as in the act of yawning. Mr. Bartlett has often seen two baboons, when first placed in the same compartment, sitting opposite to each other and thus alternately opening their mouths; and this action seems frequently to end in a real yawn. Mr. Bartlett be-

[4] *Nat. Hist. of Mammalia* (1841), p. 351.

lieves that both animals wish to show to each other that they are provided with a formidable set of teeth, as is undoubtedly the case. As I could hardly credit the reality of this yawning gesture, Mr. Bartlett insulted an old baboon and put him into a violent passion; and he almost immediately thus acted. Some species of Macacus and of Cercopithecus [5] behave in the same manner. Baboons likewise show their anger, as was observed by Brehm with those which he kept alive in Abyssinia, in another manner, namely, by striking the ground with one hand, "like an angry man striking the table with his fist." I have seen this movement with the baboons in the Zoological Gardens; but sometimes the action seems rather to represent the searching for a stone or other object in their beds of straw.

Mr. Sutton has often observed the face of the *Macacus rhesus,* when much enraged, growing red. As he was mentioning this to me, another moneky attacked a *rhesus,* and I saw its face redden as plainly as that of a man in a violent passion. In the course of a few minutes, after the battle, the face of this monkey recovered its natural tint. At the same time that the face reddened, the naked posterior part of the body, which is always red, seemed to grow still redder; but I cannot positively assert that this was the case. When the mandrill is in any way excited, the brilliantly coloured, naked parts of the skin are said to become still more vividly coloured.

With several species of baboons the ridge of the forehead projects much over the eyes, and is studded with a few long hairs, representing our eyebrows. These animals are always looking about them, and in order to look upwards they raise their eyebrows. They have thus, as it would appear, acquired the habit of frequently moving their eyebrows. However this may be, many kinds of monkeys, especially the baboons, when angered or in any way excited, rapidly and incessantly move their eyebrows up and down, as well as the hairy skin of their foreheads.[6] As we associate in the case of man the raising and lowering of the eyebrows with definite states of the mind, the almost incessant movement of the eyebrows by monkeys gives them a senseless expression. I once observed a man who had a trick of

[5] Brehm, *Thierleben,* B. I, s. 84. On baboons striking the ground, s. 61.
[6] Brehm remarks (*Thierleben,* s. 68) that the eyebrows of the *Inuus ecaudatus* are frequently moved up and down when the animal is angered.

continually raising his eyebrows without any corresponding emotion, and this gave to him a foolish appearance; so it is with some persons who keep the corners of their mouths a little drawn backwards and upwards, as if by an incipient smile, though at the time they are not amused or pleased.

A young orang, made jealous by her keeper attending to another monkey, slightly uncovered her teeth, and, uttering a peevish noise like *tish-shist,* turned her back on him. Both orangs and chimpanzees when a little more angered protrude their lips greatly, and make a harsh barking noise. A young female chimpanzee, in a violent passion, presented a curious resemblance to a child in the same state. She screamed loudly with widely open mouth, the lips being retracted so that the teeth were fully exposed. She threw her arms wildly about, sometimes clasping them over her head. She rolled on the ground, sometimes on her back, sometimes on her belly, and bit everything within reach. A young gibbon (*Hylobates syndactylus*) in a passion has been described [7] as behaving in almost exactly the same manner.

The lips of young orangs and chimpanzees are protruded, sometimes to a wonderful degree, under various circumstances. They act thus, not only when slightly angered, sulky, or disappointed, but when alarmed at anything—in one instance, at the sight of a turtle,[8]—and likewise when pleased. But neither the degree of protrusion nor the shape of the mouth is exactly the same, as I believe, in all cases; and the sounds which are then uttered are different. The accompanying drawing * represents a chimpanzee made sulky by an orange having been offered him, and then taken away. A similar protrusion or pouting of the lips, though to a much slighter degree, may be seen in sulky children.

Many years ago, in the Zoological Gardens, I placed a looking-glass on the floor before two young orangs, who, as far as it was known, had never before seen one. At first they gazed at their own images with the most steady surprise, and often changed their point of view. They then approached close and protruded their lips towards the image, as if to kiss it, in ex-

[7] G. Bennett, *Wanderings in New South Wales, &c.* (1834), Vol. II, p. 153.

[8] W. L. Martin, *Nat. Hist. of Mamm. Animals* (1841), p. 405.

* In the original edition.—S.E.H.

actly the same manner as they had previously done towards each other, when first placed, a few days before, in the same room. They next made all sorts of grimaces, and put themselves in various attitudes before the mirror; they pressed and rubbed the surface; they placed their hands at different distances behind it; looked behind it; and finally seemed almost frightened, started a little, became cross, and refused to look any longer.

When we try to perform some little action which is difficult and requires precision, for instance, to thread a needle, we generally close our lips firmly, for the sake, I presume, of not disturbing our movements by breathing; and I noticed the same action in a young orang. The poor little creature was sick, and was amusing itself by trying to kill the flies on the window-panes with its knuckles; this was difficult as the flies buzzed about, and at each attempt the lips were firmly compressed, and at the same time slightly protruded.

Although the countenances, and more especially the gestures, of orangs and chimpanzees are in some respects highly expressive, I doubt whether on the whole they are so expressive as those of some other kinds of monkeys. This may be attributed in part to their ears being immovable, and in part to the nakedness of their eyebrows, of which the movements are thus rendered less conspicuous. When, however, they raise their eyebrows their foreheads become, as with us, transversely wrinkled. In comparison with man, their faces are inexpressive, chiefly owing to their not frowning under any emotion of the mind—that is, as far as I have been able to observe, and I carefully attended to this point. Frowning, which is one of the most important of all the expressions in man, is due to the contraction of the corrugators by which the eyebrows are lowered and brought together, so that vertical furrows are formed on the forehead. Both the orang and chimpanzee are said [9] to possess this muscle, but it seems rarely brought into action, at least in a conspicuous manner. I made my hands into a sort of cage, and placing some tempting fruit within, allowed both a young orang and chimpanzee to try their utmost to get it out; but although they

[9] Prof. Owen on the Orang, *Proc. Zool. Soc.* (1830), p. 28. On the Chimpanzee, see Prof. Macalister, in *Annals and Mag. of Nat. Hist.* (1871), Vol. VII, p. 342, who states that the *corrugator supercilii* is inseparable from the *orbicularis palpebrarum*.

grew rather cross, they showed not a trace of a frown. Nor was there any frown when they were enraged. Twice I took two chimpanzees from their rather dark room suddenly into bright sunshine, which would certainly have caused us to frown; they blinked and winked their eyes, but only once did I see a very slight frown. On another occasion, I tickled the nose of a chimpanzee with a straw, and as it crumpled up its face, slight vertical furrows appeared between the eyebrows. I have never seen a frown on the forehead of the orang.

The gorilla when enraged is described as erecting its crest of hair, throwing down its under lip, dilating its nostrils, and uttering terrific yells. Messrs. Savage and Wyman [10] state that the scalp can be freely moved backwards and forwards, and that when the animal is excited it is strongly contracted; but I presume that they mean by this latter expression that the scalp is lowered; for they likewise speak of the young chimpanzee, when crying out, "as having the eyebrows strongly contracted." The great power of movement in the scalp of the gorilla, of many baboons, and other monkeys, deserves notice in relation to the power possessed by some few men, either through reversion or persistence, of voluntarily moving their scalps.[11]

Astonishment, terror. A living fresh-water turtle was placed at my request in the same compartment in the Zoological Gardens with many monkeys; and they showed unbounded astonishment, as well as some fear. This was displayed by their remaining motionless, staring intently with widely opened eyes, their eyebrows being often moved up and down. Their faces seemed somewhat lengthened. They occasionally raised themselves on their hind-legs to get a better view. They often retreated a few feet, and then turning their heads over one shoulder, again stared intently. It was curious to observe how much less afraid they were of the turtle than of a living snake which I had formerly placed in their compartment; [12] for in the course of a few minutes some of the monkeys ventured to approach and touch the turtle. On the other hand, some of the larger baboons were greatly terrified, and grinned as if on the point of screaming out.

[10] *Boston Journal of Nat. Hist.* (1845–47), Vol. v, p. 423. On the Chimpanzee, ibid. (1843–44), Vol. iv, p. 365.
[11] See on this subject, *The Descent of Man,* Vol. i, p. 20.
[12] *The Descent of Man,* Vol. i, p. 43.

When I showed a little dressed-up doll to the *Cynopithecus niger*, it stood motionless, stared intently with widely opened eyes, and advanced its ears a little forwards. But when the turtle was placed in its compartment, this monkey also moved its lips in an odd, rapid, jabbering manner, which the keeper declared was meant to conciliate or please the turtle.

I was never able clearly to perceive that the eyebrows of astonished monkeys were kept permanently raised, though they were frequently moved up and down. Attention, which precedes astonishment, is expressed by man by a slight raising of the eyebrows; and Dr. Duchenne informs me that when he gave to the monkey formerly mentioned some quite new article of food, it elevated its eyebrows a little, thus assuming an appearance of close attention. It then took the food in its fingers, and, with lowered or rectilinear eyebrows, scratched, smelt, and examined it—an expression of reflection being thus exhibited. Sometimes it would throw back its head a little, and again with suddenly raised eyebrows re-examine and finally taste the food.

In no case did any monkey keep its mouth open when it was astonished. Mr. Sutton observed for me a young orang and chimpanzee during a considerable length of time; and however much they were astonished, or whilst listening intently to some strange sound, they did not keep their mouths open. This fact is surprising, as with mankind hardly any expression is more general than a widely open mouth under the sense of astonishment. As far as I have been able to observe, monkeys breathe more freely through their nostrils than men do; and this may account for their not opening their mouths when they are astonished; for, as we shall see in a future chapter, man apparently acts in this manner when startled, at first for the sake of quickly drawing a full inspiration, and afterwards for the sake of breathing as quietly as possible.

Terror is expressed by many kinds of monkeys by the utterance of shrill screams, the lips being drawn back, so that the teeth are exposed. The hair becomes erect, especially when some anger is likewise felt. Mr. Sutton has distinctly seen the face of the *Macacus rhesus* grow pale from fear. Monkeys also tremble from fear; and sometimes they void their excretions. I have seen one which, when caught, almost fainted from an excess of terror.

The Descent of Man

CHAPTER VI

On the Affinities and Genealogy of Man

Even if it be granted that the difference between man and his nearest allies is as great in corporeal structure as some naturalists maintain, and although we must grant that the difference between them is immense in mental power, yet the facts given in the earlier chapters appear to declare, in the plainest manner, that man is descended from some lower form, notwithstanding that connecting links have not hitherto been discovered.

Man is liable to numerous, slight, and diversified variations, which are induced by the same general causes, are governed and transmitted in accordance with the same general laws, as in the lower animals. Man has multiplied so rapidly that he has necessarily been exposed to struggle for existence, and consequently to natural selection. He has given rise to many races, some of which differ so much from each other that they have often been ranked by naturalists as distinct species. His body is constructed on the same homological plan as that of other mammals. He passes through the same phases of embryological development. He retains many rudimentary and useless structures, which no doubt were once serviceable. Characters occasionally make their reappearance in him, which we have reason to believe were possessed by his early progenitors. If the origin of man had been wholly different from that of all other animals, these various appearances would be mere empty deceptions; but such an admission is incredible. These appearances, on the other hand, are intelligible, at least to a large extent, if man is the co-descendant with other mammals of some unknown and lower form.

Some naturalists, from being deeply impressed with the mental and spiritual powers of man, have divided the whole organic world into three kingdoms, the human, the animal, and the vegetable, thus giving to man a separate kingdom.[1] Spiritual

[1] Isidore Geoffroy St.-Hilaire gives a detailed account of the position assigned to man by various naturalists in their classifications: *Hist. Nat. Gén.* (1859), Tom. II, pp. 170–189.

powers cannot be compared or classed by the naturalist: but he may endeavour to show, as I have done, that the mental faculties of man and the lower animals do not differ in kind, although immensely in degree. A difference in degree, however great, does not justify us in placing man in a distinct kingdom, as will perhaps be best illustrated by comparing the mental powers of two insects, namely, a coccus or scale-insect and an ant, which undoubtedly belong to the same class. The difference is here greater than, though of a somewhat different kind from, that between man and the highest mammal. The female coccus, whilst young, attaches itself by its proboscis to a plant; sucks the sap, but never moves again; is fertilised and lays eggs; and this is its whole history. On the other hand, to describe the habits and mental powers of worker-ants would require, as Pierre Huber has shown, a large volume; I may, however, briefly specify a few points. Ants certainly communicate information to each other, and several unite for the same work, or for games of play. They recognise their fellow-ants after months of absence, and feel sympathy for each other. They build great edifices, keep them clean, close the doors in the evening, and post sentries. They make roads as well as tunnels under rivers, and temporary bridges over them, by clinging together. They collect food for the community, and when an object too large for entrance is brought to the nest, they enlarge the door, and afterwards build it up again. They store up seeds, of which they prevent the germination, and which, if damp, are brought up to the surface to dry. They keep aphids and other insects as milch-cows. They go out to battle in regular bands, and freely sacrifice their lives for the common weal. They emigrate according to a preconcerted plan. They capture slaves. They move the eggs of their aphids, as well as their own eggs and cocoons, into warm parts of the nest, in order that they may be quickly hatched; and endless similar facts could be given.[2] On the whole, the difference in mental power between an ant and a coccus is immense; yet no one has ever dreamed of placing these insects in distinct classes, much less in distinct kingdoms. No doubt the

[2] Some of the most interesting facts ever published on the habits of ants are given by Mr. Belt, in his *Naturalist in Nicaragua* (1874). See also Mr. Moggridge's admirable work, *Harvesting Ants, &c.* (1873), also "L'Instinct chez les Insectes," by M. George Pouchet, *Revue des Deux Mondes,* February 1870, p. 682.

difference is bridged over by other insects; and this is not the case with man and the higher apes. But we have every reason to believe that the breaks in the series are simply the results of many forms having become extinct.

Professor Owen, relying chiefly on the structure of the brain, has divided the mammalian series into four sub-classes. One of these he devotes to man; in another he places both the marsupials and the Monotremata; so that he makes man as distinct from all other mammals as are these two latter groups conjoined. This view has not been accepted, as far as I am aware, by any naturalist capable of forming an independent judgment, and therefore need not here be further considered.

We can understand why a classification founded on any single character or organ—even an organ so wonderfully complex and important as the brain—or on the high development of the mental faculties, is almost sure to prove unsatisfactory. This principle has indeed been tried with hymenopterous insects; but when thus classed by their habits or instincts, the arrangement proved thoroughly artificial.[3] Classifications may, of course, be based on any character whatever, as on size, colour, or the element inhabited; but naturalists have long felt a profound conviction that there is a natural system. This system, it is now generally admitted, must be, as far as possible, genealogical in arrangement—that is, the co-descendants of the same form must be kept together in one group, apart from the co-descendants of any other form; but if the parent-forms are related, so will be their descendants, and the two groups together will form a larger group. The amount of difference between the several groups—that is, the amount of modification which each has undergone—is expressed by such terms as genera, families, orders, and classes. As we have no record of the lines of descent, the pedigree can be discovered only by observing the degrees of resemblance between the beings which are to be classed. For this object numerous points of resemblance are of much more importance than the amount of similarity or dissimilarity in a few points. If two languages were found to resemble each other in a multitude of words and points of construction, they would be universally recognised as having sprung from a common source, notwithstanding that they differed greatly in some few

[3] Westwood, *Modern Class in Insects* (1840), Vol. II, p. 87.

words or points of construction. But with organic beings the points of resemblance must not consist of adaptations to similar habits of life: two animals may, for instance, have had their whole frames modified for living in the water, and yet they will not be brought any nearer to each other in the natural system. Hence we can see how it is that resemblances in several unimportant structures, in useless and rudimentary organs, or not now functionally active, or in an embryological condition, are by far the most serviceable for classification; for they can hardly be due to adaptations within a late period; and thus they reveal the old lines of descent or of true affinity.

We can further see why a great amount of modification in some one character ought not to lead us to separate widely any two organisms. A part which already differs much from the same part in other allied forms has already, according to the theory of evolution, varied much; consequently it would (as long as the organism remained exposed to the same exciting conditions) be liable to further variations of the same kind; and these, if beneficial, would be preserved, and thus be continually augmented. In many cases the continued development of a part, for instance, of the beak of a bird, or of the teeth of a mammal, would not aid the species in gaining its food, or for any other object; but with man we can see no definite limit to the continued development of the brain and mental faculties, as far as advantage is concerned. Therefore in determining the position of man in the natural or genealogical system, the extreme development of his brain ought not to outweigh a multitude of resemblances in other less important or quite unimportant points.

The greater number of naturalists who have taken into consideration the whole structure of man, including his mental faculties, have followed Blumenbach and Cuvier, and have placed man in a separate order, under the title of the Bimana, and therefore on an equality with the orders of the Quadrumana, Carnivora, &c. Recently many of our best naturalists have recurred to the view first propounded by Linnaeus, so remarkable for his sagacity, and have placed man in the same order with the Quadrumana, under the title of the Primates. The justice of this conclusion will be admitted: for in the first place, we must bear in mind the comparative insignificance for classification of the great development of the brain in man, and that the strongly marked differences between the skulls of man and the Quadrumana (lately insisted upon by Bischoff, Aeby, and others) apparently follow from their differently developed brains. In the second place, we must remember that nearly all the other and more important differences between man and the Quadrumana are manifestly adaptive in their nature, and relate chiefly to the erect position of man; such as the structure of his hand, foot, and pelvis, the curvature of his spine, and the position of his head. The family of seals offers a good illustration of the small importance of adaptive characters for classification. These animals differ from all other Carnivora in the form of their bodies and in the structure of their limbs, far more than does man from the higher apes; yet in most systems, from that of Cuvier to the most recent one by Mr. Flower,[4] seals are ranked as a mere family in the order of the Carnivora. If man had not been his own classifier, he would never have thought of founding a separate order for his own reception.

It would be beyond my limits, and quite beyond my knowledge, even to name the innumerable points of structure in which man agrees with the other Primates. Our great anatomist and philosopher, Prof. Huxley, has fully discussed this subject,[5] and concludes that man in all parts of his organisation differs less from the higher apes than these do from the lower members of the same group. Consequently there "is no justification for placing man in a distinct order."

In an early part of this work I brought forward various facts showing how closely man agrees in constitution with the higher mammals; and this agreement must depend on our close similarity in minute structure and chemical composition. I gave, as instances, our liability to the same diseases, and to the attacks of allied parasites; our tastes in common for the same stimulants, and the similar effects produced by them. as well as by various drugs, and other such facts.

As small unimportant points of resemblance between man and the Quadrumana are not commonly noticed in systematic works, and as, when numerous, they clearly reveal our relationship, I will specify a few such points. The relative position of our features is manifestly the same; and the various emotions are

---

[4] *Proc. Zoolog. Soc.* (1863), p. 4.
[5] *Evidence as to Man's Place in Nature* (1863), p. 70, *et passim*.

displayed by nearly similar movements of the muscles and skin, chiefly above the eyebrows and round the mouth. Some few expressions are, indeed, almost the same, as in the weeping of certain kinds of monkeys and in the laughing noise made by others, during which the corners of the mouth are drawn backwards, and the lower eyelids wrinkled. The external ears are curiously alike. In man the nose is much more prominent than in most monkeys; but we may trace the commencement of an aquiline curvature in the nose of the Hoolock gibbon; and this in the *Semnopithecus nasica* is carried to a ridiculous extreme.

The faces of many monkeys are ornamented with beards, whiskers, or moustaches. The hair on the head grows to a great length in some species of Semnopithecus; [6] and in the Bonnet monkey (*Macacus radiatus*) it radiates from a point on the crown, with a parting down the middle. It is commonly said that the forehead gives to man his noble and intellectual appearance; but the thick hair on the head of the Bonnet monkey terminates downwards abruptly, and is succeeded by hair so short and fine that at a little distance the forehead, with the exception of the eyebrows, appears quite naked. It has been erroneously asserted that eyebrows are not present in any monkey. In the species just named the degree of nakedness of the forehead differs in different individuals; and Eschricht states [7] that in our children the limit between the hairy scalp and the naked forehead is sometimes not well defined; so that here we seem to have a trifling case of reversion to a progenitor, in whom the forehead had not as yet become quite naked.

It is well known that the hair on our arms tends to converge from above and below to a point at the elbow. This curious arrangement, so unlike that in most of the lower mammals, is common to the gorilla, chimpanzee, orang, some species of Hylobates, and even to some few American monkeys. But in *Hylobates agilis* the hair on the fore-arm is directed downwards or towards the wrist in the ordinary manner; and in *H. lar* it is nearly erect, with only a very slight forward inclination; so that in this latter species it is in a transitional state. It can hardly be doubted that with most mammals the thickness of the hair on

[6] Isid. Geoffroy St.-Hilaire, *Hist. Nat. Gén.* (1859), Tom. II, p. 217.
[7] "Ueber die Richtung der Haare," &c., Muller's *Archiv für Anat. und Phys.* (1837), s. 51.

the back and its direction is adapted to throw off the rain; even the transverse hairs on the fore-legs of a dog may serve for this end when he is coiled up asleep. Mr. Wallace, who has carefully studied the habits of the orang, remarks that the convergence of the hair towards the elbow on the arms of the orang may be explained as serving to throw off the rain, for this animal during rainy weather sits with its arms bent, and with the hands clasped round a branch or over its head. According to Livingstone, the gorilla also "sits in pelting rain with his hands over his head." [8] If the above explanation is correct, as seems probable, the direction of the hair on our own arms offers a curious record of our former state; for no one supposes that it is now of any use in throwing off the rain; nor, in our present erect condition, is it properly directed for this purpose.

It would, however, be rash to trust too much to the principle of adaptation in regard to the direction of the hair in man or his early progenitors; for it is impossible to study the figures given by Eschricht of the arrangement of the hair on the human foetus (this being the same as in the adult) and not agree with this excellent observer that other and more complex causes have intervened. The points of convergence seem to stand in some relation to those points in the embryo which are last closed in during development. There appears, also, to exist some relation between the arrangement of the hair on the limbs, and the course of the medullary arteries.[9]

It must not be supposed that the resemblances between man and certain apes in the above and many other points—such as in having a naked forehead, long tresses on the head, &c.—are all necessarily the result of unbroken inheritance from a common progenitor, or of subsequent reversion. Many of these resemblances are more probably due to analogous variation, which follows, as I have elsewhere attempted to show,[10] from co-descended organisms having a similar constitution, and having

[8] Quoted by Reade, *The African Sketch Book* (1873), Vol. I, p. 152.
[9] On the hair in Hylobates, see *Nat. Hist. of Mammals,* by C. L. Martin (1841), p. 415. Also, Isid. Geoffroy St.-Hilaire on the American monkeys and other kinds, *Hist. Nat. Gén.* (1859), Vol. II, pp. 216, 243. Eschricht, ibid. s. 46, 55, 61. Owen, *Anat. of Vertebrates,* Vol. III, p. 619. Wallace, *Contributions to the Theory of Natural Selection* (1870), p. 344.
[10] *Origin of Species,* 5th edit. (1869), p. 194. *The Variation of Animals and Plants under Domestication* (1868), Vol. II, p. 348.

been acted on by like causes inducing similar modifications. With respect to the similar direction of the hair on the fore-arms of man and certain monkeys, as this character is common to almost all the anthropomorphous apes, it may probably be attributed to inheritance; but this is not certain, as some very distinct American monkeys are thus characterised.

Although, as we have now seen, man has no just right to form a separate order for his own reception, he may perhaps claim a distinct sub-order or family. Prof. Huxley, in his last work,[11] divides the Primates into three sub-orders; namely, the Anthropidae with man alone, the Simiadae including monkeys of all kinds, and the Lemuridae with the diversified genera of lemurs. As far as differences in certain important points of structure are concerned, man may no doubt rightly claim the rank of a sub-order; and this rank is too low, if we look chiefly to his mental faculties. Nevertheless, from a genealogical point of view it appears that this rank is too high, and that man ought to form merely a family, or possibly even only a sub-family. If we imagine three lines of descent proceeding from a common stock, it is quite conceivable that two of them might after the lapse of ages be so slightly changed as still to remain as species of the same genus, whilst the third line might become so greatly modified as to deserve to rank as a distinct sub-family, family, or even order. But in this case it is almost certain that the third line would still retain through inheritance numerous small points of resemblance with the other two. Here, then, would occur the difficulty, at present insoluble, how much weight we ought to assign in our classifications to strongly marked differences in some few points—that is, to the amount of modification undergone; and how much to close resemblance in numerous unimportant points, as indicating the lines of descent or genealogy. To attach much weight to the few but strong differences is the most obvious and perhaps the safest course, though it appears more correct to pay great attention to the many small resemblances, as giving a truly natural classification.

In forming a judgment on this head with reference to man, we must glance at the classification of the Simiadae. This family is divided by almost all naturalists into the Catarhine group, or

[11] *An Introduction to the Classification of Animals* (1869), p. 99.

Old World monkeys, all of which are characterised (as their name expresses) by the peculiar structure of their nostrils, and by having four premolars in each jaw; and into the Platyrhine group or New World monkeys (including two very distinct sub-groups), all of which are characterised by differently constructed nostrils, and by having six premolars in each jaw. Some other small differences might be mentioned. Now man unquestionably belongs in his dentition, in the structure of his nostrils, and some other respects to the Catarhine or Old World division; nor does he resemble the Platyrhines more closely than the Catarhines in any characters, excepting in a few of not much importance and apparently of an adaptive nature. It is therefore against all probability that some New World species should have formerly varied and produced a man-like creature, with all the distinctive characters proper to the Old World division; losing at the same time all its own distinctive characters. There can, consequently, hardly be a doubt that man is an off-shoot from the Old World Simian stem; and that under a genealogical point of view, he must be classed with the Catarhine division.[12]

The anthropomorphous apes, namely, the gorilla, chimpanzee, orang, and hylobates, are by most naturalists separated from the other Old World monkeys, as a distinct sub-group. I am aware that Gratiolet, relying on the structure of the brain, does not admit the existence of this sub-group, and no doubt it is a broken one. Thus the orang, as Mr. St. G. Mivart remarks,[13] "is one of the most peculiar and aberrant forms to be found in the order." The remaining non-anthropomorphous Old World monkeys are again divided by some naturalists into two or three smaller sub-groups; the genus Semnopithecus, with its peculiar sacculated stomach, being the type of one such sub-group. But it appears from M. Gaudry's wonderful discoveries in Attica that during the Miocene period a form existed there which connected

---

[12] This is nearly the same classification as that provisionally adopted by St. George Mivart (*Transact. Phil. Soc.* [1867], p. 300), who, after separating the Lemuridae, divides the remainder of the Primates into the Hominidae, the Simiadae which answer to the Catarhines, the Cebidae, and the Hapalidae—these two latter groups answering to the Platyrhines. Mr. Mivart still abides by the same view; see *Nature* (1871), p. 481.

[13] *Transact. Zoolog. Soc.* (1867), Vol. vi, p. 214.

Semnophithecus and Macacus; and this probably illustrates the manner in which the other and higher groups were once blended together.

If the anthropomorphous apes be admitted to form a natural sub-group, then as man agrees with them, not only in all those characters which he possesses in common with the whole Catarhine group, but in other peculiar characters, such as the absence of a tail and of callosities, and in general appearance, we may infer that some ancient member of the anthropomorphous sub-group gave birth to man. It is not probable that, through the law of analogous variation, a member of one of the other lower sub-groups should have given rise to a man-like creature, resembling the higher anthropomorphous apes in so many respects. No doubt man, in comparison with most of his allies, has undergone an extraordinary amount of modification, chiefly in consequence of the great development of his brain and his erect position; nevertheless, we should bear in mind that he "is but one of several exceptional forms of Primates." [14]

Every naturalist who believes in the principle of evolution will grant that the two main divisions of the Simiadae, namely the Catarhine and Platyrhine monkeys, with their sub-groups, have all proceeded from some one extremely ancient progenitor. The early descendants of this progenitor, before they had diverged to any considerable extent from each other, would still have formed a single natural group; but some of the species or incipient genera would have already begun to indicate by their diverging characters the future distinctive marks of the Catarhine and Platyrhine divisions. Hence the members of this supposed ancient group would not have been so uniform in their dentition, or in the structure of their nostrils, as are the existing Catarhine monkeys in one way and the Platyrhines in another way, but would have resembled in this respect the allied Lemuridae, which differ greatly from each other in the form of their muzzles,[15] and to an extraordinary degree in their dentition.

The Catarhine and Platyrhine monkeys agree in a multitude of characters, as is shown by their unquestionably belonging to one and the same order. The many characters which they

[14] St. G. Mivart, *Transact. Phil. Soc.* (1867), p. 410.
[15] Messrs. Murie and Mivart on the Lemuroidea, *Transact. Zoolog. Soc.* (1869), Vol. VII, p. 5.

possess in common can hardly have been independently acquired by so many distinct species; so that these characters must have been inherited. But a naturalist would undoubtedly have ranked as an ape or a monkey an ancient form which possessed many characters common to the Catarhine and Platyrhine monkeys, other characters in an intermediate condition, and some few, perhaps, distinct from those now found in either group. And as man from a genealogical point of view belongs to the Catarhine or Old World stock, we must conclude, however much the conclusion may revolt our pride, that our early progenitors would have been properly thus designated.[16] But we must not fall into the error of supposing that the early progenitor of the whole simian stock, including man, was identical with, or even closely resembled, any existing ape or monkey.

On the Birthplace and Antiquity of Man. We are naturally led to inquire, where was the birthplace of man at that stage of descent when our progenitors diverged from the Catarhine stock? The fact that they belonged to this stock clearly shews that they inhabited the Old World; but not Australia nor any oceanic island, as we may infer from the laws of geographical distribution. In each great region of the world the living mammals are closely related to the extinct species of the same region. It is therefore probable that Africa was formerly inhabited by extinct apes closely allied to the gorilla and chimpanzee; and as these two species are now man's nearest allies, it is somewhat more probable that our early progenitors lived on the African continent than elsewhere. But it is useless to speculate on this subject; for two or three anthropomorphous apes, one the Dryopithecus [17] of Lartet, nearly as large as a man, and closely allied to Hylobates, existed in Europe during the Miocene age; and since so remote a period the earth has certainly undergone many great revolutions, and there has been ample time for migration on the largest scale.

At the period and place, whenever and wherever it was, when

[16] Häckel has come to this same conclusion. See "Ueber die Entstehung des Menschengeschlechts," in Virchow's *Sammlung. gemein. wissen. Vorträge* (1868), s. 61. Also his *Natürliche Schöpfungsgeschichte* (1868), in which he gives in detail his views on the genealogy of man.
[17] Dr. C. Forsyth Major, "Sur les Singes Fossiles trouvés en Italie," *Soc. Ital. des Sc. Nat.* (1872), Tom. xv.

man first lost his hairy covering, he probably inhabited a hot country, a circumstance favourable for the frugiferous diet on which, judging from analogy, he subsisted. We are far from knowing how long ago it was when man first diverged from the Catarhine stock; but it may have occurred at an epoch as remote as the Eocene period; for that the higher apes had diverged from the lower apes as early as the Upper Miocene period is shown by the existence of the Dryopithecus. We are also quite ignorant at how rapid a rate organisms, whether high or low in the scale, may be modified under favourable circumstances; we know, however, that some have retained the same form during an enormous lapse of time. From what we see going on under domestication, we learn that some of the co-descendants of the same species may be not at all, some a little, and some greatly changed, all within the same period. Thus it may have been with man, who has undergone a great amount of modification in certain characters in comparison with the higher apes.

The great break in the organic chain between man and his nearest allies, which cannot be bridged over by any extinct or living species, has often been advanced as a grave objection to the belief that man is descended from some lower form; but this objection will not appear of much weight to those who, from general reasons, believe in the general principle of evolution. Breaks often occur in all parts of the series, some being wide, sharp, and defined, others less so in various degrees; as between the orang and its nearest allies, between the Tarsius and the other Lemuridae, between the elephant, and in a more striking manner between the Ornithorhynchus or Echidna, and all other mammals. But these breaks depend merely on the number of related forms which have become extinct. At some future period, not very distant as measured by centuries, the civilised races of man will almost certainly exterminate, and replace, the savage races throughout the world. At the same time the anthropomorphous apes, as Prof. Schaaffhausen has remarked,[18] will no doubt be exterminated. The break between man and his nearest allies will then be wider, for it will intervene between man in a more civilised state, as we may hope, even than the Caucasian, and some ape as low as a baboon, instead of as now between the Negro or Australian and the gorilla.

[18] *Anthropological Review*, April 1867, p. 236.

With respect to the absence of fossil remains serving to connect man with his ape-like progenitors, no one will lay much stress on this fact who reads Sir C. Lyell's discussion,[19] where he shows that in all the vertebrate classes the discovery of fossil remains has been a very slow and fortuitous process. Nor should it be forgotten that those regions which are the most likely to afford remains connecting man with some extinct ape-like creature have not as yet been searched by geologists.

Lower Stages in the Genealogy of Man. We have seen that man appears to have diverged from the Catarhine or Old World division of the Simiadae, after these had diverged from the New World division. We will now endeavour to follow the remote traces of his genealogy, trusting principally to the mutual affinities between the various classes and orders, with some slight reference to the periods, as far as ascertained, of their successive appearance on the earth. The Lemuridae stand below and near to the Simiadae, and constitute a very distinct family of the Primates, or, according to Häckel and others, a distinct order. This group is diversified and broken to an extraordinary degree, and includes many aberrant forms. It has, therefore, probably suffered much extinction. Most of the remnants survive on islands, such as Madagascar and the Malayan archipelago, where they have not been exposed to so severe a competition as they would have been on well-stocked continents. This group likewise presents many gradations, leading, as Huxley remarks,[20] "insensibly from the crown and summit of the animal creation down to creatures from which there is but a step, as it seems, to the lowest, smallest, and least intelligent of the placental mammalia." From these various considerations it is probable that the Simiadae were originally developed from the progenitors of the existing Lemuridae; and these in their turn from forms standing very low in the mammalian series.

The marsupials stand in many important characters below the placental mammals. They appeared at an earlier geological period, and their range was formerly much more extensive than at present. Hence the Placentata are generally supposed to have

[19] *Elements of Geology* (1865), pp. 583–585. *Antiquity of Man,* (1863), p. 145.
[20] *Man's Place in Nature,* p. 105.

been derived from the Implacentata or marsupials; not, however, from forms closely resembling the existing marsupials, but from their early progenitors. The Monotremata are plainly allied to the marsupials, forming a third and still lower division in the great mammalian series. They are represented at the present day solely by the Ornithorhynchus and Echidna; and these two forms may be safely considered as relics of a much larger group representatives of which have been preserved in Australia through some favourable concurrence of circumstances. The Monotremata are eminently interesting, as leading in several important points of structure towards the class of reptiles.

In attempting to trace the genealogy of the Mammalia, and therefore of man, lower down in the series, we become involved in greater and greater obscurity; but as a most capable judge, Mr. Parker, has remarked, we have good reason to believe that no true bird or reptile intervenes in the direct line of descent. He who wishes to see what ingenuity and knowledge can effect, may consult Prof. Häckel's works.[21] I will content myself with a few general remarks. Every evolutionist will admit that the five great vertebrate classes, namely, mammals, birds, reptiles, amphibians, and fishes, are descended from some one prototype; for they have much in common, especially during their embryonic state. As the class of fishes is the most lowly organised, and appeared before the others, we may conclude that all the members of the vetebrate kingdom are derived from some fish-like animal. The belief that animals so distinct as a monkey, an elephant, a humming-bird, a snake, a frog, and a fish, &c., could all have sprung from the same parents will appear monstrous to those who have not attended to the recent progress of natural history. For this belief implies the former existence of links binding closely together all these forms, now so utterly unlike.

Nevertheless, it is certain that groups of animals have existed, or do now exist, which serve to connect several of the great vertebrate classes more or less closely. We have seen that the

[21] Elaborate tables are given in his *Generelle Morphologie* (B. II., s. cliii. and s. 425); and with more especial reference to man in his *Natürliche Schöpfungsgeschichte* (1868). Prof. Huxley, in reviewing this latter work (*The Academy* [1869], p. 42) says that he considers the phylum or lines of descent of the Vertebrata to be admirably discussed by Häckel, although he differs on some points. He expresses, also, his high estimate of the general tenor and spirit of the whole work.

Ornithorhynchus graduates towards reptiles; and Prof. Huxley has discovered, and is confirmed by Mr. Cope and others, that the Dinosaurians are in many important characters intermediate between certain reptiles and certain birds—the birds referred to being the ostrich-tribe (itself evidently a widely diffused remnant of a larger group) and the Archeopteryx, that strange Secondary bird, with a long lizard-like tail. Again, according to Prof. Owen,[22] the Ichthyosaurians—great sea-lizards furnished with paddles—present many affinities with fishes, or rather, according to Huxley, with amphibians; a class which, including in its highest division frogs and toads, is plainly allied to the fishes. These latter fishes swarmed during the earlier geological periods, and were constructed on what is called a generalised type, that is, they presented diversified affinities with other groups of organisms. The Lepidosiren is also so closely allied to amphibians and fishes, that naturalists long disputed in which of these two classes to rank it; it, and also some few Ganoid fishes have been preserved from utter extinction by inhabiting rivers, which are harbours of refuge, and are related to the great waters of the ocean in the same way that islands are to continents.

Lastly, one single member of the immense and diversified class of fishes, namely, the lancelet or amphioxus, is so different from all other fishes that Häckel maintains that it ought to form a distinct class in the vertebrate kingdom. This fish is remarkable for its negative characters; it can hardly be said to possess a brain, vertebral column, or heart, &c.; so that it was classed by the older naturalists amongst the worms. Many years ago Prof. Goodsir perceived that the lancelet presented some affinities with the Ascidians, which are invertebrate, hermaphrodite, marine creatures permanently attached to a support. They hardly appear like animals, and consist of a simple, tough, leathery sack, with two small projecting orifices. They belong to the Molluscoida of Huxley—a lower division of the great kingdom of the Mollusca; but they have recently been placed by some naturalists amongst the Vermes or worms. Their larvae somewhat resemble tadpoles in shape,[23] and have the power of swimming freely about. M.

[22] *Palaeontology* (1860), p. 199.
[23] At the Falkland Islands I had the satisfaction of seeing, in April 1833, and therefore some years before any other naturalist, the locomo-

Kovalevsky [24] has lately observed that the larvae of Ascidians
are related to the Vertebrata, in their manner of development, in
the relative position of the nervous system, and in possessing a
structure closely like the *chorda dorsalis* of vertebrate animals;
and in this he has been since confirmed by Prof. Kupffer. M.
Kovalevsky writes to me from Naples that he has now carried
these observations yet further; and should his results be well
established, the whole will form a discovery of the very greatest
value. Thus, if we may rely on embryology, ever the safest guide
in classification, it seems that we have at last gained a clue to the
source whence the Vertebrata were derived.[25] We should then
be justified in believing that at an extremely remote period a
group of animals existed, resembling in many respects the larvae
of our present Ascidians, which diverged into two great branches
—the one retrograding in development and producing the present
class of Ascidians, the other rising to the crown and summit of
the animal kingdom by giving birth to the Vertebrata.

We have thus far endeavoured rudely to trace the genealogy
of the Vertebrata by the aid of their mutual affinities. We will
now look to man as he exists; and we shall, I think, be able
partially to restore the structure of our early progenitors, during
successive periods, but not in due order of time. This can be
effected by means of the rudiments which man still retains, by
the characters which occasionally make their appearance in him

---

tive larvae of a compound Ascidian, closely allied to Synoicum, but apparently generically distinct from it. The tail was about five times as
long as the oblong head, and terminated in a very fine filament. It was,
as sketched by me under a simple microscope, plainly divided by transverse opaque partitions, which I presume represent the great cells figured
by Kovalevsky. At an early stage of development the tail was closely
coiled round the head of the larva.

[24] *Mémoires de l'Acad. des Sciences de St. Pétersbourg* (1866), Tom.
x, No. 15.

[25] But I am bound to add that some competent judges dispute this
conclusion; for instance, M. Giard, in a series of papers in the *Archives
de Zoologie Expérimentale,* for 1872. Nevertheless, this naturalist remarks, p. 281, "L'organisation de la larve ascidienne en dehors de toute
hypothèse et de toute théorie, nous montre comment la nature peut
produire la disposition fondamentale du type vertébré (l'existence d'une
corde dorsale) chez un invertébré par la seule condition vitale de l'adaptation, et cette simple possibilité du passage supprime l'abîme entre les
deux sous-règnes, encore bien qu'on ignore par où le passage s'est fait en
réalité."

through reversion, and by the aid of the principles of morphology and embryology. The various facts, to which I shall here allude, have been given in the previous chapters.

The early progenitors of man must have been once covered with hair, both sexes having beards; their ears were probably pointed, and capable of movement; and their bodies were provided with a tail, having the proper muscles. Their limbs and bodies were also acted on by many muscles which now only occasionally reappear, but are normally present in the Quadrumana. At this or some earlier period, the great artery and nerve of the humerus ran through a supra-condyloid foramen. The intestine gave forth a much larger diverticulum or caecum than that now existing. The foot was then prehensile, judging from the condition of the great toe in the foetus; and our progenitors, no doubt, were arboreal in their habits, and frequented some warm, forest-clad land. The males had great canine teeth, which served them as formidable weapons. At a much earlier period the uterus was double; the excreta were voided through a cloaca; and the eye was protected by a third eyelid or nictitating membrane. At a still earlier period the progenitors of man must have been aquatic in their habits; for morphology plainly tells us that our lungs consist of a modified swim-bladder, which once served as a float. The clefs on the neck in the embryo of man show where the branchiae once existed. In the lunar or weekly recurrent periods of some of our functions we apparently still retain traces of our primordial birthplace, a shore washed by the tides. At about this same early period the true kidneys were replaced by the corpora wolffiana. The heart existed as a simple pulsating vessel; and the chorda dorsalis took the place of a vertebral column. These early ancestors of man, thus seen in the dim recesses of time, must have been as simply, or even still more simply, organised than the lancelet or amphioxus.

There is one other point deserving a fuller notice. It has long been known that in the vertebrate kingdom one sex bears rudiments of various accessory parts, appertaining to the reproductive system, which properly belong to the opposite sex; and it has now been ascertained that at a very early embryonic period both sexes possess true male and female glands. Hence some remote progenitor of the whole vertebrate kingdom appears to have been hermaphrodite or androgynous.[26] But here we encounter a singular difficulty. In the manmmalian class the males possess rudiments of a uterus with the adjacent passage, in their vesiculae prostaticae; they bear also rudiments of mammae, and some male marsupials have traces of a marsupial sack.[27] Other analogous facts could be added. Are we, then, to suppose that some extremely ancient mammal continued androgynous, after it had acquired the chief distinctions of its class, and therefore after it had diverged from the lower classes of the vertebrate kingdom? This seems very improbable, for we have to look to fishes, the lowest of all the classes, to find any still existent androgynous forms.[28] That various accessory parts proper to each sex are found in a rudimentary condition in the opposite sex may be explained by such organs having been gradually acquired by the one sex, and then transmitted in a more or less imperfect state to the other. When we treat of sexual selection, we shall meet with innumerable instances of this form of transmission—as in the case of the spurs, plumes, and brilliant colours acquired for battle or ornament by male birds, and inherited by the females in an imperfect or rudimentary condition.

The possession by male mammals of functionally imperfect mammary organs is, in some respects, especially curious. The Monotremata have the proper milk-secreting glands with orifices,

[26] This is the conclusion of Prof. Gegenbaur, one of the highest authorities in comparative anatomy; see *Grundzüge der vergleich. Anat.* (1870), s. 876. The result has been arrived at chiefly from the study of the Amphibia; but it appears from the researches of Waldeyer (as quoted in *Journal of Anat. and Phys.* [1869], p. 161), that the sexual organs of even "the higher vertebrata are, in their early condition, hermaphrodite." Similar views have long been held by some authors, though until recently without a firm basis.

[27] The male Thylacinus offers the best instance. Owen, *Anatomy of Vertebrates,* Vol. III, p. 771.

[28] Hermaphroditism has been observed in several species of Serranus, as well as in some other fishes, where it is either normal and symmetrical, or abnormal and unilateral. Dr. Zouteveen has given me references on this subject, more especially to a paper by Prof. Halbertsma, in the *Transact. of the Dutch Acad. of Sciences,* Vol. XVI. Dr. Günther doubts the fact, but it has now been recorded by too many good observers to be any longer disputed. Dr. M. Lessona writes to me that he has verified the observatioɴs made by Cavolini on Serranus. Prof. Ercolani has recently shown (*Accad. delle Scienze,* Bologna, Dec. 28, 1871) that eels are androgynous.

but no nipples; and as these animals stand at the very base of the mammalian series, it is probable that the progenitors of the class also had milk-secreting glands, but no nipples. This conclusion is supported by what is known of their manner of development; for Prof. Turner informs me, on the authority of Kölliker and Langer, that in the embryo the mammary glands can be distinctly traced before the nipples are in the least visible; and the development of successive parts in the individual generally represents and accords with the development of successive beings in the same line of descent. The marsupials differ from the Monotremata by possessing nipples; so that probably these organs were first acquired by the marsupials, after they had diverged from, and risen above, the Monotremata, and were then transmitted to the placental mammals.[29] No one will suppose that the marsupials still remained androgynous after they had approximately acquired their present structure. How then are we to account for male mammals possessing mammae? It is possible that they were first developed in the females and then transferred to the males; but from what follows this is hardly probable.

It may be suggested, as another view, that long after the progenitors of the whole mammalian class had ceased to be androgynous, both sexes yielded milk, and thus nourished their young; and in the case of the marsupials, that both sexes carried their young in marsupial sacks. This will not appear altogether improbable, if we reflect that the males of existing syngnathous fishes receive the eggs of the females in their abdominal pouches, hatch them, and afterwards, as some believe, nourish the young; [30] that certain other male fishes hatch the eggs within

[29] Prof. Gegenbaur has shown (*Jenaische Zeitschrift,* Bd. VII, p. 212) that two distinct types of nipples prevail throughout the several mammalian orders, but that it is quite intelligible how both could have been derived from the nipples of the marsupials, and the latter from those of the Monotremata. See, also, a memoir by Dr. Max Huss, on the mammary glands, ibid. B. VIII, p. 176.

[30] Mr. Lockwood believes (as quoted in *Quart. Journal of Science,* April 1868, p. 269), from what he has observed of the development of Hippocampus, that the walls of the abdominal pouch of the male in some way afford nourishment. On male fishes hatching the ova in their mouths, see a very interesting paper by Prof. Wyman, in *Proc. Boston Soc. of Nat. Hist.,* Sept. 15, 1857; also Prof. Turner, in *Journal of Anat. and Phys.,* Nov. 1, 1866, p. 78. Dr. Günther has likewise described similar cases.

their mouths or branchial cavities; that certain male toads take the chaplets of eggs from the females, and wind them round their own thighs, keeping them there until the tadpoles are born; that certain male birds undertake the whole duty of incubation, and that male pigeons, as well as the females, feed their nestlings with a secretion from their crops. But the above suggestion first occurred to me from the mammary glands of male mammals being so much more perfectly developed than the rudiments of the other accessory reproductive parts which are found in the one sex though proper to the other. The mammary glands and nipples, as they exist in male mammals, can indeed hardly be called rudimentary; they are merely not fully developed, and not functionally active. They are sympathetically affected under the influence of certain diseases, like the same organs in the female. They often secrete a few drops of milk at birth and at puberty: this latter fact occurred in the curious case, before referred to, where a young man possessed two pairs of mammae. In man and some other male mammals these organs have been known occasionally to become so well developed during maturity as to yield a fair supply of milk. Now if we suppose that during a former prolonged period male mammals aided the females in nursing their offspring,[31] and that afterwards from some cause (as from the production of a smaller number of young) the males ceased to give this aid, disuse of the organs during maturity would lead to their becoming inactive; and from two well-known principles of inheritance, this state of inactivity would probably be transmitted to the males at the corresponding age of maturity. But at an earlier age these organs would be left unaffected, so that they would be almost equally well developed in the young of both sexes.

Conclusion. Von Baer has defined advancement or progress in the organic scale better than anyone else, as resting on the amount of differentiation and specialisation of the several parts of a being—when arrived at maturity, as I should be inclined to add. Now as organisms have become slowly adapted to diversified lines of life by means of natural selection, their parts will have become more and more differentiated and specialised for

[31] Mlle. C. Royer has suggested a similar view in her *Origine de l'Homme*, &c. (1870).

various functions, from the advantage gained by the division of physiological labour. The same part appears often to have been modified first for one purpose, and then long afterwards for some other and quite distinct purpose; and thus all the parts are rendered more and more complex. But each organism still retains the general type of structure of the progenitor from which it was aboriginally derived. In accordance with this view it seems, if we turn to geological evidence, that organisation on the whole has advanced throughout the world by slow and interrupted steps. In the great kingdom of the Vertebrata it has culminated in man. It must not, however, be supposed that groups of organic beings are always supplanted and disappear as soon as they have given birth to other and more perfect groups. The latter, though victorious over their predecessors, may not have become better adapted for all places in the economy of nature. Some old forms appear to have survived from inhabiting protected sites, where they have not been exposed to very severe competition; and these often aid us in constructing our genealogies, by giving us a fair idea of former and lost populations. But we must not fall into the error of looking at the existing members of any lowly organised group as perfect representatives of their ancient predecessors.

The most ancient progenitors in the kingdom of the Vertebrata, at which we are able to obtain an obscure glance, apparently consisted of a group of marine animals,[32] resembling the

---

[32] The inhabitants of the sea-shore must be greatly affected by the tides; animals living either about the *mean* high-water mark, or about the *mean* low-water mark, pass through a complete cycle of tidal changes in a fortnight. Consequently, their food supply will undergo marked changes week by week. The vital functions of such animals, living under these conditions for many generations, can hardly fail to run their course in regular weekly periods. Now it is a mysterious fact that in the higher and now terrestrial Vertebrata, as well as in other classes, many normal and abnormal processes have one or more whole weeks as their periods; this would be rendered intelligible if the Vertebrata are descended from an animal allied to the existing tidal Ascidians. Many instances of such periodic processes might be given, as the gestation of mammals, the duration of fevers, &c. The hatching of eggs affords also a good example, for, according to Mr. Bartlett (*Land and Water,* Jan. 7, 1871), the eggs of the pigeon are hatched in two weeks; those of the fowl in three; those of the duck in four; those of the goose in five; and those of the ostrich in seven weeks. As far as we can judge, a recurrent period, if approximately of the right duration for any process or function, would not, when once gained, be liable to change;

larvae of existing Ascidians. These animals probably gave rise to a group of fishes, as lowly organised as the lancelet; and from these the Ganoids, and other fishes like the Lepidosiren, must have been developed. From such fish a very small advance would carry us on to the amphibians. We have seen that birds and reptiles were once intimately connected together; and the Monotremata now connect mammals with reptiles in a slight degree. But no one can at present say by what line of descent the three higher and related classes, namely, mammals, birds, and reptiles, were derived from the two lower vertebrate classes, namely, amphibians and fishes. In the class of mammals the steps are not difficult to conceive which led from the ancient Monotremata to the ancient Marsupials; and from these to the early progenitors of the placental mammals. We may thus ascend to the Lemuridae; and the interval is not very wide from these to the Simiadae. The Simiadae then branched off into two great stems, the New World and Old World monkeys; and from the latter, at a remote period, Man, the wonder and glory of the Universe, proceeded.

Thus we have given to man a pedigree of prodigious length, but not, it may be said, of noble quality. The world, it has often been remarked, appears as if it had long been preparing for the advent of man: and this, in one sense, is strictly true, for he owes his birth to a long line of progenitors. If any single link in this chain had never existed, man would not have been exactly what he now is. Unless we wilfully close our eyes, we may, with our present knowledge, approximately recognise our parentage; nor need we feel ashamed of it. The most humble organism is something much higher than the inorganic dust under our feet; and no one with an unbiassed mind can study any living creature, however humble, without being struck with enthusiasm at its marvellous structure and properties.

---

consequently it might be thus transmitted through almost any number of generations. But if the function changed, the period would have to change, and would be apt to change almost abruptly by a whole week. This conclusion, if sound, is highly remarkable; for the period of gestation in each mammal, and the hatching of each bird's eggs, and many other vital processes, thus betray to us the primordial birthplace of these animals.

Insectivorous Plants

CHAPTER I

Drosera Rotundifolia, or the Common Sun-Dew

During the summer of 1860, I was surprised by finding how large a number of insects were caught by the leaves of the common sun-dew (*Drosera rotundifolia*) on a heath in Sussex. I had heard that insects were thus caught, but knew nothing further on the subject.[1] I gathered by chance a dozen plants, bearing fifty-six fully expanded leaves, and on thirty-one of these

[1] As Dr. Nitschke has given (*Bot. Zeitung* [1860], p. 229) the bibliography of Drosera, I need not here go into details. Most of the notices published before 1860 are brief and unimportant. The oldest paper seems to have been one of the most valuable, namely, by Dr. Roth in 1782. There is also an interesting though short account of the habits of Drosera by Dr. Milde, in the *Bot. Zeitung* (1852), p. 540. In 1855, in the *Annales des Sc. nat. bot.*, Tom. III, pp. 297 and 304, MM. Groenland and Trécul each published papers, with figures, on the structure of the leaves; but M. Trécul went so far as to doubt whether they possessed any power of movement. Dr. Nitschke's papers in the *Bot. Zeitung* for 1860 and 1861 are by far the most important ones which have been published, both on the habits and structure of this plant; and I shall frequently have occasion to quote from them. His discussions on several points, for instance on the transmission of an excitement from one part of the leaf to another, are excellent. On Dec. 11, 1862, Mr. J. Scott read a paper before the Botanical Society of Edinburgh, which was published in the *Gardener's Chronicle* (1863), p. 30. Mr. Scott shows that gentle irritation of the hairs, as well as insects placed on the disc of the leaf, cause the hairs to bend inwards. Mr. A. W. Bennett also gave another interesting account of the movements of the leaves before the British Association for 1873. In this same year Dr. Warming published an essay, in which he describes the structure of the so-called hairs, entitled "Sur la Différence entre les Trichomes, &c." extracted from the *Proceedings of the Soc. d'Hist. Nat. de Copenhague*. I shall also have occasion hereafter to refer to a paper by Mrs. Treat, of New Jersey, on some American species of Drosera. Dr. Burdon Sanderson delivered a lecture on Dionaea, before the Royal Institution (published in *Nature*, June 14, 1874), in which a short account of my observations on the power of true digestion possessed by Drosera and Dionaea first appeared. Prof. Asa Gray has done good service by calling attention to Drosera, and to other plants having similar habits, in *The Nation* (1874), pp. 261 and 232, and in other publications. Dr. Hooker, also, in his important address on Carnivorous Plants (Brit. Assoc., Belfast, 1874), has given a history of the subject.

dead insects or remnants of them adhered; and, no doubt, many more would have been caught afterwards by these same leaves, and still more by those as yet not expanded. On one plant all six leaves had caught their prey; and on several plants very many leaves had caught more than a single insect. On one large leaf I found the remains of thirteen distinct insects. Flies (Diptera) are captured much oftener than other insects. The largest kind which I have seen caught was a small butterfly (*Caenonympha pamphilus*); but the Rev. H. M. Wilkinson informs me that he found a large living dragon-fly with its body firmly held by two leaves. As this plant is extremely common in some districts, the number of insects thus annually slaughtered must be prodigious. Many plants cause the death of insects, for instance the sticky buds of the horse-chestnut (*Aesculus hippocastanum*), without thereby receiving, as far as we can perceive, any advantage; but it was soon evident that Drosera was excellently adapted for the special purpose of catching insects, so that the subject seemed well worthy of investigation.

The results have proved highly remarkable: the more important ones being, firstly, the extraordinary sensitiveness of the glands to slight pressure and to minute doses of certain nitrogenous fluids, as shown by the movements of the so-called hairs or tentacles [2]; secondly, the power possessed by the leaves of rendering soluble or digesting nitrogenous substances, and of afterwards absorbing them; thirdly, the changes which take place within the cells of the tentacles when the glands are excited in various ways.

It is necessary, in the first place, to describe briefly the plant. It bears from two or three to five or six leaves, generally extended more or less horizontally, but sometimes standing vertically upwards. The shape and general appearance of a leaf is shown, as seen from above, in fig. 1, and as seen laterally, in fig. 2. The leaves are commonly a little broader than long, but this was not the case in the one here figured. The whole upper surface is covered with gland-bearing filaments, or tentacles, as I shall call them, from their manner of acting. The glands were

[2] The drawings of Drosera and Dionaea, given in this work [in the original edition—S.E.H.], were made for me by my son George Darwin; those of Aldrovanda, and of the several species of Utricularia, by my son Francis. They have been excellently reproduced on wood by Mr. Cooper, 188 Strand.

counted on thirty-one leaves, but many of these were of unusually large size, and the average number was one hundred and ninety-two, the greatest number being two hundred and sixty, and the least one hundred and thirty. The glands are each surrounded by large drops of extremely viscid secretion, which, glittering in the sun, have given rise to the plant's poetical name of the sun-dew.

The tentacles on the central part of the leaf or disc are short and stand upright, and their pedicels are green. Towards the margin they become longer and longer and more inclined outwards, with their pedicels of a purple colour. Those on the extreme margin project in the same plane with the leaf, or more commonly (see fig. 2) are considerably reflexed. A few tentacles spring from the base of the footstalk or petiole, and these are the longest of all, being sometimes nearly a quarter of an inch in length. On a leaf bearing altogether two hundred and fifty-two tentacles, the short ones on the disc, having green pedicels, were in number to the longer submarginal and marginal tentacles, having purple pedicels, as nine to sixteen.

A tentacle consists of a thin, straight, hair-like pedicel, carrying a gland on the summit. The pedicel is somewhat flattened, and is formed of several rows of elongated cells, filled with purple fluid or granular matter.[3] There is, however, a narrow zone close beneath the glands of the longer tentacles, and a broader zone near their bases, of a green tint. Spiral vessels, accompanied by simple vascular tissue, branch off from the vascular bundles in the blade of the leaf, and run up all the tentacles into the glands.

Several eminent physiologists have discussed the homological nature of these appendages or tentacles, that is, whether they ought to be considered as hairs (trichomes) or prolongations of the leaf. Nitschke has shown that they include all the elements proper to the blade of a leaf; and the fact of their including vascular tissue was formerly thought to prove that they were prolongations of the leaf, but it is now known that vessels sometimes enter true hairs.[4] The

[3] According to Nitschke (*Bot. Zeitung* [1861], p. 224) the purple fluid results from the metamorphosis of chlorophyll. Mr. Sorby examined the colouring matter with the spectroscope, and informs me that it consists of the commonest species of erythrophyll, "which is often met with in leaves with low vitality, and in parts, like the petioles, which carry on leaf-functions in a very imperfect manner. All that can be said, therefore, is that the hairs (or tentacles) are coloured like parts of a leaf which do not fulfil their proper office."

[4] Dr. Nitschke has discussed this subject in *Bot. Zeitung* (1861), p. 241, &c. See also Dr. Warming, "Sur la Différence entre les Trichomes,

power of movement which they possess is a strong argument against their being viewed as hairs. The conclusion which seems to me the most probable will be given in Chap. XV, namely, that they existed primordially as glandular hairs, or mere epidermic formations, and that their upper part should still be so considered; but that their lower part, which alone is capable of movement, consists of a prolongation of the leaf; the spiral vessels being extended from this to the uppermost part. We shall hereafter see that the terminal tentacles of the divided leaves of Roridula are still in an intermediate condition.

The glands, with the exception of those borne by the extreme marginal tentacles, are oval, and of nearly uniform size, viz., about four five-hundredths of an inch in length. Their structure is remarkable, and their functions complex, for they secrete, absorb, and are acted on by various stimulants. They consist of an outer layer of small polygonal cells, containing purple granular matter or fluid, and with the walls thicker than those of the pedicels. Within this layer of cells there is an inner one of differently shaped ones, likewise filled with purple fluid, but of a slightly different tint, and differently affected by chloride of gold. These two layers are sometimes well seen when a gland has been crushed or boiled in caustic potash. According to Dr. Warming, there is still another layer of much more elongated cells, as shown in the accompanying section (fig. 3) copied from his work; but these cells were not seen by Nitschke, nor by me. In the centre there is a group of elongated, cylindrical cells of unequal lengths, bluntly pointed at their upper ends, truncated or rounded at their lower ends, closely pressed together, and remarkable from being surrounded by a spiral line, which can be separated as a distinct fibre.

These latter cells are filled with limpid fluid, which after long immersion in alcohol deposits much brown matter. I presume that they are actually connected with the spiral vessels which run up the tentacles, for on several occasions the latter were seen to divide into two or three excessively thin branches, which could be traced close up to the spiriferous cells. Their development has been described by Dr. Warming. Cells of the same kind have been observed in other plants, as I hear from Dr. Hooker, and were seen by me in the margins of the leaves of Pinguicula. Whatever their function may be, they are not necessary for the secretion of the digestive fluid, or for absorption, or for the communication of a motor impulse to

&c." (1873), who gives references to various publications. See also Groenland and Trécul, *Annal. des Sc. nat. bot.* (4th series, 1855), Tom. III, pp. 297 and 303.

other parts of the leaf, as we may infer from the structure of the glands in some other genera of the Droseraceae.

The extreme marginal tentacles differ slightly from the others. Their bases are broader, and besides their own vessels, they receive a fine branch from those which enter the tentacles on each side. Their glands are much elongated, and lie embedded on the upper surface of the pedicel, instead of standing at the apex. In other respects they do not differ essentially from the oval ones, and in one specimen I found every possible transition between the two states. In another specimen there were no long-headed glands. These marginal tentacles lose their irritability earlier than the others; and when a stimulus is applied to the centre of the leaf, they are excited into action after the others. When cut-off leaves are immersed in water, they alone often become inflected.

The purple fluid or granular matter which fills the cells of the glands differs to a certain extent from that within the cells of the pedicels. For when a leaf is placed in hot water or in certain acids, the glands become quite white and opaque, whereas the cells of the pedicels are rendered of a bright red, with the exception of those close beneath the glands. These latter cells lose their pale red tint; and the green matter which they, as well as the basal cells, contain becomes of a brighter green. The petioles bear many multicellular hairs, some of which near the blade are surmounted, according to Nitschke, by a few rounded cells, which appear to be rudimentary glands. Both surfaces of the leaf, the pedicels of the tentacles, especially the lower sides of the outer ones, and the petioles, are studded with minute papillae (hairs or trichomes), having a conical basis, and bearing on their summits two, and occasionally three or even four, rounded cells, containing much protoplasm. These papillae are generally colourless, but sometimes include a little purple fluid. They vary in development, and graduate, as Nitschke [5] states, and as I repeatedly observed, into the long multicellular hairs. The latter, as well as the papillae, are probably rudiments of formerly existing tentacles.

I may here add, in order not to recur to the papillae, that they do not secrete, but are easily permeated by various fluids: thus when living or dead leaves are immersed in a solution of one part of chloride of gold, or of nitrate of silver, to four hundred and thirty-seven of water, they are quickly blackened, and the discoloration soon spreads to the surrounding tissue. The long multicellular hairs are not so quickly affected. After a leaf had been left in a weak in-

[5] Nitschke has elaborately described and figured these papillae, *Bot. Zeitung* (1861), pp. 234, 253, 254.

fusion of raw meat for ten hours, the cells of the papillae had evidently absorbed animal matter, for instead of limpid fluid they now contained small aggregated masses of protoplasm, which slowly and incessantly changed their forms. A similar result followed from an immersion of only fifteen minutes in a solution of one part of carbonate of ammonia to two hundred and eighteen of water, and the adjoining cells of the tentacles, on which the papillae were seated, now likewise contained aggregated masses of protoplasm. We may therefore conclude that when a leaf has closely clasped a captured insect in the manner immediately to be described, the papillae, which project from the upper surface of the leaf and of the tentacles, probably absorb some of the animal matter dissolved in the secretion; but this cannot be the case with the papillae on the backs of the leaves or on the petioles.

Preliminary Sketch of the Action of the Several Parts, and of the Manner in Which Insects Are Captured

If a small organic or inorganic object be placed on the glands in the centre of a leaf, these transmit a motor impulse to the marginal tentacles. The nearer ones are first affected and slowly bend towards the centre, and then those farther off, until at last all become closely inflected over the object. This takes place in from one hour to four or five or more hours. The difference in the time required depends on many circumstances; namely, on the size of the object and on its nature, that is, whether it contains soluble matter of the proper kind; on the vigour and age of the leaf; whether it has lately been in action; and, according to Nitschke,[6] on the temperature of the day, as likewise seemed to me to be the case. A living insect is a more efficient object than a dead one, as in struggling it presses against the glands of many tentacles. An insect, such as a fly, with thin integuments, through which animal matter in solution can readily pass into the surrounding dense secretion, is more efficient in causing prolonged inflection than an insect with a thick coat, such as a beetle. The inflection of the tentacles takes place indifferently in the light and darkness; and the plant is not subject to any nocturnal movement of so-called sleep.

If the glands on the disc are repeatedly touched or brushed,

[6] *Bot. Zeitung* (1860), p. 246.

although no object is left on them, the marginal tentacles curve inwards. So again, if drops of various fluids, for instance of saliva or of a solution of any salt of ammonia, are placed on the central glands, the same result quickly follows, sometimes in under half an hour.

The tentacles in the act of inflection sweep through a wide space; thus a marginal tentacle, extended in the same plane with the blade, moves through an angle of 180°; and I have seen the much reflected tentacles of a leaf which stood upright move through an angle of not less than 270°. The bending part is almost confined to a short space near the base; but a rather larger portion of the elongated exterior tentacles becomes slightly incurved, the distal half in all cases remaining straight. The short tentacles in the centre of the disc when directly excited do not become inflected; but they are capable of inflection if excited by a motor impulse received from other glands at a distance. Thus, if a leaf is immersed in an infusion of raw meat, or in a weak solution of ammonia (if the solution is at all strong, the leaf is paralysed), all the exterior tentacles bend inwards (see fig. 4), excepting those near the centre, which remain upright; but these bend towards any exciting object placed on one side of the disc, as shown in fig. 5. The glands in fig. 4 may be seen to form a dark ring round the centre; and this follows from the exterior tentacles increasing in length in due proportion, as they stand nearer to the circumference.

The kind of inflection which the tentacles undergo is best shown when the gland of one of the long exterior tentacles is in any way excited; for the surrounding ones remain unaffected. In the accompanying outline (fig. 6) we see one tentacle, on which a particle of meat had been placed, thus bent towards the centre of the leaf, with two others retaining their original position. A gland may be excited by being simply touched three or four times, or by prolonged contact with organic or inorganic objects, and various fluids. I have distinctly seen, through a lens, a tentacle beginning to bend in ten seconds, after an object had been placed on its gland; and I have often seen strongly pronounced inflection in under one minute. It is surprising how minute a particle of any substance, such as a bit of thread or hair or splinter of glass, if placed in actual contact with the surface of a gland, suffices to cause the tentacle to bend. If the object, which has been carried by this movement to the centre, be not very small, or if it contains soluble nitrogenous matter, it acts on the central glands; and these transmit a motor impulse to the exterior tentacles, causing them to bend inwards.

Not only the tentacles, but the blade of the leaf often, but by no means always, becomes much incurved, when any strongly exciting substance or fluid is placed on the disc. Drops of milk and of a solution of nitrate of ammonia or soda are particularly apt to produce this effect. The blade is thus converted into a little cup. The manner in which it bends varies greatly. Sometimes the apex alone, sometimes one side, and sometimes both sides become incurved. For instance, I placed bits of hard-boiled egg on three leaves; one had the apex bent towards the base; the second had both distal margins much incurved, so that it became almost triangular in outline, and this perhaps is the commonest case; whilst the third blade was not at all affected, though the tentacles were as closely inflected as in the two previous cases. The whole blade also generally rises or bends upwards, and thus forms a smaller angle with the footstalk than it did before. This appears at first sight a distinct kind of movement, but it results from the incurvation of that part of the margin which is attached to the footstalk, causing the blade, as a whole, to curve or move upwards.

The length of time during which the tentacles as well as the blade remain inflected over an object placed on the disc depends on various circumstances; namely, on the vigour and age of the leaf, and, according to Dr. Nitschke, on the temperature, for during cold weather when the leaves are inactive, they re-expand at an earlier period than when the weather is warm. But the nature of the object is by far the most important circumstance; I have repeatedly found that the tentacles remain clasped for a much longer average time over objects which yield soluble nitrogenous matter than over those, whether organic or inorganic, which yield no such matter. After a period varying from one to seven days, the tentacles and blade re-expand, and are then ready to act again. I have seen the same leaf inflected three successive times over insects placed on the disc; and it would probably have acted a greater number of times.

The secretion from the glands is extremely viscid, so that it can be drawn out into long threads. It appears colourless, but stains little balls of paper pale pink. An object of any kind placed on a gland always causes it, as I believe, to secrete more freely; but the mere presence of the object renders this difficult to ascertain. In some cases, however, the effect was strongly marked, as when particles of sugar were added; but the result in this case is probably due merely to exosmose. Particles of carbonate and phosphate of ammonia and of some other salts, for instance sulphate of zinc, likewise increase the secretion. Immersion in a solution of one part of chloride of gold, or of some other salts, to four hundred and thirty-seven of water, excites the glands to largely increased secretion; on the other hand, tartrate of antimony produces no such effect. Immersion in many acids (of the strength of one part to four hundred and thirty-seven of water) likewise causes a wonderful amount of secretion, so that when the leaves are lifted out long ropes of extremely viscid fluid hang from them. Some acids, on the other hand, do not act in this manner. Increased secretion is not necessarily dependent on the inflection of the tentacle, for particles of sugar and of sulphate of zinc cause no movement.

It is a much more remarkable fact that when an object, such as a bit of meat or an insect, is placed on the disc of a leaf, as soon as the surrounding tentacles become considerably inflected their glands pour forth an increased amount of secretion. I ascertained this by selecting leaves with equal-sized drops on the two sides, and by placing bits of meat on one side of the disc; and as soon as the tentacles on this side became much inflected, but before the glands touched the meat, the drops of secretion became larger. This was repeatedly observed, but a record was kept of only thirteen cases, in nine of which increased secretion was plainly observed; the four failures being due either to the leaves being rather torpid, or to the bits of meat being too small to cause much inflection. We must therefore conclude that the central glands, when strongly excited, transmit some influence to the glands of the circumferential tentacles, causing them to secrete more copiously.

It is a still more important fact (as we shall see more fully when we treat of the digestive power of the secretion) that when the tentacles become inflected, owing to the central glands having been stimulated mechanically, or by contact with animal matter, the secretion not only increases in quantity, but changes its nature and becomes acid; and this occurs before the glands have touched the object on the centre of the leaf. This acid is of a different nature from that contained in the tissue of the leaves. As long as the tentacles remain closely inflected, the glands continue to secrete, and the secretion is acid; so that, if neutralised by carbonate of soda, it again becomes acid after a few hours. I have observed the same leaf with the tentacles closely inflected over rather indigestible substances, such as chemically prepared casein, pouring forth acid secretion for eight successive days, and over bits of bone for ten successive days.

The secretion seems to possess, like the gastric juice of the higher animals, some antiseptic power. During very warm weather I placed close together two equal-sized bits of raw meat, one on a leaf of the Drosera, and the other surrounded by wet moss. They were thus left for forty-eight hours, and then examined. The bit on the moss swarmed with infusoria, and was so much decayed that the transverse striae on the muscular fibres could no longer be clearly distinguished; whilst the bit on the leaf, which was bathed by the secretion, was free from infusoria, and its striae were perfectly distinct in the central and undissolved portion. In like manner small cubes of albumen and cheese placed on wet moss became threaded with filaments of mould, and had their surfaces slightly discoloured and disintegrated; whilst those on the leaves of Drosera remained clean, the albumen being changed into transparent fluid.

As soon as tentacles, which have remained closely inflected during several days over an object, begin to re-expand, their glands secrete less freely, or cease to secrete, and are left dry. In this state they are covered with a film of whitish, semi-fibrous matter, which was held in solution by the secretion. The drying of the glands during the act of re-expansion is of some little service to the plant; for I have often observed that objects adhering to the leaves could then be blown away by a breath of air, the leaves being thus left unencumbered and free for future action. Nevertheless, it often happens that all the glands do not become completely dry; and in this case delicate objects, such as fragile insects, are sometimes torn by the re-expansion of the tentacles into fragments, which remain scattered all over the leaf. After the re-expansion is complete, the glands quickly begin to re-secrete, and as soon as full-sized drops are formed the tentacles are ready to clasp a new object.

When an insect alights on the central disc, it is instantly entangled by the viscid secretion, and the surrounding tentacles after a time begin to bend, and ultimately clasp it on all sides. Insects are generally killed, according to Dr. Nitschke, in about a quarter of an hour, owing to their tracheae being closed by the secretion. If an insect adheres to only a few of the glands of the exterior tentacles, these soon become inflected and carry their prey to the tentacles next succeeding them inwards; these then bend inwards, and so onwards, until the insect is ultimately carried by a curious sort of rolling movement to the centre of the leaf. Then, after an interval, the tentacles on all sides become inflected and bathe their prey with their secretion, in the same manner as if the insect had first alighted on the central disc. It is surprising how minute an insect suffices to cause this action: for instance, I have seen one of the smallest species of gnats (Culex), which had just settled with its excessively delicate feet on the glands of the outermost tentacles, and these were already beginning to curve inwards, though not a single gland had as yet touched the body of the insect. Had I not interfered, this minute gnat would assuredly have been carried to the centre of the leaf and been securely clasped on all sides. We shall hereafter see what excessively small doses of certain organic fluids and saline solutions cause strongly marked inflection.

Whether insects alight on the leaves by mere chance, as a resting-place, or are attracted by the odour of the secretion, I know not. I suspect from the number of insects caught by the English species of Drosera, and from what I have observed with some exotic species kept in my greenhouse, that the odour is attractive. In this latter case the leaves may be compared with a baited trap; in the former case with a trap laid in a run frequented by game, but without any bait.

That the glands possess the power of aborption is shown by their almost instantaneously becoming dark-coloured when given a minute quantity of carbonate of ammonia; the change of colour being chiefly or exclusively due to the rapid aggregation of their contents. When certain other fluids are added, they become pale-coloured. Their power of absorption is, however, best shown by the widely different results which follow, from placing drops of various nitrogenous and non-nitrogenous fluids of the same density on the glands of the disc, or on a single marginal gland; and likewise by the very different lengths of time during which the tentacles remain inflected over objects, which yield or do not yield soluble nitrogenous matter. This same conclusion might indeed have been inferred from the structure and movements of the leaves, which are so admirably adapted for capturing insects.

The absorption of animal matter from captured insects explains how Drosera can flourish in extremely poor peaty soil—in some cases where nothing but sphagnum moss grows, and mosses depend altogether on the atmosphere for their nourishment. Although the leaves at a hasty glance do not appear green, owing to the purple colour of the tentacles, yet the upper and lower surfaces of the blade, the pedicels of the central tentacles, and the petioles contain chlorophyll, so that, no doubt, the plant obtains and assimilates carbonic acid from the air. Nevertheless, considering the nature of the soil where it grows, the supply of nitrogen would be extremely limited, or quite deficient, unless the plant had the power of obtaining this important element from captured insects. We can thus understand how it is that the roots are so poorly developed. These usually consist of only two or three slightly divided branches, from half to one inch in length, furnished with absorbent hairs. It appears, therefore, that the roots serve only to imbibe water; though, no doubt, they would absorb nutritious matter if present in the soil; for as we shall hereafter see, they absorb a weak solution of carbonate of ammonia. A plant of Drosera, with the edges of its leaves curled inwards, so as to form a temporary stomach, with the glands of the closely inflected tentacles pouring forth their acid secretion, which dissolves animal matter, afterwards to be absorbed, may be said to feed like an animal. But, differently from an animal, it drinks by means of its roots; and it must drink largely, so as to retain many drops of viscid fluid round the glands, sometimes as many as two hundred and sixty, exposed during the whole day to a glaring sun.

CHAPTER VI

The Digestive Power of the Secretion of Drosera

As we have seen that nitrogenous fluids act very differently on the leaves of Drosera from non-nitrogenous fluids, and as the leaves remain clasped for a much longer time over various organic bodies than over inorganic bodies, such as bits of glass, cinder, wood, &c., it becomes an interesting inquiry whether they can only absorb matter already in solution, or render it soluble —that is, have the power of digestion. We shall immediately see that they certainly have this power, and that they act on albuminous compounds in exactly the same manner as does the gastric juice of mammals, the digested matter being afterwards absorbed. This fact, which will be clearly proved, is a wonderful one in the physiology of plants. I must here state that I have been aided throughout all my later experiments by many valuable suggestions and assistance given me with the greatest kindness by Dr. Burdon Sanderson.

It may be well to premise for the sake of any reader who knows nothing about the digestion of albuminous compounds by animals that this is effected by means of a ferment, pepsin, together with weak hydrochloric acid, though almost any acid will serve. Yet neither pepsin nor an acid by itself has any such power.[1] We have seen that when the glands of the disc are excited by the contact of any object, especially of one containing nitrogenous matter, the outer tentacles and often the blade become inflected; the leaf being thus converted into a temporary cup or stomach. At the same time the discal glands secrete more copiously, and the secretion becomes acid. Moreover, they transmit some influence to the glands of the exterior tentacles, causing them to pour forth a more copious secretion, which also becomes acid or more acid than it was before.

As this result is an important one, I will give the evidence. The secretion of many glands on thirty leaves, which had not been in any way excited, was tested with litmus paper; and the

---

[1] It appears, however, according to Schiff, and contrary to the opinion of some physiologists, that weak hydrochloric dissolves, though slowly, a very minute quantity of coagulated albumen. Schiff, *Phys. de la Digestion* (1867), Tom. II, p. 25.

secretion of twenty-two of these leaves did not in the least affect the colour, whereas that of eight caused an exceedingly feeble and sometimes doubtful tinge of red. Two other old leaves, however, which appeared to have been inflected several times, acted much more decidedly on the paper. Particles of clean glass were then placed on five of the leaves, cubes of albumen on six, and bits of raw meat on three, on none of which was the secretion at this time in the least acid. After in interval of 24 hrs., when almost all the tentacles on these fourteen leaves had become more or less inflected, I again tested the secretion, selecting glands which had not as yet reached the centre or touched any object, and it was now plainly acid. The degree of acidity of the secretion varied somewhat on the glands of the same leaf. On some leaves, a few tentacles did not, from some unknown cause, become inflected, as often happens; and in five instances their secretion was found not to be in the least acid; whilst the secretion of the adjoining and inflected tentacles on the same leaf was decidedly acid. With leaves excited by particles of glass placed on the central glands, the secretion which collects on the disc beneath them was much more strongly acid than that poured forth from the exterior tentacles, which were as yet only moderately inflected. When bits of albumen (and this is naturally alkaline), or bits of meat were placed on the disc, the secretion collected beneath them was likewise strongly acid. As raw meat moistened with water is slightly acid, I compared its action on litmus paper before it was placed on the leaves, and afterwards when bathed in the secretion; and there could not be the least doubt that the latter was very much more acid. I have indeed tried hundreds of times the state of the secretion on the discs of leaves which were inflected over various objects, and never failed to find it acid. We may, therefore, conclude that the secretion from unexcited leaves, though extremely viscid, is not acid or only slightly so, but that it becomes acid, or much more strongly so, after the tentacles have begun to bend over any inorganic or organic object; and still more strongly acid after the tentacles have remained for some time closely clasped over any object.

I may here remind the reader that the secretion appears to be to a certain extent antiseptic, as it checks the appearance of mould and infusoria, thus preventing for a time the discoloration and decay of such substances as the white of an egg, cheese, &c. It therefore acts like the gastric juice of the higher animals, which is known to arrest putrefaction by destroying the microzymes.

As I was anxious to learn what acid the secretion contained, four hundred and forty-five leaves were washed in distilled water given me by Prof. Frankland; but the secretion is so viscid that it is scarcely possible to scrape or wash off the whole. The conditions were also unfavourable, as it was late in the year and the leaves were small. Prof. Frankland with great kindness undertook to test the fluid thus collected. The leaves were excited by clean particles of glass placed on them twenty-four hours previously. No doubt much more acid would have been secreted had the leaves been excited by animal matter, but this would have rendered the analysis more difficult. Prof. Frankland informs me that the fluid contained no trace of hydrochloric, sulphuric, tartaric, oxalic, or formic acids. This having been ascertained, the remainder of the fluid was evaporated nearly to dryness, and acidified with sulphuric acid; it then evolved volatile acid vapour, which was condensed and digested with carbonate of silver. "The weight of the silver salt thus produced was only .37 gr., much too small a quantity for the accurate determination of the molecular weight of the acid. The number obtained, however, corresponded nearly with that of propionic acid; and I believe that this, or a mixture of acetic and butyric acids, were present in the liquid. The acid doubtless belongs to the acetic or fatty series."

Prof. Frankland, as well as his assistant, observed (and this is an important fact) that the fluid, "when acidified with sulphuric acid, emitted a powerful odour like that of pepsin." The leaves from which the secretion had been washed were also sent to Prof. Frankland; they were macerated for some hours, then acidified with sulphuric acid and distilled, but no acid passed over. Therefore the acid which fresh leaves contain, as shown by their discolouring litmus paper when crushed, must be of a different nature from that present in the secretion. Nor was any odour of pepsin emitted by them.

Although it has long been known that pepsin with acetic acid has the power of digesting albuminous compounds, it appeared advisable to ascertain whether acetic acid could be replaced, without the loss of digestive power, by the allied acids which are believed to occur in the secretion of Drosera, namely, propionic, butyric, or valerianic. Dr. Burdon Sanderson was so kind as to make for me the following experiments, the results of which are valuable, independently of the present inquiry. Prof. Frankland supplied the acids.

"1. The purpose of the following experiments was to determine the digestive activity of liquids containing pepsin, when acidulated with certain volatile acids belonging to the acetic series, in comparison with liquids acidulated with hydrochloric acid, in proportion similar to that in which it exists in gastric juice.

"2. It has been determined empirically that the best results are obtained in artificial digestion when a liquid containing two per thousand of hydrochloric acid gas by weight is used. This corresponds to about 6.25 cubic centimetres per litre of ordinary strong hydrochloric acid. The quantities of propionic, butyric, and valerianic acids respectively which are required to neutralise as much base as 6.25 cubic centimetres of HCl, are in grammes 4.04 of propionic acid, 4.82 of butyric acid, and 5.68 of valerianic acid. It was therefore judged expedient, in comparing the digestive powers of these acids with that of hydrochloric acid, to use them in these proportions.

"3. Five hundred cub. cent. of a liquid containing about eight cub. cent. of a glycerine extract of the mucous membrane of the stomach of a dog killed during digestion having been prepared, ten cub. cent. of it were evaporated and dried at 110°. This quantity yielded 0.0031 of residue.

"4. Of this liquid four quantities were taken which were severally acidulated with hydrochloric, propionic, butyric, and valerianic acids, in the proportions above indicated. Each liquid was then placed in a tube, which was allowed to float in a water bath, containing a thermometer which indicated a temperature of 38° to 40° Cent. Into each, a quantity of unboiled fibrin was introduced, and the whole allowed to stand for four hours, the temperature being maintained during the whole time, and care being taken that each contained throughout an excess of fibrin. At the end of the period each liquid was filtered. Of the filtrate, which of course contained as much of the fibrin as had been digested during the four hours, 10 cub. cent. were measured out and evaporated, and dried at 110° as before. The residues were respectively—

In the liquid containing hydrochloric acid	0.4079
In the liquid containing propionic acid	0.0601
In the liquid containing butyric acid	0.1468
In the liquid containing valerianic acid	0.1254

"Hence, deducting from each of these the above-mentioned residue, left when the digestive liquid itself was evaporated, viz., 0.0031, we have:

For propionic acid 0.0570
For butyric acid 0.1437
For valerianic acid 0.1223

as compared with 0.4048 for hydrochloric acid; these several numbers expressing the quantities of fibrin by weight digested in presence of equivalent quantities of the respective acids under identical conditions.

"The results of the experiment may be stated thus: If 100 represent the digestive power of a liquid containing pepsin with the usual proportion of hydrochloric acid, 14.0, 35.4, and 30.2, will represent respectively the digestive powers of the three acids under investigation.

"5. In a second experiment in which the procedure was in every respect the same, excepting that all the tubes were plunged into the same water-bath, and the residues dried at 115° C., the results were as follows—

"Quantity of fibrin dissolved in four hours by ten cub. cent. of the liquid:

Propionic acid 0.0563
Butyric acid 0.0835
Valerianic acid 0.0615

"The quantity digested by a similar liquid containing hydrochloric acid was 0.3376. Hence, taking this as 100, the following numbers represent the relative quantities digested by the other acids:

Propionic acid 16.5
Butyric acid 24.7
Valerianic acid 16.1

"6. A third experiment of the same kind gave—

"Quantity of fibrin digested in four hours by ten cub. cent. of the liquid:

Hydrochloric acid 0.2915
Propionic acid 0.1490
Butyric acid 0.1044
Valerianic acid 0.0520

"Comparing, as before, the three last numbers with the first taken as 100, the digestive power of propionic acid is represented by 16.8; that of butyric acid by 35.8; and that of valerianic by 17.8.

"The mean of these three sets of observations (hydrochloric acid being taken as 100) gives for:

Propionic acid 15.8
Butyric acid 32.0
Valerianic acid 21.4

"7. A further experiment was made to ascertain whether the digestive activity of butyric acid (which was selected as being apparently the most efficacious) was relatively greater at ordinary temperatures than at the temperature of the body. It was found that whereas ten cub. cent. of a liquid containing the ordinary proportion of hydrochloric acid digested 0.1311 gramme, a similar liquid prepared with butyric acid digested 0.0455 gramme of fibrin.

"Hence, taking the quantities digested with hydrochloric acid at the temperature of the body as 100, we have the digestive power of hydrochloric acid at the temperature of 16° to 18° Cent. represented by 44.9; that of butyric acid at the same temperature being 15.6."

We here see that at the lower of these two temperatures, hydrochloric acid with pepsin digests, within the same time, rather less than half the quantity of fibrin compared with what it digests at the higher temperature; and the power of butyric acid is reduced in the same proportion under similar conditions and temperatures. We have also seen that butyric acid, which is much more efficacious than propionic or valerianic acids, digests with pepsin at the higher temperature less than a third of the fibrin which is digested at the same temperature by hydrochloric acid.

I will now give in detail my experiments on the digestive power of the secretion of Drosera, dividing the substances tried into two series, namely those which are digested more or less completely, and those which are not digested. We shall presently see that all these substances are acted on by the gastric juice of the higher animals in the same manner. I beg leave to call attention to the experiments under the head albumen, showing that the secretion loses its power when neutralised by an alkali, and recovers it when an acid is added.

Substances Which Are Completely or Partially Digested by the Secretion of Drosera

Albumen. After having tried various substances, Dr. Burdon Sanderson suggested to me the use of cubes of coagulated albumen or hard-boiled egg. I may premise that five cubes of the same size as those used in the following experiments were placed for the sake of comparison at the same time on wet moss close to the plants of Drosera. The weather was hot, and after four days some of the cubes were discoloured and mouldy, with their angles a little rounded; but they were not surrounded by a zone of transparent fluid as in the case of those undergoing digestion. Other cubes retained their angles and white colour. After eight days all were somewhat reduced in size, discoloured, with their angles much rounded. Nevertheless in four out of the five specimens, the central parts were still white and opaque. So that their state differed widely, as we shall see, from that of the cubes subjected to the action of the secretion.

Experiment 1. Rather large cubes of albumen were first tried the tentacles were well inflected in twenty-four hours; after an additional day the angles of the cubes were dissolved and rounded [2]; but the cubes were too large, so that the leaves were injured, and after seven days one died and the others were dying. Albumen which has been kept for four or five days, and which, it may be presumed, has begun to decay slightly, seems to act more quickly than freshly boiled eggs. As the latter were generally used, I often moistened them with a little saliva, to make the tentacles close more quickly.

Experiment 2. A cube of one tenth of an inch (i.e. with each side one tenth of an inch, or 2.54 mm., in length) was placed on a leaf, and after fifty hours it was converted into a sphere about three fortieths of an inch (1.905 mm.) in diameter, surrounded by perfectly transparent fluid. After ten days the leaf re-expanded, but there was still left on the disc a minute bit of albumen now rendered transparent. More albumen had been given to this leaf than could be dissolved or digested.

Experiment 3. Two cubes of albumen of one twentieth of an inch (1.27 mm.) were placed on two leaves. After forty-six hours every atom of one was dissolved, and most of the liquefied matter was absorbed, the fluid which remained being in this, as in all other cases, very acid and viscid. The other cube was acted on at a rather slower rate.

Experiment 4. Two cubes of albumen of the same size as the last

---

[2] In all my numerous experiments on the digestion of cubes of albumen, the angles and edges were invariably first rounded. Now, Schiff states (*Leçons phys. de la Digestion* [1867], Vol. ii, p. 149) that this is characteristic of the digestion of albumen by the gastric juice of animals. On the other hand, he remarks, "Les dissolutions, en chimie, ont lieu sur *toute* la surface des corps en contact avec l'agent dissolvant."

were placed on two leaves, and were converted in fifty hours into two large drops of transparent fluid; but when these were removed from beneath the inflected tentacles, and viewed by reflected light under the microscope, fine streaks of white opaque matter could be seen in the one, and traces of similar streaks in the other. The drops were replaced on the leaves, which re-expanded after ten days; and now nothing was left except a very little transparent acid fluid.

Experiment 5. This experiment was slightly varied, so that the albumen might be more quickly exposed to the action of the secretion. Two cubes, each of about one fortieth of an inch (.635 mm.), were placed on the same leaf, and two similar cubes on another leaf. These were examined after twenty-one hours and thirty minutes, and all four were found rounded. After forty-six hours the two cubes on the one leaf were completely liquefied, the fluid being perfectly transparent; on the other leaf some opaque white streaks could still be seen in the midst of the fluid. After seventy-two hours these streaks disappeared, but there was still a little viscid fluid left on the disc; whereas it was almost all absorbed on the first leaf. Both leaves were now beginning to re-expand.

The best and almost sole test of the presence of some ferment analogous to pepsin in the secretion appeared to be to neutralise the acid of the secretion with an alkali, and to observe whether the process of digestion ceased; and then to add a little acid and observe whether the process recommenced. This was done, and, as we shall see, with success, but it was necessary first to try two control experiments; namely, whether the addition of minute drops of water of the same size as those of the dissolved alkalies to be used would stop the process of digestion; and, secondly, whether minute drops of weak hydrochloric acid, of the same strength and size as those to be used, would injure the leaves. The two following experiments were therefore tried:

Experiment 6. Small cubes of albumen were put on three leaves, and minute drops of distilled water on the head of a pin were added two or three times daily. These did not in the least delay the process; for, after forty-eight hours, the cubes were completely dissolved on all three leaves. On the third day the leaves began to re-expand, and on the fourth day all the fluid was absorbed.

Experiment 7. Small cubes of albumen were put on two leaves, and minute drops of hydrochloric acid, of the strength of one part to four hundred and thirty-seven of water, were added two or three times. This did not in the least delay, but seemed rather to hasten, the process of digestion; for every trace of the albumen disappeared in twenty-four hours and thirty minutes. After three days the leaves partially re-expanded, and by this time almost all the viscid fluid on their discs was absorbed. It is almost superfluous to state that cubes of albumen of the same size as those above used, left for seven days in a little hydrochloric acid of the above strength, retained all their angles as perfect as ever.

Experiment 8. Cubes of albumen (of one twentieth of an inch, or 2.54 mm.) were placed on five leaves, and minute drops of a solution of one part of carbonate of soda to four hundred and thirty-seven of water were added at intervals to three of them, and drops of carbonate of potash of the same strength to the other two. The drops were given on the head of a rather large pin, and I ascertained that each was equal to about one tenth of a minim (.0059 ml.), so that each contained only $\frac{1}{4800}$ of a grain (.0135 mg.) of the alkali. This was not sufficient, for after forty-six hours all five cubes were dissolved.

Experiment 9. The last experiment was repeated on four leaves, with this difference, that drops of the same solution of carbonate of soda were added rather oftener, as often as the secretion became acid, so that it was much more effectually neutralised. And now after twenty-four hours the angles of three of the cubes were not in the least rounded, those of the fourth being so in a very slight degree. Drops of extremely weak hydrochloric acid (viz., one part to eight hundred and forty-seven of water) were then added, just enough to neutralise the alkali which was still present; and now digestion immediately recommenced, so that after twenty-three hours and thirty minutes three of the cubes were completely dissolved, whilst the fourth was converted into a minute sphere, surrounded by transparent fluid; and this sphere next day disappeared.

Experiment 10. Stronger solutions of carbonate of soda and of potash were next used, viz. one part to one hundred and nine of water; and as the same-sized drops were given as before, each drop contained $\frac{1}{1200}$ of a grain (.0539 mg.) of either salt. Two cubes of albumen (each about one fortieth of an inch, or .635 mm.) were placed on the same leaf, and two on another. Each leaf received, as soon as the secretion became slightly acid (and this occurred four times within twenty-four hours), drops either of the soda or potash, and the acid was thus effectually neutralised. The experiment now succeeded perfectly, for after twenty-two hours the angles of the cubes were as sharp as they were at first, and we know from Experiment 5 that such small cubes would have been completely rounded within this time by the secretion in its natural state. Some of the fluid was now removed with blotting-paper from the discs of the leaves, and minute drops of hydrochloric acid of the strength of one part to two hundred of water was added. Acid of this greater strength was used as the solutions of the alkalies were stronger. The process of digestion now commenced, so that within forty-eight hours from the time when the acid was given the four cubes were not only completely dissolved, but much of the liquefied albumen was absorbed.

Experiment 11. Two cubes of albumen (one fortieth of an inch, or .635 mm.) were placed on two leaves, and were treated with alkalies as in the last experiment, and with the same result; for after twenty-two hours they had their angles perfectly sharp, showing that the digestive process had been completely arrested. I then wished to ascertain what would be the effect of using stronger hydrochloric acid; so I added minute drops of the strength of one per cent. This proved rather too strong, for after forty-eight hours from the time when the acid was added one cube was still almost perfect, and the other only very slightly rounded, and both were stained slightly pink. This latter fact shows that the leaves were injured,[3] for during the normal process of digestion the albumen is not thus coloured, and we can thus understand why the cubes were not dissolved.

From these experiments we clearly see that the secretion has the power of dissolving albumen, and we further see that if an alkali is added, the process of digestion is stopped, but immediately recommences as soon as the alkali is neutralised by weak hydrochloric acid. Even if I had tried no other experiments than these, they would have almost sufficed to prove that the glands of Drosera secrete some ferment analogous to pepsin, which in presence of an acid gives to the secretion its power of dissolving albuminous compounds.

Splinters of clean glass were scattered on a large number of leaves, and these became moderately inflected. They were cut off and divided into three lots; two of them, after being left for some time in a little distilled water, were strained, and some discoloured, viscid, slightly acid fluid was thus obtained. The third lot was well soaked in a few drops of glycerine, which is well known to dissolve pepsin. Cubes of albumen (one twentieth of an inch) were now placed in the three fluids in watch-glasses, some of which were kept for several days at about 90° Fahr.

---

[3] Sachs remarks (*Traité de Bot.* [1874], p. 774) that cells which are killed by freezing, by too great heat, or by chemical agents, allow all their colouring matter to escape into the surrounding water.

(32°.2 Cent.), and others at the temperature of my room; but none of the cubes were dissolved, the angles remaining as sharp as ever. This fact probably indicates that the ferment is not secreted until the glands are excited by the absorption of a minute quantity of already soluble animal matter—a conclusion which is supported by what we shall hereafter see with respect to Dionaea. Dr. Hooker likewise found that, although the fluid within the pitchers of Nepenthes possesses extraordinary power of digestion, yet when removed from the pitchers before they have been excited and placed in a vessel, it has no such power, although it is already acid; and we can account for this fact only on the supposition that the proper ferment is not secreted until some exciting matter is absorbed.

On three other occasions eight leaves were strongly excited with albumen moistened with saliva; they were then cut off, and allowed to soak for several hours or for a whole day in a few drops of glycerine. Some of this extract was added to a little hydrochloric acid of various strengths (generally one to four hundred of water), and minute cubes of albumen were placed in the mixture.[4] In two of these trials the cubes were not in the least acted on; but in the third the experiment was successful. For in a vessel containing two cubes, both were reduced in size in three hours; and after twenty-four hours mere streaks of un-dissolved albumen were left. In a second vessel, containing two minute ragged bits of albumen, both were likewise reduced in size in three hours, and after twenty-four hours completely dis-appeared. I then added a little weak hydrochloric acid to both vessels, and placed fresh cubes of albumen in them; but these were not acted on. This latter fact is intelligible according to the high authority of Schiff,[5] who has demonstrated, as he believes, in opposition to the view held by some physiologists, that a cer-tain small amount of pepsin is destroyed during the act of di-gestion. So that if my solution contained, as is probable, an ex-tremely small amount of the ferment, this would have been consumed by the dissolution of the cubes of albumen first given; none being left when the hydrochloric acid was added. The de-struction of the ferment during the process of digestion, or its

[4] As a control experiment bits of albumen were placed in the same glycerine with hydrochloric acid of the same strength; and the albumen, as might have been expected, was not in the least affected after two days.
[5] *Leçons phys. de la Digestion* (1867), Tom. II, pp. 114–126.

absorption after the albumen had been converted into a peptone, will also account for only one out of the three latter sets of experiments having been successful.

Digestion of Roast Meat. Cubes of about one twentieth of an inch (1.27 mm.) of moderately roasted meat were placed on five leaves which became in twelve hours closely inflected. After forty-eight hours I gently opened one leaf, and the meat now consisted of a minute central sphere, partially digested and surrounded by a thick envelope of transparent viscid fluid. The whole, without being much disturbed, was removed and placed under the microscope. In the central part the transverse striae on the muscular fibres were quite distinct; and it was interesting to observe how gradually they disappeared, when the same fibre was traced into the surrounding fluid. They disappeared by the striae being replaced by transverse lines formed of excessively minute dark points, which towards the exterior could be seen only under a very high power; and ultimately these points were lost. When I made these observations, I had not read Schiff's account [6] of the digestion of meat by gastric juice, and I did not understand the meaning of the dark points. But this is explained in the following statement, and we further see how closely similar is the process of digestion by gastric juice and by the secretion of Drosera.

"On a dit que le suc gastrique faisait perdre à la fibre musculaire ses stries transversales. Ainsi énoncée, cette proposition pourrait donner lieu à une équivoque, car ce qui se perd, ce n'est que *l'aspect* extérieur de la striature et non les éléments anatomiques qui la composent. On sait que les stries qui donnent un aspect si caractéristique à la fibre musculaire, sont le résultat de la juxtaposition et du parallélisme des corpuscules élémentaires, placés, à distances égales, dans l'intérieur des fibrilles contiguës. Or, dès que le tissu connectif qui relie entre elles les fibrilles élémentaires vient à se gonfler et à se dissoudre, et que les fibrilles elles-mêmes se dissocient, ce parallélisme est détruit et avec lui l'aspect, le phénomène optique des stries. Si, après la désagrégation des fibres, on examine au microscope les fibrilles élémentaires, on distingue encore très-nettement à leur intérieur les corpuscules, et on continue à les voir, de plus en plus pâles, jusqu'au moment où les fibrilles elles-mêmes se liquéfient et disparaissent dans le suc gastrique. Ce qui constitue la striature,

---

[6] *Leçons phys. de la Digestion,* Tom. II, p. 145.

à proprement parler, n'est donc pas détruit, avant la liquéfaction de
la fibre charnue elle-même."

In the viscid fluid surrounding the central sphere of undigested
meat there were globules of fat and little bits of fibro-elastic
tissue; neither of which were in the least digested. There were
also little free parallelograms of yellowish, highly translucent
matter. Schiff, in speaking of the digestion of meat by gastric
juice, alludes to such parallelograms, and says:

"Le gonflement par lequel commence la digestion de la viande,
résulte de l'action du suc gastrique acide sur le tissu connectif qui
se dissout d'abord, et qui, par sa liquéfaction, désagrége les fibrilles.
Celles-ci se dissolvent ensuite en grande partie, mais, avant de passer
à l'état liquide, elles tendent à se briser en petits fragments trans-
versaux. Les '*sarcous elements*' de Bowman, qui ne sont autre chose
que les produits de cette division transversale des fibrilles élémentaires, peuvent être préparés et isolés à l'aide du suc gastrique,
pourvu qu'on n'attend pas jusqu'à la liquéfaction complète du
muscle."

After an interval of seventy-two hours, from the time when the
five cubes were placed on the leaves, I opened the four remaining
ones. On two nothing could be seen but little masses of transparent viscid fluid; but when these were examined under a high
power, fat-globules, bits of fibro-elastic tissue, and some few
parallelograms of sarcous matter, could be distinguished, but not
a vestige of transverse striae. On the other two leaves there were
minute spheres of only partially digested meat in the centre of
much transparent fluid.

Fibrin. Bits of fibrin were left in water during four days, whilst
the following experiments were tried, but they were not in the
least acted on. The fibrin which I first used was not pure, and
included dark particles: it had either not been well prepared or
had subsequently undergone some change. Thin portions, about
one twentieth of an inch square, were placed on several leaves,
and though the fibrin was soon liquefied, the whole was never
dissolved. Smaller particles were then placed on four leaves, and
minute drops of hydrochloric acid (one part to four hundred and
thirty-seven of water) were added; this seemed to hasten the
process of digestion, for on one leaf all was liquefied and ab-

sorbed after twenty hours; but on the three other leaves some undissolved residue was left after forty-eight hours. It is remarkable that in all the above and following experiments, as well as when much larger bits of fibrin were used, the leaves were very little excited; and it was sometimes necessary to add a little saliva to induce complete inflection. The leaves, moreover, began to re-expand after only forty-eight hours, whereas they would have remained inflected for a much longer time had insects, meat, cartilage, albumen, &c., been placed on them.

I then tried some pure white fibrin, sent me by Dr. Burdon Sanderson.

Experiment 1. Two particles, barely one twentieth of an inch (1.27 mm.) square, were placed on opposite sides of the same leaf. One of these did not excite the surrounding tentacles, and the gland on which it rested soon dried. The other particle caused a few of the short adjoining tentacles to be inflected, the more distant ones not being affected. After twenty-four hours both were almost, and after seventy-two hours completely, dissolved.

Experiment 2. The same experiment with the same result, only one of the two bits of fibrin exciting the short surrounding tentacles. This bit was so slowly acted on that after a day I pushed it on to some fresh glands. In three days from the time when it was first placed on the leaf it was completely dissolved.

Experiment 3. Bits of fibrin of about the same size as before were placed on the disks of two leaves; these caused very little inflection in twenty-three hours, but after forty-eight hours both were well clasped by the surrounding short tentacles, and after an additional twenty-four hours were completely dissolved. On the disc of one of these leaves much clear acid fluid was left.

Experiment 4. Similar bits of fibrin were placed on the discs of two leaves; as after two hours the glands seemed rather dry, they were freely moistened with saliva; this soon caused strong inflection both of the tentacles and blades, with copious secretion from the glands. In eighteen hours the fibrin was completely liquefied, but undigested atoms still floated in the liquid; these, however, disappeared in under two additional days.

From these experiments it is clear that the secretion completely dissolves pure fibrin. The rate of dissolution is rather slow; but this depends merely on this substance not exciting the leaves sufficiently, so that only the immediately adjoining tentacles are inflected, and the supply of secretion is small.

Syntonin. This substance, extracted from muscle, was kindly prepared for me by Dr. Moore. Very differently from fibrin, it acts quickly and energetically. Small portions placed on the discs of three leaves caused their tentacles and blades to be strongly inflected within eight hours; but no further observations were made. It is probably due to the presence of this substance that raw meat is too powerful a stimulant, often injuring or even killing the leaves.

Areolar Tissue. Small portions of this tissue from a sheep were placed on the discs of three leaves; these became moderately well inflected in twenty-four hours, but began to re-expand after forty-eight hours, and were fully re-expanded in seventy-two hours, always reckoning from the time when the bits were first given. This substance, therefore, like fibrin, excites the leaves for only a short time. The residue left on the leaves, after they were fully re-expanded, was examined under a high power and found much altered, but, owing to the presence of a quantity of elastic tissue, which is never acted on, could hardly be said to be in a liquefied condition.

Some areolar tissue free from elastic tissue was next procured from the visceral cavity of a toad, and moderately sized, as well as very small, bits were placed on five leaves. After twenty-four hours two of the bits were completely liquefied; two others were rendered transparent, but not quite liquefied; whilst the fifth was but little affected. Several glands on the three latter leaves were now moistened with a little saliva, which soon caused much inflection and secretion, with the result that in the course of twelve additional hours one leaf alone showed a remnant of undigested tissue. On the discs of the four other leaves (to one of which a rather large bit had been given) nothing was left except some transparent viscid fluid. I may add that some of this tissue included points of black pigment, and these were not at all affected. As a control experiment, small portions of this tissue were left in water and on wet moss for the same length of time, and remained white and opaque. From these facts it is clear that areolar tissue is easily and quickly digested by the secretion; but that it does not greatly excite the leaves.

Cartilage. Three cubes (one twentieth of an inch, or 1.27 mm.) of white, translucent, extremely tough cartilage were cut from the end of a slightly roasted leg-bone of a sheep. These were placed on three leaves, borne by poor, small plants in my greenhouse during November; and it seemed in the highest degree improbable that so hard a substance would be digested under such unfavourable circumstances. Nevertheless, after forty-eight hours, the cubes were largely dissolved and converted into minute spheres, surrounded by transparent, very acid fluid. Two of these spheres were completely softened to their centres; whilst the third still contained a very small irregularly shaped core of solid cartilage. Their surfaces were seen under the microscope to be curiously marked by prominent ridges, showing that the cartilage had been unequally corroded by the secretion. I need hardly say that cubes of the same cartilage, kept in water for the same length of time, were not in the least affected.

During a more favourable season, moderately sized bits of the skinned ear of a cat, which includes cartilage, areolar and elastic tissue, were placed on three leaves. Some of the glands were touched with saliva, which caused prompt inflection. Two of the leaves began to re-expand after three days, and the third on the fifth day. The fluid residue left on their discs was now examined, and consisted in one case of perfectly transparent, viscid matter; in the other two cases, it contained some elastic tissue and apparently remnants of half digested areolar tissue.

Fibro-cartilage (from between the vertebrae of the tail of a sheep). Moderately sized and small bits (the latter about one twentieth of an inch) were placed on nine leaves. Some of these were well and some very little inflected. In the latter case the bits were dragged over the discs, so that they were well bedaubed with the secretion, and many glands thus irritated. All the leaves re-expanded after only two days; so that they were but little excited by this substance. The bits were not liquefied, but were certainly in an altered condition, being swollen, much more transparent, and so tender as to disintegrate very easily. My son Francis prepared some artificial gastric juice, which was proved efficient by quickly dissolving fibrin, and suspended portions of the fibro-cartilage in it. These swelled and became hyaline, exactly like those exposed to the secretion of Drosera, but were not dissolved. This result surprised me much, as two physiologists were of opinion that fibro-cartilage would be easily digested by gastric juice. I therefore asked Dr. Klein to examine the specimens; and he reports that the two which had been subjected to

artificial gastric juice were "in that state of digestion in which
we find connective tissue when treated with an acid, viz., swollen,
more or less hyaline, the fibrillar bundles having become homo-
geneous and lost their fibrillar structure." In the specimens which
had been left on the leaves of Drosera, until they re-expanded,
"parts were altered, though only slightly so, in the same manner
as those subjected to the gastric juice, as they had become more
transparent, almost hyaline, with the fibrillation of the bundles
indistinct." Fibro-cartilage is therefore acted on in nearly the
same manner by gastric juice and by the secretion of Drosera.

Bone. Small smooth bits of the dried hyoidal bone of a fowl
moistened with saliva were placed on two leaves, and a similarly
moistened splinter of an extremely hard, broiled mutton-chop
bone on a third leaf. These leaves soon became strongly inflected,
and remained so for an unusual length of time; namely, one leaf
for ten and the other two for nine days. The bits of bone were
surrounded all the time by acid secretion. When examined under
a weak power, they were found quite softened, so that they were
readily penetrated by a blunt needle, torn into fibres, or compressed. Dr. Klein was so kind as to make sections of both bones
and examine them. He informs me that both presented the
normal appearance of decalcified bone, with traces of the earthy
salts occasionally left. The corpuscles with their processes were
very distinct in most parts; but in some parts, especially near the
periphery of the hyoidal bone, none could be seen. Other parts
again appeared amorphous, with even the longitudinal striation
of bone not distinguishable. This amorphous structure, as Dr.
Klein thinks, may be the result either of the incipient digestion
of the fibrous basis or of all the animal matter having been
removed, the corpuscles being thus rendered invisible. A hard,
brittle, yellowish substance occupied the position of the medulla
in the fragments of the hyoidal bone.

As the angles and little projections of the fibrous basis were
not in the least rounded or corroded, two of the bits were placed
on fresh leaves. These by the next morning were closely inflected,
and remained so—the one for six and the other for seven days—
therefore for not so long a time as on the first occasion, but for a
much longer time than ever occurs with leaves inflected over
inorganic or even over many organic bodies. The secretion during the whole time coloured litmus paper of a bright red; but this

may have been due to the presence of the acid superphosphate of lime. When the leaves re-expanded, the angles and projections of the fibrous basis were as sharp as ever. I therefore concluded, falsely as we shall presently see, that the secretion cannot touch the fibrous basis of bone. The more probable explanation is that the acid was all consumed in decomposing the phosphate of lime which still remained; so that none was left in a free state to act in conjunction with the ferment on the fibrous basis.

Enamel and Dentine. As the secretion decalcified ordinary bone, I determined to try whether it would act on enamel and dentine, but did not expect that it would succeed with so hard a substance as enamel. Dr. Klein gave me some thin transverse slices of the canine tooth of a dog; small angular fragments of which were placed on four leaves; and these were examined each succeeding day at the same hour. The results are, I think, worth giving in detail.

Experiment 1. May 1st, fragment placed on leaf; 3rd, tentacles but little inflected, so a little saliva was added; 6th, as the tentacles were not strongly inflected, the fragment was transferred to another leaf, which acted at first slowly, but by the 9th closely embraced it. On the 11th this second leaf began to re-expand; the fragment was manifestly softened, and Dr. Klein reports, "a great deal of enamel and the greater part of the dentine decalcified."

Experiment 2. May 1st, fragment placed on leaf; 2nd, tentacles fairly well inflected, with much secretion on the disc, and remained so until the 7th, when the leaf re-expanded. The fragment was now transferred to a fresh leaf, which next day (8th) was inflected in the strongest manner, and thus remained until the 11th, when it re-expanded. Dr. Klein reports, "A great deal of enamel and the greater part of the dentine decalcified."

Experiment 3. May 1st, fragment moistened with saliva and placed on a leaf, which remained well inflected until 5th, when it re-expanded. The enamel was not at all, and the dentine only slightly, softened. The fragment was now transferred to a fresh leaf, which next morning (6th) was strongly inflected, and remained so until the 11th. The enamel and dentine both now somewhat softened; and Dr. Klein reports, "Less than half the enamel, but the greater part of the dentine, decalcified."

Experiment 4. May 1st, a minute and thin bit of dentine, moistened with saliva, was placed on a leaf, which was soon inflected, and re-expanded on the 5th. The dentine had become as flexible as thin paper. It was then transferred to a fresh leaf, which next morning (6th) was strongly inflected, and reopened on the 10th. The decalcified dentine was now so tender that it was torn into shreds merely by the force of the re-expanding tentacles.

From these experiments it appears that enamel is attacked by the secretion with more difficulty than dentine, as might have been expected from its extreme hardness; and both with more difficulty than ordinary bone. After the process of dissolution has once commenced, it is carried on with greater ease; this may be inferred from the leaves, to which the fragments were transferred, becoming in all four cases strongly inflected in the course of a single day; whereas the first set of leaves acted much less quickly and energetically. The angles or projections of the fibrous basis of the enamel and dentine (except, perhaps, in No. 4, which could not be well observed) were not in the least rounded; and Dr. Klein remarks that their microscopical structure was not altered. But this could not have been expected, as the decalcification was not complete in the three specimens which were carefully examined.

Fibrous Basis of Bone. I at first concluded, as already stated, that the secretion could not digest this substance. I therefore asked Dr. Burdon Sanderson to try bone, enamel, and dentine, in artificial gastric juice, and he found that they were after a considerable time completely dissolved. Dr. Klein examined some of the small lamellae, into which part of the skull of a cat became broken up after about a week's immersion in the fluid, and he found that towards the edges the "matrix appeared rarefied, thus producing the appearance as if the canaliculi of the bone-corpuscles had become larger. Otherwise the corpuscles and their canaliculi were very distinct." So that with bone subjected to artificial gastric juice complete decalcification precedes the dissolution of the fibrous basis. Dr. Burdon Sanderson suggested to me that the failure of Drosera to digest the fibrous basis of bone, enamel, and dentine, might be due to the acid being consumed in the decomposition of the earthy salts, so that there was none left for the work of digestion. Accordingly, my son thoroughly decalcified the bone of a sheep with weak hydrochloric acid; and seven minute fragments of the fibrous basis were placed on so many leaves, four of the fragments being first damped with saliva to aid prompt inflection. All seven leaves became inflected, but only very moderately, in the course of a day. They quickly began to re-expand; five of them on the second day, and the other two on the third day. On all seven leaves the fibrous tissue was converted into perfectly transparent, viscid, more or less liquefied little masses. In the middle, however, of one, my son saw under a high power a few corpuscles, with traces of fibrillation in the surrounding transparent matter. From these facts it is clear that the leaves are very little excited by the fibrous basis of bone, but that the secretion easily and quickly liquefies it, if thoroughly decalcified. The glands which had remained in contact for two or three days with the viscid masses were not discoloured, and apparently had absorbed little of the liquefied tissue, or had been little affected by it.

Phosphate of Lime. As we have seen that the tentacles of the first set of leaves remained clasped for nine or ten days over minute fragments of bone, and the tentacles of the second set for six or seven days over the same fragments, I was led to suppose that it was the phosphate of lime, and not any included animal matter, which caused such long continued inflection. It is at least certain from what has just been shown that this cannot have been due to the presence of the fibrous basis. With enamel and dentine (the former of which contains only four per cent of organic matter) the tentacles of two successive sets of leaves remained inflected altogether for eleven days. In order to test my belief in the potency of phosphate of lime, I procured some from Prof. Frankland absolutely free of animal matter and of any acid. A small quantity moistened with water was placed on the discs of two leaves. One of these was only slightly affected; the other remained closely inflected for ten days, when a few of the tentacles began to re-expand, the rest being much injured or killed. I repeated the experiment, but moistened the phosphate with saliva to insure prompt inflection; one leaf remained inflected for six days (the little saliva used would not have acted for nearly so long a time) and then died; the other leaf tried to re-expand on the sixth day, but after nine days failed to do so, and likewise died. Although the quantity of phosphate given to the above four leaves was extremely small, much was left in every case undissolved. A larger quantity wetted with water was next placed on the discs of three leaves; and these became most strongly inflected in the course of twenty-four hours. They never re-expanded; on the fourth day they looked sickly, and on the sixth were almost dead. Large drops of not very viscid fluid hung from their edges during the six days. This fluid was tested each day with litmus paper, but never coloured it; and this circumstance I do not understand, as the superphosphate of lime is acid. I suppose that some superphosphate must have been formed by the acid of the secretion acting on the phosphate, but that it was all absorbed and injured the leaves; the large drops which hung from their edges being an abnormal and dropsical secretion. Anyhow, it is manifest that the phosphate of lime is a most powerful stimulant. Even small doses are more or less poisonous, probably on the same principle that raw meat and other nutritious substances, given in excess, kill the leaves. Hence the conclusion, that the long continued inflection of the tentacles over fragments of bone, enamel, and dentine, is caused by the presence of phosphate of lime, and not of any included animal matter, is no doubt correct.

Gelatine. I used pure gelatine in thin sheets given me by Prof. Hoffmann. For comparison, squares of the same size as those placed on the leaves were left close by on wet moss. These soon swelled, but retained their angles for three days; after five days they formed rounded, softened masses, but even on the eighth day a trace of gelatine could still be detected. Other squares were immersed in water, and these, though much swollen, retained their angles for six days. Squares of one tenth of an inch (2.54 mm.), just moistened with water, were placed on two leaves; and after two or three days nothing was left on them but some acid viscid fluid, which in this and other cases never showed any tendency to regelatinise; so that the secretion must act on the gelatine differently to what water does, and apparently in the same manner as gastric juice.[7] Four squares of the same size as before were then soaked for three days in water, and placed on large leaves; the gelatine was liquefied and rendered acid in two days, but did not excite much inflection. The leaves began to re-expand after four or five days, much viscid fluid being left on their discs, as if but little had been absorbed. One of these leaves, as soon as it re-expanded, caught a small fly, and after twenty-four hours was closely inflected, showing how much more potent than gelatine is the animal matter absorbed from an insect. Some

[7] Dr. Lauder Brunton, *Handbook for the Phys. Laboratory* (1873), pp. 477, 487; Schiff, *Leçons phys. de la Digestion* (1867), p. 249.

larger pieces of gelatine, soaked for five days in water, were next placed on three leaves, but these did not become much inflected the fourth day. On this day one leaf began to re-expand; the second on the fifth; and third on the sixth. These several facts prove that gelatine is far from acting energetically on Drosera.

In the last chapter it was shown that a solution of isinglass of commerce, as thick as milk or cream, induces strong inflection. I therefore wished to compare its action with that of pure gelatine. Solutions of one part of both substances to two hundred and eighteen of water were made; and half-minim drops (.0296 ml.) were placed on the discs of eight leaves, so that each received $\frac{1}{480}$ of a grain, or .135 mg. The four with the isinglass were much more strongly inflected than the other four. I conclude therefore that isinglass contains some, though perhaps very little, soluble albuminous matter. As soon as these eight leaves re-expanded, they were given bits of roast meat, and in some hours all became greatly inflected; again showing how much more meat excites Drosera than does gelatine or isinglass. This is an interesting fact, as it is well known that gelatine by itself has little power of nourishing animals.[8]

Chondrin. This was sent me by Dr. Moore in a gelatinous state. Some was slowly dried, and a small chip was placed on a leaf, and a much larger chip on a second leaf. The first was liquefied in a day; the larger piece was much swollen and softened, but was not completely liquefied until the third day. The undried jelly was next tried, and as a control experiment small cubes were left in water for four days and retained their angles. Cubes of the same size were placed on two leaves, and larger cubes on two other leaves. The tentacles and laminae of the latter were closely inflected after twenty-two hours, but those of the two leaves with the smaller cubes only to a moderate degree. The jelly on all four was by this time liquefied, and rendered very acid. The glands were blackened from the aggregation of their protoplasmic contents. In forty-six hours from the time when the jelly was given, the leaves had almost re-expanded, and completely so after seventy hours; and now only a little slightly adhesive fluid was left unabsorbed on their discs.

[8] Dr. Lauder Brunton gives in the *Medical Record,* January 1873, p. 36, an account of Voit's view of the indirect part which gelatine plays in nutrition.

One part of chondrin jelly was dissolved in two hundred and eighteen parts of boiling water, and half-minim drops were given to four leaves; so that each received about $\frac{1}{480}$ of a grain (.135 mg.) of the jelly; and, of course, much less of dry chondrin. This acted most powerfully, for after only three hours and thirty minutes all four leaves were strongly inflected. Three of them began to re-expand after twenty-four hours, and in forty-eight hours were completely open; but the fourth had only partially re-expanded. All the liquefied chondrin was by this time absorbed. Hence a solution of chondrin seems to act far more quickly and energetically than pure gelatine or isinglass; but I am assured by good authorities that it is most difficult, or impossible, to know whether chondrin is pure, and if it contained any albuminous compound, this would have produced the above effects. Nevertheless, I have thought these facts worth giving, as there is so much doubt on the nutritious value of gelatine; and Dr. Lauder Brunton does not know of any experiments with respect to animals on the relative value of gelatine and chondrin.

Milk. We have seen in the last chapter that milk acts most powerfully on the leaves; but whether this is due to the contained casein or albumen, I know not. Rather large drops of milk excite so much secretion (which is very acid) that it sometimes trickles down from the leaves, and this is likewise characteristic of chemically·prepared casein. Minute drops of milk, placed on leaves, were coagulated in about ten minutes. Schiff denies [9] that the coagulation of milk by gastric juice is exclusively due to the acid which is present, but attributes it in part to the pepsin; and it seems doubtful whether with Drosera the coagulation can be wholly due to the acid, as the secretion does not commonly colour litmus paper until the tentacles have become well inflected; whereas the coagulation commences, as we have seen, in about ten minutes. Minute drops of skimmed milk were placed on the discs of five leaves; and a large proportion of the coagulated matter or curd was dissolved in six hours and still more completely in eight hours. These leaves re-expanded after two days, and the viscid fluid left on their discs was then carefully scraped off and examined. It seemed at first sight as if all the casein had not been dissolved, for a little matter was left which appeared of a whitish colour by reflected light. But this

[9] *Leçons, &c.,* Tom. II, p. 151.

matter, when examined under a high power, and when compared with a minute drop of skimmed milk coagulated by acetic acid, was seen to consist exclusively of oil-globules, more or less aggregated together, with no trace of casein. As I was not familiar with the microscopical appearance of milk, I asked Dr. Lauder Brunton to examine the slides, and he tested the globules with ether, and found that they were dissolved. We may, therefore, conclude that the secretion quickly dissolves casein, in the state in which it exists in milk.

Chemically Prepared Casein. This substance, which is insoluble in water, is supposed by many chemists to differ from the casein of fresh milk. I procured some, consisting of hard globules, from Messrs. Hopkins and Williams, and tried many experiments with it. Small particles and the powder, both in a dry state and moistened with water, caused the leaves on which they were placed to be inflected very slowly, generally not until two days had elapsed. Other particles, wetted with weak hydrochloric acid (one part to four hundred and thirty-seven of water), acted in a single day, as did some casein freshly prepared for me by Dr. Moore. The tentacles commonly remained inflected for from seven to nine days; and during the whole of this time the secretion was strongly acid. Even on the eleventh day some secretion left on the disc of a fully re-expanded leaf was strongly acid. The acid seems to be secreted quickly, for in one case the secretion from the discal glands, on which a little powdered casein had been strewed, coloured litmus paper, before any of the exterior tentacles were inflected.

Small cubes of hard casein, moistened with water, were placed on two leaves; after three days one cube had its angles a little rounded, and after seven days both consisted of rounded softened masses, in the midst of much viscid and acid secretion; but it must not be inferred from this fact that the angles were dissolved, for cubes immersed in water were similarly acted on. After nine days these leaves began to re-expand, but in this and other cases the casein did not appear, as far as could be judged by the eye, much, if at all, reduced in bulk. According to Hoppe-Seyler and Lubavin [10] casein consists of an albuminous, with a non-albuminous, substance; and the absorption of a very small quan-

[10] Dr. Lauder Brunton, *Handbook for Phys. Lab.*, p. 529.

tity of the former would excite the leaves, and yet not decrease the casein to a perceptible degree. Schiff asserts [11]—and this is an important fact for us—that "la caséine purifiée des chimistes est un corps presque complètement inattaquable par le suc gastrique." So that here we have another point of accordance between the secretion of Drosera and gastric juice, as both act so differently on the fresh casein of milk, and on that prepared by chemists.

A few trials were made with cheese; cubes of one twentieth of an inch (1.27 mm.) were placed on four leaves, and these after one or two days became well inflected, their glands pouring forth much acid secretion. After five days they began to re-expand, but one died, and some of the glands on the other leaves were injured. Judging by the eye, the softened and subsided masses of cheese, left on the discs, were very little or not at all reduced in bulk. We may, however, infer from the time during which the tentacles remained inflected—from the changed colour of some of the glands—and from the injury done to others, that matter had been absorbed from the cheese.

Legumin. I did not procure this substance in a separate state; but there can hardly be a doubt that it would be easily digested, judging from the powerful effect produced by drops of a decoction of green peas, as described in the last chapter. Thin slices of a dried pea, after being soaked in water, were placed on two leaves; these became somewhat inflected in the course of a single hour, and most strongly so in twenty-one hours. They re-expanded after three or four days. The slices were not liquefied, for the walls of the cells, composed of cellulose, are not in the least acted on by the secretion.

Pollen. A little fresh pollen from the common pea was placed on the discs of five leaves, which soon became closely inflected, and remained so for two or three days.

The grains being then removed, and examined under the microscope, were found discoloured, with the oil-globules remarkably aggregated. Many had their contents much shrunk, and some were almost empty. In only a few cases were the pollen-tubes emitted. There could be no doubt that the secretion had penetrated the outer coats of the grains, and had partially di-

[11] *Leçons, &c.,* Tom. II, p. 153.

gested their contents. So it must be with the gastric juice of the insects which feed on pollen, without masticating it.[12] Drosera in a state of nature cannot fail to profit to a certain extent by this power of digesting pollen, as innumerable grains from the carices, grasses, rumices, fir-trees, and other wind-fertilised plants, which commonly grow in the same neighbourhood, will be inevitably caught by the viscid secretion surrounding the many glands.

Gluten. This substance is composed of two albuminoids, one soluble, the other insoluble in alcohol.[13] Some was prepared by merely washing wheaten flour in water. A provisional trial was made with rather large pieces placed on two leaves; these, after twenty-one hours, were closely inflected, and remained so for four days, when one was killed and the other had its glands extremely blackened, but was not afterwards observed. Smaller bits were placed on two leaves; these were only slightly inflected in two days, but afterwards became much more so. Their secretion was not so strongly acid as that of leaves excited by casein. The bits of gluten, after lying for three days on the leaves, were more transparent than other bits left for the same time in water. After seven days both leaves re-expanded, but the gluten seemed hardly at all reduced in bulk. The glands which had been in contact with it were extremely black. Still smaller bits of half putrid gluten were now tried on two leaves; these were well inflected in twenty-four hours, and thoroughly in four days, the glands in contact being much blackened. After five days one leaf began to re-expand, and after eight days both were fully re-expanded, some gluten being still left on their discs. Four little chips of dried gluten, just dipped in water, were next tried, and these acted rather differently from fresh gluten. One leaf was almost fully re-expanded in three days, and the other three leaves in four days. The chips were greatly softened, almost liquefied, but not nearly all dissolved. The glands which had been in contact with them, instead of being much blackened, were of a very pale colour, and many of them were evidently killed.

In not one of these ten cases was the whole of the gluten dissolved, even when very small bits were given. I therefore asked

[12] Mr. A. W. Bennett found the undigested coats of the grains in the intestinal canal of pollen-eating Diptera; see *Journal of Hort. Soc. of London* (1874), Vol. IV, p. 158.

[13] Watts's *Dict. of Chemistry* (1872), Vol. II, p. 873.

Dr. Burdon Sanderson to try gluten in artificial digestive fluid of pepsin with hydrochloric acid; and this dissolved the whole. The gluten, however, was acted on much more slowly than fibrin; the proportion dissolved within four hours being as 40.8 of gluten to 100 of fibrin. Gluten was also tried in two other digestive fluids, in which hydrochloric acid was replaced by propionic and butyric acids, and it was completely dissolved by these fluids at the ordinary temperature of a room. Here, then, at last, we have a case in which it appears that there exists an essential difference in digestive power between the secretion of Drosera and gastric juice; the difference being confined to the ferment, for, as we have just seen, pepsin in combination with acids of the acetic series acts perfectly on gluten. I believe that the explanation lies simply in the fact that gluten is too powerful a stimulant (like raw meat, or phosphate of lime, or even too large a piece of albumen), and that it injures or kills the glands before they have had time to pour forth a sufficient supply of the proper secretion. That some matter is absorbed from the gluten, we have clear evidence in the length of time during which the tentacles remain inflected, and in the greatly changed colour of the glands.

At the suggestion of Dr. Sanderson, some gluten was left for fifteen hours in weak hydrochloric acid (.02 per cent), in order to remove the starch. It became colourless, more transparent, and swollen. Small portions were washed and placed on five leaves, which were soon closely inflected, but to my surprise re-expanded completely in forty-eight hours. A mere vestige of gluten was left on two of the leaves, and not a vestige on the other three. The viscid and acid secretion, which remained on the discs of the three latter leaves, was scraped off and examined by my son under a high power; but nothing could be seen except a little dirt, and a good many starch grains which had not been dissolved by the hydrochloric acid. Some of the glands were rather pale. We thus learn that gluten, treated with weak hydrochloric acid, is not so powerful or so enduring a stimulant as fresh gluten, and does not much injure the glands; and we further learn that it can be digested quickly and completely by the secretion.

Globulin or *Crystallin*. This substance was kindly prepared for me from the lens of the eye by Dr. Moore, and consisted of hard, colourless, transparent fragments. It is said [14] that globulin ought to "swell up in water and dissolve, for the most part forming a gummy liquid"; but this did not occur with the above fragments, though kept in water for four days. Particles, some moistened with water, others with weak hydrochloric acid, others soaked in water for one or two days, were placed on nineteen leaves. Most of these leaves, especially those with the long soaked particles, became strongly inflected in a few hours. The greater number re-expanded after three or four days; but three of the leaves remained inflected during one, two, or three additional days. Hence some exciting matter must have been absorbed; but the fragments, though perhaps softened in a greater degree than those kept for the same time in water, retained all their angles as sharp as ever. As globulin is an albuminous substance, I was astonished at this result; and my object being to compare the action of the secretion with that of gastric juice, I asked Dr. Burdon Sanderson to try some of the globulin used by me. He reports that "it was subjected to a liquid containing 0.2 per cent. of hydrochloric acid, and about 1 per cent. of glycerine extract of the stomach of a dog. It was then ascertained that this liquid was capable of digesting 1.31 of its weight of unboiled fibrin in one hour; whereas, during the hour, only 0.141 of the above globulin was dissolved. In both cases an excess of the substance to be digested was subjected to the liquid." [15] We thus see that within the same time less than one-ninth by weight of globulin than of fibrin was dissolved; and bearing in mind that pepsin with acids of the acetic series has only about one-third of the digestive power of pepsin with hydrochloric acid, it is not surprising that the fragments of globulin were not corroded or rounded by the secretion of Drosera, though some soluble matter was certainly extracted from them and absorbed by the glands.

Haematin. Some dark red granules, prepared from bullock's blood, were given me; these were found by Dr. Sanderson to be insoluble in water, acids, and alcohol, so that they were probably haematin, together with other bodies derived from the blood. Particles with little drops of water were placed on four leaves, three of which were pretty closely inflected in two days; the fourth only moderately so. On the third day the glands in contact with the haematin were black-

[14] Watts's *Dict. of Chemistry,* Vol. ii, p. 874.

[15] I may add that Dr. Sanderson prepared some fresh globulin by Schmidt's method, and of this 0.865 was dissolved within the same time, namely, one hour; so that it was far more soluble than that which I used, though less soluble than fibrin, of which, as we have seen, 1.31 was dissolved. I wish that I had tried on Drosera globulin prepared by this method.

ened, and some of the tentacles seemed injured. After five days two leaves died, and the third was dying; the fourth was beginning to re-expand, but many of its glands were blackened and injured. It is therefore clear that matter had been absorbed which was either actually poisonous or of too stimulating a nature. The particles were much more softened than those kept for the same time in water, but, judging by the eye, very little reduced in bulk. Dr. Sanderson tried this substance with artificial digestive fluid, in the manner described under globulin, and found that whilst 1.31 of fibrin, only 0.456 of the haematin was dissolved in an hour; but the dissolution by the secretion of even a less amount would account for its action on Drosera. The residue left by the artificial digestive fluid at first yielded nothing more to it during several succeeding days.

Substances Which Are Not Digested by the Secretion

All the substances hitherto mentioned cause prolonged inflection of the tentacles, and are either completely or at least partially dissolved by the secretion. But there are many other substances, some of them containing nitrogen, which are not in the least acted on by the secretion, and do not induce inflection for a longer time than do inorganic and insoluble objects. These un-exciting and indigestible substances are, as far as I have observed, epidermic productions (such as bits of human nails, balls of hair, the quills of feathers), fibro-elastic tissue, mucin, pepsin, urea, chitine, chlorophyll, cellulose, gun-cotton, fat, oil, and starch.

To these may be added dissolved sugar and gum, diluted alcohol, and vegetable infusions not containing albumen, for none of these, as shown in the last chapter, excite inflection. Now, it is a remarkable fact, which affords additional and important evidence, that the ferment of Drosera is closely similar to or identical with pepsin, that none of these same substances are, as far as it is known, digested by the gastric juice of animals, though some of them are acted on by the other secretions of the alimentary canal. Nothing more need be said about some of the above enumerated substances, excepting that they were repeatedly tried on the leaves of Drosera, and were not in the least affected by the secretion. About the others it will be advisable to give my experiments.

Fibro-elastic Tissue. We have already seen that when little cubes of meat, &c., were placed on leaves, the muscles, areolar tissue, and cartilage were completely dissolved, but the fibro-elastic tissue, even the most delicate threads, were left without the least signs of having been attacked. And it is well known that this tissue cannot be digested by the gastric juice of animals.[16]

Mucin. As this substance contains about seven per cent of nitrogen, I expected that it would have excited the leaves greatly and been digested by the secretion, but in this I was mistaken. From what is stated in chemical works, it appears extremely doubtful whether mucin can be prepared as a pure principle. That which I used (prepared by Dr. Moore) was dry and hard. Particles moistened with water were placed on four leaves, but after two days there was only a trace of inflection in the immediately adjoining tentacles. These leaves were then tried with bits of meat, and all four soon became strongly inflected. Some of the dried mucin was then soaked in water for two days, and little cubes of the proper size were placed on three leaves. After four days the tentacles round the margins of the discs were a little inflected, and the secretion collected on the disc was acid, but the exterior tentacles were not affected. One leaf began to re-expand on the fourth day, and all were fully re-expanded on the sixth. The glands which had been in contact with the mucin were a little darkened. We may therefore conclude that a small amount of some impurity of a moderately exciting nature had been absorbed. That the mucin employed by me did contain some soluble matter was proved by Dr. Sanderson, who on subjecting it to artificial gastric juice found that in one hour some was dissolved, but only in the proportion of 23 to 100 of fibrin during the same time. The cubes, though perhaps rather softer than those left in water for the same time, retained their angles as sharp as ever. We may therefore infer that the mucin itself was not dissolved or digested. Nor is it digested by the gastric juice of living animals, and according to Schiff [17] it is, a layer of this substance which protects the coats of the stomach from being corroded during digestion.

Pepsin. My experiments are hardly worth giving, as it is scarcely possible to prepare pepsin free from other albuminoids; but I was curious to ascertain, as far as that was possible, whether the ferment of the secretion of Drosera would act on the ferment of the gastric juice of animals. I first used the common pepsin sold for medicinal purposes, and afterwards some which was much purer,

[16] See, for instance, Schiff, *Phys. de la Digestion* (1867), Tom. II, p. 38.

[17] *Leçons phys. de la Digestion* (1867), Tom. II, p. 304.

prepared for me by Dr. Moore. Five leaves to which a considerable quantity of the former was given remained inflected for five days; four of them then died, apparently from too great stimulation. I then tried Dr. Moore's pepsin, making it into a paste with water, and placing such small particles on the discs of five leaves that all would have been quickly dissolved had it been meat or albumen. The leaves were soon inflected; two of them began to re-expand after only twenty hours, and the other three were almost completely re-expanded after forty-four hours. Some of the glands which had been in contact with the particles of pepsin, or with the acid secretion surrounding them, were singularly pale, whereas others were singularly dark-coloured. Some of the secretion was scraped off and examined under a high power; and it abounded with granules undistinguishable from those of pepsin left in water for the same length of time. We may therefore infer, as highly probable (remembering what small quantities were given), that the ferment of Drosera does not act on or digest pepsin, but absorbs from it some albuminous impurity which induces inflection, and which in large quantity is highly injurious. Dr. Lauder Brunton at my request endeavoured to ascertain whether pepsin with hydrochloric acid would digest pepsin, and as far as he could judge, it had no such power. Gastric juice, therefore, apparently agrees in this respect with the secretion of Drosera.

Urea. It seemed to me an interesting inquiry whether this refuse of the living body, which contains much nitrogen, would, like so many other animal fluids and substances, be absorbed by the glands of Drosera and cause inflection. Half-minim drops of a solution of one part to four hundred and thirty-seven of water were placed on the discs of four leaves, each drop containing the quantity usually employed by me, namely, $\frac{1}{960}$ of a grain, or .0674 mg.; but the leaves were hardly at all affected. They were then tested with bits of meat, and soon became closely inflected. I repeated the same experiment on four leaves with some fresh urea prepared by Dr. Moore; after two days there was no inflection; I then gave them another dose, but still there was no inflection. These leaves were afterwards tested with similarly sized drops of an infusion of raw meat, and in six hours there was considerable inflection, which became excessive in twenty-four hours. But the urea apparently was not quite pure, for when four leaves were immersed in two dr. (7.1 ml.) of the solution, so that all the glands, instead of merely those on the disc, were enabled to absorb any small amount of impurity in solution, there was considerable inflection after twenty-four hours, certainly more than would have followed from a similar immersion in pure water. That the urea, which was not perfectly white, should have contained a sufficient quantity of albuminous matter, or of some salt of ammonia, to have caused the above effect, is far from surprising, for, as we shall see in the next chapter, astonishingly small doses of ammonia are highly efficient. We may therefore conclude that urea itself is not exciting or nutritious to Drosera; nor is it modified by the secretion, so as to be rendered nutritious, for, had this been the case, all the leaves with drops on their discs assuredly would have been well inflected. Dr. Lauder Brunton informs me that from experiments made at my request at St. Bartholomew's Hospital it appears that urea is not acted on by artificial gastric juice, that is by pepsin with hydrochloric acid.

Chitine. The chitinous coats of insects naturally captured by the leaves do not appear in the least corroded. Small square pieces of the delicate wing and of the elytron of a Staphylinus were placed on some leaves, and after these had re-expanded, the pieces were carefully examined. Their angles were as sharp as ever, and they did not differ in appearance from the other wing and elytron of the same insect which had been left in water. The elytron, however, had evidently yielded some nutritious matter, for the leaf remained clasped over it for four days; whereas the leaves with bits of the true wing re-expanded on the second day. Any one who will examine the excrement of insect-eating animals will see how powerless their gastric juice is on chitine.

Cellulose. I did not obtain this substance in a separate state, but tried angular bits of dry wood, cork, sphagnum moss, linen, and cotton thread. None of these bodies were in the least attacked by the secretion, and they caused only that moderate amount of inflection which is common to all inorganic objects. Gun-cotton, which consists of cellulose, with the hydrogen replaced by nitrogen, was tried with the same result. We have seen that a decoction of cabbage-leaves excites the most powerful inflection. I therefore placed two little square bits of the blade of a cabbage-leaf, and four little cubes cut from the midrib, on six leaves of Drosera. These became well inflected in twelve hours, and remained so for between two and four days; the bits of cabbage being bathed all the time by acid secretion. This shows that some exciting matter, to which I shall presently refer, had been absorbed; but the angles of the squares and cubes remained as sharp as ever, proving that the framework of cellulose had not been attacked. Small square bits of spinach-leaves were tried with the same result; the glands pouring forth a moderate supply of acid secretion, and the tentacles remaining inflected for three days. We have also seen that the delicate coats of pollen grains are not dissolved by the secretion. It is well known that the gastric juice of animals does not attack cellulose.

Chlorophyll. This substance was tried, as it contains nitrogen. Dr. Moore sent me some preserved in alcohol; it was dried, but soon deliquesced. Particles were placed on four leaves; after three hours the secretion was acid; after eight hours there was a good deal of inflection, which in twenty-four hours became fairly well marked. After four days two of the leaves began to open, and the other two were then almost fully re-expanded. It is therefore clear that this chlorophyll contained matter which excited the leaves to a moderate degree; but judging by the eye, little or none was dissolved; so that in a pure state it would not probably have been attacked by the secretion. Dr. Sanderson tried that which I used, as well as some freshly prepared, with artificial digestive liquid, and found that it was not digested. Dr. Lauder Brunton likewise tried some prepared by the process given in the British Pharmacopoeia, and exposed it for five days at the temperature of 37° Cent. to digestive liquid, but it was not diminished in bulk, though the fluid acquired a slightly brown colour. It was also tried with the glycerine extract of pancreas with a negative result. Nor does chlorophyll seem affected by the intestinal secretions of various animals, judging by the colour of their excrement.

It must not be supposed from these facts that the grains of chlorophyll, as they exist in living plants, cannot be attacked by the secretion; for these grains consist of protoplasm merely coloured by chlorophyll. My son Francis placed a thin slice of spinach leaf, moistened with saliva, on a leaf of Drosera, and other slices on damp cotton-wool, all exposed to the same temperature. After nineteen hours the slice on the leaf of Drosera was bathed in much secretion from the inflected tentacles, and was now examined under the microscope. No perfect grains of chlorophyll could be distinguished; some were shrunken, of a yellowish-green colour, and collected in the middle of the cells; others were disintegrated and formed a yellowish mass, likewise in the middle of the cells. On the other hand, in the slices surrounded by damp cotton-wool, the grains of chlorophyll were green and as perfect as ever. My son also placed some slices in artificial gastric juice, and these were acted on in nearly the same manner as by the secretion. We have seen that bits of fresh cabbage and spinach leaves cause the tentacles to be inflected and the glands to pour forth much acid secretion; and there can be little doubt that it is the protoplasm forming the grains of chlorophyll, as well as that lining the walls of the cells, which excites the leaves.

Fat and Oil. Cubes of almost pure uncooked fat, placed on several leaves, did not have their angles in the least rounded. We have also seen that the oil-globules in milk are not digested. Nor does olive oil dropped on the discs of leaves cause any inflection; but when they are immersed in olive oil, they become strongly inflected; but to this subject I shall have to recur. Oily substances are not digested by the gastric juice of animals.

Starch. Rather large bits of dry starch caused well-marked inflection, and the leaves did not re-expand until the fourth day; but I have no doubt that this was due to the prolonged irritation of the glands, as the starch continued to absorb the secretion. The particles were not in the least reduced in size; and we know that leaves immersed in an emulsion of starch are not at all affected. I need hardly say that starch is not digested by the gastric juice of animals.

Action of the Secretion on Living Seeds

The results of some experiments on living seeds, selected by hazard, may here be given, though they bear only indirectly on our present subject of digestion.

Seven cabbage seeds of the previous year were placed on the same number of leaves. Some of these leaves were moderately, but the greater number only slightly inflected, and most of them re-expanded on the third day. One, however, remained clasped till the fourth, and another till the fifth day. These leaves therefore were excited somewhat more by the seeds than by inorganic objects of the same size. After they re-expanded, the seeds were placed under favourable conditions on damp sand; other seeds of the same lot being tried at the same time in the same manner, and found to germinate well. Of the seven seeds which had been exposed to the secretion, only three germinated; and one of the three seedlings soon perished, the tip of its radicle being from the first decayed, and the edges of its cotyledons of a dark brown colour; so that altogether five out of the seven seeds ultimately perished.

Radish seeds (*Raphanus sativus*) of the previous year were placed on three leaves, which became moderately inflected, and re-expanded on the third or fourth day. Two of these seeds were transferred to damp sand; only one germinated, and that very slowly. This seedling had an extremely short, crooked, diseased radicle, with no absorbent hairs; and the cotyledons were oddly mottled with purple, with the edges blackened and partly withered.

Cress seeds (*Lepidum sativum*) of the previous year were placed on four leaves; two of these next morning were moderately and two strongly inflected, and remained so for four, five, and even six days. Soon after these seeds were placed on the leaves and had become damp, they secreted in the usual manner a layer of tenacious mucus; and to ascertain whether it was the absorption of this substance by the glands which caused so much inflection, two seeds were put into water, and as much of the mucus as possible scraped off. They were then placed on leaves, which became very strongly inflected in the course of three hours, and were still closely inflected on the third day; so that it evidently was not the mucus which excited so much inflection; on the contrary, this served to a certain extent as a protection to the seeds. Two of the six seeds germinated whilst still lying on the leaves, but the seedlings, when transferred to damp sand, soon died; of the other four seeds, only one germinated.

Two seeds of mustard (*Sinapis nigra*), two of celery (*Apium graveolens*)—both of the previous year, two seeds well soaked of caraway (*Carum carui*), and two of wheat did not excite the leaves more than inorganic objects often do. Five seeds, hardly ripe, of a buttercup (Ranunculus), and two fresh seeds of *Anemone nemorosa* induced only a little more effect. On the other hand, four seeds, perhaps not quite ripe, of *Carex sylvatica* caused the leaves on which they were placed to be very strongly inflected; and these only began to re-expand on the third day, one remaining inflected for seven days.

It follows from these few facts that different kinds of seeds excite the leaves in very different degrees; whether this is solely due to the nature of their coats is not clear. In the case of the cress seeds, the partial removal of the layer of mucus hastened the inflection of the tentacles. Whenever the leaves remain inflected during several days over seeds, it is clear that they absorb some matter from them. That the secretion penetrates their coats is also evident from the large proportion of cabbage, radish, and cress seeds which were killed, and from several of the seedlings being greatly injured. This injury to the seeds and seedlings may, however, be due solely to the acid of the secretion, and not to any process of digestion; for Mr. Traherne Moggridge has shown that very weak acids of the acetic series are highly injurious to seeds. It never occurred to me to observe whether seeds are often blown on to the viscid leaves of plants growing in a state of nature; but this can hardly fail sometimes to occur, as we shall hereafter see in the case of Pinguicula. If so, Drosera will profit to a slight degree by absorbing matter from such seeds.

Summary and Concluding Remarks on the Digestive Power of Drosera

When the glands on the disc are excited either by the absorption of nitrogenous matter or by mechanical irritation, their secretion increases in quantity and becomes acid. They likewise transmit some influence to the glands of the exterior tentacles, causing them to secrete more copiously; and their secretion likewise becomes acid. With animals, according to Schiff,[18] mechanical irritation excites the glands of the stomach to secrete an acid, but not pepsin. Now, I have every reason to believe (though the fact is not fully established), that although the glands of Drosera are continually secreting viscid fluid to replace that lost by evaporation, yet they do not secrete the ferment proper for digestion when mechanically irritated, but only after absorbing certain matter, probably of a nitrogenous nature. I infer that this is the case, as the secretion from a large number of leaves which had been irritated by particles of glass placed on their discs did not digest albumen; and more especially from the analogy of Dionaea and Nepenthes. In like manner, the glands of the stomach of animals secrete pepsin, as Schiff asserts, only after they have absorbed certain soluble substances, which he designates as peptogenes. There is, therefore, a remarkable parallelism between the glands of Drosera and those of the stomach in the secretion of their proper acid and ferment.

The secretion, as we have seen, completely dissolves albumen, muscle, fibrin, areolar tissue, cartilage, and fibrous basis of bone, gelatin, chondrin, casein in the state in which it exists in milk, and gluten which has been subjected to weak hydrochloric acid. Syntonin and legumin excite the leaves so powerfully and quickly that there can hardly be a doubt that both would be dissolved by the secretion. The secretion failed to digest fresh gluten, apparently from its injuring the glands, though some was absorbed. Raw meat, unless in very small bits, and large pieces of albumen, &c., likewise injure the leaves, which seem to suffer, like animals, from a surfeit. I know not whether the analogy is a real one, but it is worth notice that a decoction of cabbage leaves is far more

[18] *Phys. de la Digestion* (1867), Tom. II, pp. 188, 245.

exciting and probably nutritious to Drosera than an infusion made with tepid water; and boiled cabbages are far more nutritious, at least to man, than the uncooked leaves. The most striking of all the cases, though not really more remarkable than many others, is the digestion of so hard and tough a substance as cartilage. The dissolution of pure phosphate of lime, of bone, dentine, and especially enamel seems wonderful; but it depends merely on the long-continued secretion of an acid; and this is secreted for a longer time under these circumstances than under any others. It was interesting to observe that as long as the acid was consumed in dissolving the phosphate of lime, no true digestion occurred; but that as soon as the bone was completely decalcified, the fibrous basis was attacked and liquefied with the greatest ease. The twelve substances above enumerated, which are completely dissolved by the secretion, are likewise dissolved by the gastric juice of the higher animals; and they are acted on in the same manner, as shown by the rounding of the angles of albumen, and more especially by the manner in which the transverse striae of the fibres of muscle disappear.

The secretion of Drosera and gastric juice were both able to dissolve some element or impurity out of the globulin and haematin employed by me. The secretion also dissolved something out of chemically prepared casein, which is said to consist of two substances; and although Schiff asserts that casein in this state is not attacked by gastric juice, he might easily have overlooked a minute quantity of some albuminous matter, which Drosera would detect and absorb. Again, fibro-cartilage, though not properly dissolved, is acted on in the same manner, both by the secretion of Drosera and gastric juice. But this substance, as well as the so-called haematin used by me, ought perhaps to have been classed with indigestible substances.

That gastric juice acts by means of its ferment, pepsin, solely in the presence of an acid, is well established; and we have excellent evidence that a ferment is present in the secretion of Drosera, which likewise acts only in the presence of an acid; for we have seen that when the secretion is neutralised by minute drops of the solution of an alkali, the digestion of albumen is completely stopped, and that on the addition of a minute dose of hydrochloric acid it immediately recommences.

The nine following substances, or classes of substances,

namely, epidermic productions, fibro-elastic tissue, mucin, pepsin, urea, chitine, cellulose, gun-cotton, chlorophyll, starch, fat and oil, are not acted on by the secretion of Drosera; nor are they, as far as is known, by the gastric juice of animals. Some soluble matter, however, was extracted from the mucin, pepsin, and chlorophyll, used by me, both by the secretion and by artificial gastric juice.

The several substances, which are completely dissolved by the secretion, and which are afterwards absorbed by the glands, affect the leaves rather differently. They induce inflection at very different rates and in very different degrees; and the tentacles remain inflected for very different periods of time. Quick inflection depends partly on the quantity of the substance given, so that many glands are simultaneously affected, partly on the facility with which it is penetrated and liquefied by the secretion, partly on its nature, but chiefly on the presence of exciting matter already in solution. Thus saliva, or a weak solution of raw meat, acts much more quickly than even a strong solution of gelatine. So again leaves which have re-expanded, after absorbing drops of a solution of pure gelatine or isinglass (the latter being the more powerful of the two), if given bits of meat, are inflected much more energetically and quickly than they were before, notwithstanding that some rest is generally requisite between two acts of inflection. We probably see the influence of texture in gelatine and globulin when softened by having been soaked in water acting more quickly than when merely wetted. It may be partly due to changed texture, and partly to changed chemical nature, that albumen, which has been kept for some time, and gluten which has been subjected to weak hydrochloric acid, act more quickly than these substances in their fresh state.

The length of time during which the tentacles remain inflected largely depends on the quantity of the substance given, partly on the facility with which it is penetrated or acted on by the secretion, and partly on its essential nature. The tentacles always remain inflected much longer over large bits or large drops than over small bits or drops. Texture probably plays a part in determining the extraordinary length of time during which the tentacles remain inflected over the hard grains of chemically prepared casein. But the tentacles remain inflected for an equally long time over finely powdered, precipitated phosphate of lime;

phosphorus in this latter case evidently being the attraction, and animal matter in the case of casein. The leaves remain long inflected over insects, but it is doubtful how far this is due to the protection afforded by their chitinous integuments; for animal matter is soon extracted from insects (probably by exosmose from their bodies into the dense surrounding secretion), as shown by the prompt inflection of the leaves. We see the influence of the nature of different substances in bits of meat, albumen, and fresh gluten acting very differently from equalsized bits of gelatine, areolar tissue, and the fibrous basis of bone. The former cause not only far more prompt and energetic, but more prolonged, inflection than do the latter. Hence we are, I think, justified in believing that gelatine, areolar tissue, and the fibrous basis of bone, would be far less nutritious to Drosera than such substances as insects, meat, albumen, &c. This is an interesting conclusion, as it is known that gelatine affords but little nutriment to animals; and so, probably, would areolar tissue and the fibrous basis of bone. The chondrin which I used acted more powerfully than gelatine, but then I do not know that it was pure. It is a more remarkable fact that fibrin, which belongs to the great class of Proteids,[19] including albumen in one of its sub-groups, does not excite the tentacles in a greater degree, or keep them inflected for a longer time, than does gelatine, or areolar tissue, or the fibrous basis of bone. It is not known how long an animal would survive if fed on fibrin alone, but Dr. Sanderson has no doubt longer than on gelatine, and it would be hardly rash to predict, judging from the effects on Drosera, that albumen would be found more nutritious than fibrin. Globulin likewise belongs to the Proteids, forming another sub-group, and this substance, though containing some matter which excited Drosera rather strongly, was hardly attacked by the secretion, and was very little or very slowly attacked by gastric juice. How far globulin would be nutritious to animals is not known. We thus see how differently the above specified several digestible substances act on Drosera; and we may infer, as highly probable, that they would in like manner be nutritious in very different degrees both to Drosera and to animals.

The glands of Drosera absorb matter from living seeds, which

[19] See the classification adopted by Dr. Michael Foster in Watts's *Dict. of Chemistry,* Supplement (1872), p. 969.

are injured or killed by the secretion. They likewise absorb matter from pollen, and from fresh leaves; and this is notoriously the case with the stomachs of vegetable-feeding animals. Drosera is properly an insectivorous plant; but as pollen cannot fail to be often blown on to the glands, as will occasionally the seeds and leaves of surrounding plants, Drosera is, to a certain extent, a vegetable-feeder.

Finally, the experiments recorded in this chapter show us that there is a remarkable accordance in the power of digestion between the gastric juice of animals with its pepsin and hydrochloric acid and the secretion of Drosera with its ferment and acid belonging to the acetic series. We can, therefore, hardly doubt that the ferment in both cases is closely similar, if not identically the same. That a plant and an animal should pour forth the same, or nearly the same, complex secretion, adapted for the same purpose of digestion, is a new and wonderful fact in physiology. But I shall have to recur to this subject in the fifteenth chapter, in my concluding remarks on the Droseraceae.

The Autobiography

May 31, 1876: Recollections of the Development of My Mind and Character

A German editor having written to me to ask for an account of the development of my mind and character with some sketch of my autobiography, I have thought that the attempt would amuse me, and might possibly interest my children or their children. I know that it would have interested me greatly to have read even so short and dull a sketch of the mind of my grandfather written by himself, and what he thought and did and how he worked. I have attempted to write the following account of myself as if I were a dead man in another world looking back at my own life. Nor have I found this difficult, for life is nearly over with me. I have taken no pains about my style of writing.

I was born at Shrewsbury on February 12, 1809. I have heard my father say that he believed that persons with powerful minds generally had memories extending far back to a very early period of life. This is not my case, for my earliest recollection goes back only to when I was a few months over four years old, when we went to near Abergele for sea-bathing, and I recollect some events and places there with some little distinctness.

My mother died in July 1817, when I was a little over eight years old, and it is odd that I can remember hardly anything about her except her death-bed, her black velvet gown, and her curiously constructed work-table. I believe that my forgetfulness is partly due to my sisters, owing to their great grief, never being able to speak about her or mention her name; and partly to her previous invalid state. In the spring of this same year I was sent to a day-school in Shrewsbury,[1] where I stayed a year. Be-

[1] Kept by Rev. G. Case, minister of the Unitarian Chapel in the High Street. Mrs. Darwin was a Unitarian and attended Mr. Case's chapel, and my father as a little boy went there with his elder sisters. But both he and his brother were christened and intended to belong to the Church of England; and after his early boyhood he seems usually to have gone to church and not to Mr. Case's. It appears (*St. James's Gazette*, Dec. 15, 1883) that a mural tablet has been erected to his memory in the chapel, which is now known as the "Free Christian Church." [Notes to the *Autobiography* are by Francis Darwin.—S.E.H.]

fore going to school I was educated by my sister Caroline, but I doubt whether this plan answered. I have been told that I was much slower in learning than my younger sister Catherine, and I believe that I was in many ways a naughty boy. Caroline was extremely kind, clever and zealous; but she was too zealous in trying to improve me; for I clearly remember, after this long interval of years, saying to myself when about to enter a room where she was—"What will she blame me for now?" and I made myself dogged so as not to care what she might say.

By the time I went to this day-school my taste for natural history, and more especially for collecting, was well developed. I tried to make out the names of plants and collected all sorts of things, shells, seals, franks, coins, and minerals. The passion for collecting, which leads a man to be a systematic naturalist, a virtuoso, or a miser, was very strong in me and was clearly innate, as none of my sisters or brothers ever had this taste.

One little event during this year has fixed itself very firmly in my mind, and I hope that it has done so from my conscience having been afterwards sorely troubled by it; it is curious as showing that apparently I was interested at this early age in the variability of plants! I told another little boy (I believe it was Leighton,[2] who afterwards became a well-known lichenologist and botanist) that I could produce variously coloured polyanthuses and primroses by watering them with certain coloured fluids, which was of course a monstrous fable, and had never been tried by me. I may here also confess that as a little boy I was much given to inventing deliberate falsehoods, and this was always done for the sake of causing excitement. For instance, I once gathered much valuable fruit from my father's trees and hid them in the shrubbery, and then ran in breathless haste to spread the news that I had discovered a hoard of stolen fruit.[3]

About this time, or as I hope at a somewhat earlier age, I

[2] Rev. W. A. Leighton, who was a schoolfellow of my father's at Mr. Case's school, remembers his bringing a flower to school and saying that his mother had taught him how by looking at the inside of the blossom the name of the plant could be discovered. Mr. Leighton goes on, "This greatly roused my attention and curiosity, and I inquired of him repeatedly how this could be done?"—but his lesson was naturally enough not transmissible.

[3] His father wisely treated this tendency not by making crimes of the fibs, but by making light of the discoveries.

sometimes stole fruit for the sake of eating it; and one of my schemes was ingenious. The kitchen garden was kept locked in the evening and was surrounded by a high wall, but by the aid of neighbouring trees I could easily get on the coping. I then fixed a long stick into the hole at the bottom of a rather large flower-pot, and by dragging this upwards pulled off peaches and plums, which fell into the pot, and the prizes were thus secured. When a very little boy I remember stealing apples from the orchard, for the sake of giving them away to some boys and young men who lived in a cottage not far off, but before I gave them the fruit I showed off how quickly I could run, and it is wonderful that I did not perceive that the surprise and admiration which they expressed at my powers of running was given for the sake of the apples. But I well remember that I was delighted at them declaring that they had never seen a boy run so fast!

I remember clearly only one other incident during the years whilst at Mr. Case's daily school—namely, the burial of a dragoon-soldier; and it is surprising how clearly I can still see the horse with the man's empty boots and carbine suspended to the saddle, and the firing over the grave. This scene deeply stirred whatever poetic fancy there was in me.[4]

In the summer of 1818 I went to Dr. Butler's great school in Shrewsbury and remained there for seven years till mid-summer 1825, when I was sixteen years old. I boarded at this school, so that I had the great advantage of living the life of a true school-boy; but as the distance was hardly more than a mile to my home, I very often ran there in the longer intervals between the callings over and before locking up at night. This I think was in many ways advantageous to me by keeping up home affections and interests. I remember in the early part of my school life that I often had to run very quickly to be in time, and from being a fleet runner was generally successful; but when in doubt I prayed earnestly to God to help me, and I well remember that

[4] It is curious that another Shrewsbury boy should have been impressed by this military funeral: Mr. Gretton, in his *Memory's Harkback*, says that the scene is so strongly impressed on his mind that he could "walk straight to the spot in St. Chad's churchyard where the poor fellow was buried." The soldier was an Inniskilling Dragoon, and the officer in command had been recently wounded at Waterloo, where his corps did good service against the French Cuirassiers.

I attributed my success to the prayers and not to my quick running, and marvelled how generally I was aided.

I have heard my father and elder sisters say that I had, as a very young boy, a strong taste for long solitary walks; but what I thought about I know not. I often became quite absorbed and once, whilst returning to school on the summit of the old fortifications round Shrewsbury, which had been converted into a public foot-path with no parapet on one side, I walked off and fell to the ground, but the height was only seven or eight feet. Nevertheless, the number of thoughts which passed through my mind during this very short but sudden and wholly unexpected fall was astonishing, and seem hardly compatible with what physiologists have, I believe, proved about each thought requiring quite an appreciable amount of time.

I must have been a very simple little fellow when I first went to the school. A boy of the name of Garnett took me into a cake-shop one day and bought some cakes for which he did not pay, as the shopman trusted him. When we came out I asked him why he did not pay for them, and he instantly answered, "Why, do you not know that my uncle left a great sum of money to the Town on condition that every tradesman should give whatever was wanted without payment to anyone who wore his old hat and moved it in a particular manner?" And he then showed me how it was moved. He then went into another shop where he was trusted and asked for some small article, moving his hat in the proper manner, and of course obtained it without payment. When we came out he said, "Now if you like to go by yourself into that cake-shop [how well I remember its exact position], I will lend you my hat, and you can get whatever you like if you move the hat on your head properly." I gladly accepted the generous offer and went in and asked for some cakes, moved the old hat, and was walking out of the shop when the shopman made a rush at me, so I dropped the cakes and ran away for dear life, and was astonished by being greeted with shouts of laughter by my false friend Garnett.

I can say in my own favour that I was as a boy humane, but I owed this entirely to the instruction and example of my sisters. I doubt indeed whether humanity is a natural or innate quality. I was very fond of collecting eggs, but I never took more than a single egg out of a bird's nest, except on one single occasion, when I took all, not for their value but from a sort of bravado.

I had a strong taste for angling and would sit for any number of hours on the bank of a river or pond watching the float. When at Maer [5] I was told that I could kill the worms with salt and water, and from that day I never spitted a living worm, though at the expense, probably, of some loss of success.

Once as a very little boy, whilst at the day-school or before that time, I acted cruelly, for I beat a puppy, I believe simply from enjoying the sense of power; but the beating could not have been severe, for the puppy did not howl, of which I feel sure as the spot was near to the house. This act lay heavily on my conscience, as is shown by my remembering the exact spot where the crime was committed. It probably lay all the heavier from my love of dogs being then, and for a long time afterwards, a passion. Dogs seemed to know this, for I was an adept in robbing their love from their masters.

Nothing could have been worse for the development of my mind than Dr. Butler's school, as it was strictly classical, nothing else being taught except a little ancient geography and history. The school as a means of education to me was simply a blank. During my whole life I have been singularly incapable of mastering any language. Especial attention was paid to verse-making, and this I could never do well. I had many friends and got together a grand collection of old verses, which by patching together, sometimes aided by other boys, I could work into any subject. Much attention was paid to learning by heart the lessons of the previous day. This I could effect with great facility, learning forty or fifty lines of Virgil or Homer whilst I was in morning chapel; but this exercise was utterly useless, for every verse was forgotten in forty-eight hours. I was not idle and, with the exception of versification, generally worked conscientiously at my classics, not using cribs. The sole pleasure I ever received from such studies was from some of the odes of Horace, which I admired greatly. When I left the school I was for my age neither high nor low in it; and I believe that I was considered by all my masters and by my father as a very ordinary boy, rather below the common standard in intellect. To my deep mortification my father once said to me, "You care for nothing but shooting, dogs, and rat-catching, and you will be a disgrace to

[5] The house of his uncle, Josiah Wedgwood the younger.

yourself and all your family." But my father, who was the kindest man I ever knew, and whose memory I love with all my heart, must have been angry and somewhat unjust when he used such words.

I may here add a few pages about my father, who was in many ways a remarkable man.

He was about six feet two inches in height, with broad shoulders, and very corpulent, so that he was the largest man whom I ever saw. When he last weighed himself he was twenty-four stone, but afterwards increased much in weight. His chief mental characteristics were his powers of observation and his sympathy, neither of which have I ever seen exceeded or even equalled. His sympathy was not only with the distresses of others but in a greater degree with the pleasures of all around him. This led him to be always scheming to give pleasure to others and, though hating extravagance, to perform many generous actions. For instance, Mr. B——, a small manufacturer in Shrewsbury, came to him one day and said he should be bankrupt unless he could at once borrow ten thousand pounds, but that he was unable to give any legal security. My father heard his reasons for believing that he could ultimately repay the money, and from my father's intuitive perception of character felt sure that he was to be trusted. So he advanced this sum, which was a very large one for him while young, and was after a time repaid.

I suppose that it was his sympathy which gave him unbounded power of winning confidence and as a consequence made him highly successful as a physician. He began to practise before he was twenty-one years old, and his fees during the first year paid for the keep of two horses and a servant. On the following year his practice was larger and so continued for above sixty years, when he ceased to attend on anyone. His great success as a doctor was the more remarkable as he told me that he at first hated his profession so much that if he had been sure of the smallest pittance, or if his father had given him any choice, nothing should have induced him to follow it. To the end of his life, the thought of an operation almost sickened him, and he could scarcely endure to see a person bled—a horror which he has transmitted to me—and I remember the horror which I felt as a schoolboy in reading about Pliny (I think) bleeding to death in a warm bath. My father told me two odd stories about bleeding. One was that as a very young man he became a Freemason. A friend of his who was a Freemason, and who pretended not to know about his strong feeling with respect to blood, remarked casually to him as they walked to the meeting, "I suppose that you do not care about losing a few drops of blood?" It seems that when he was received as a member, his eyes were bandaged and his coat-sleeves turned up. Whether any such ceremony is now performed I know not, but my father mentioned the case as an excellent instance of the power of imagination, for he distinctly felt the blood trickling down his arm and could hardly believe his own eyes when he afterwards could not find the smallest prick on his arm.

A great slaughtering butcher from London once consulted my grandfather, when another man very ill was brought in, and my grandfather wished to have him instantly bled by the accompanying apothecary. The butcher was asked to hold the patient's arm, but he made some excuse and left the room. Afterwards he explained to my grandfather that although he believed that he had killed with his own hands more animals than any other man in London, yet absurd as it might seem he assuredly should have fainted if he had seen the patient bled.

Owing to my father's power of winning confidence, many patients, especially ladies, consulted him when suffering from any misery, as a sort of Father-Confessor. He told me that they always began by complaining in a vague manner about their health, and by practice he soon guessed what was really the matter. He then suggested that they had been suffering in their minds, and now they would pour out their troubles, and he heard nothing more about the body. Family quarrels were a common subject. When gentlemen complained to him about their wives, and the quarrel seemed serious, my father advised them to act in the following manner; and his advice always succeeded if the gentleman followed it to the letter, which was not always the case. The husband was to say to the wife that he was very sorry that they could not live happily together, that he felt sure that she would be happier if separated from him, that he did not blame her in the least (this was the point on which the man oftenest failed), that he would not blame her to any of her relations or friends, and lastly that he would settle on her as large a provision as he could afford. She was then asked to deliberate on this proposal. As no fault had been found, her temper was unruffled, and she soon felt what an awkward position she would be in with no accusation to rebut and with her husband and not herself proposing a separation. Invariably the lady begged her husband not to think of separation and usually behaved much better ever afterwards.

Owing to my father's skill in winning confidence he received many strange confessions of misery and guilt. He often remarked how many miserable wives he had known. In several instances husbands and wives had gone on pretty well together for between twenty and thirty years and then hated each other bitterly; this he attributed to their having lost a common bond in their young children having grown up.

But the most remarkable power which my father possessed was that of reading the characters, and even the thoughts, of those whom he saw even for a short time. We had many instances of this power, some of which seemed almost supernatural. It saved my father from ever making (with one exception, and the character of this man was soon discovered) an unworthy friend. A strange clergyman came to Shrewsbury and seemed to be a rich man; everbody called on him, and he was invited to many houses. My father called and on his return home told my sisters on no account to invite him or his family to our house, for he felt sure that the man was not to be trusted. After a few months he suddenly bolted, being heavily in debt, and was found out to be little better than an habitual swindler. Here is a case of trustfulness which not many men would have ventured on. An Irish gentleman, a complete stranger, called on my father one day and said that he had lost his purse and that it would be a serious inconvenience to him to wait in Shrewsbury until he could receive a remittance from Ireland. He then asked my father to lend him twenty pounds, which was immediately done, as my father felt certain that the story was a true one. As soon as a letter could arrive from Ireland, one came with the most profuse thanks and enclosing, as he said, a twenty-pound Bank of England note; but no note was enclosed. I asked my father whether this did not stagger him, but he answered, "Not in the least." On the next day another letter came with many apologies for having forgotten (like a true Irishman) to put the note into his letter of the day before.

A connection of my father's consulted him about his son who was strangely idle and would settle to no work. My father said, "I believe that the foolish young man thinks that I shall bequeath him a large sum of money. Tell him that I have declared to you that I shall not leave him a penny." The father of the youth owned with shame that this preposterous idea had taken possession of his son's mind, and he asked my father how he could possibly have discovered it. But my father said he did not in the least know.

The Earl of —— brought his nephew, who was insane but quite gentle, to my father; and the young man's insanity led him to accuse himself of all the crimes under heaven. When my father afterwards talked about the case with the uncle, he said, "I am sure that your nephew is really guilty of . . . a heinous crime." Whereupon the Earl of —— exclaimed, "Good God, Dr. Darwin, who told you? We thought that no human being knew the fact except ourselves!" My father told me the story many years after the event, and I asked him how he distinguished the true from the false self-accusations; and it was very characteristic of my father that he said he could not explain how it was.

The following story shows what good guesses my father could make. Lord Sherburn, afterwards the first Marquis of Lansdowne, was famous (as Macaulay somewhere remarks) for his knowledge of the affairs of Europe, on which he greatly prided himself. He consulted my father medically and afterwards harangued him on the state of Holland. My father had studied medicine at Leyden and one day went a long walk into the country with a friend, who took him to the house of a clergyman (we will say the Rev. Mr. A——, for I have forgotten his name), who had married an Englishwoman. My father was very hungry, and there was little for luncheon except cheese, which he could never eat. The old lady was surprised and grieved at this and assured my father that it was an excellent cheese and had been sent her from Bowood, the seat of Lord Sherburn. My father wondered why a cheese should be sent her from Bowood but thought nothing more about it until it flashed across his mind many years afterwards, whilst Lord Sherburn was talking about

Holland. So he answered, "I should think from what I saw of the Rev. Mr. A—— that he was a very able man and well acquainted with the state of Holland." My father saw that the Earl, who immediately changed the conversation, was much startled. On the next morning my father received a note from the Earl saying that he had delayed starting on his journey and wished particularly to see my father. When he called, the Earl said, "Dr. Darwin, it is of the utmost importance to me and to the Rev. Mr. A—— to learn how you have discovered that he is the source of my information about Holland." So my father had to explain the state of the case, and he supposed that Lord Sherburn was much struck with his diplomatic skill in guessing, for during many years afterwards he received many kind messages from him through various friends. I think that he must have told the story to his children; for Sir C. Lyell asked me many years ago why the Marquis of Lansdowne (the son or grandson of the first marquis) felt so much interest about me, whom he had never seen, and my family. When forty new members (the forty thieves, as they were then called) were added to the Athenaeum Club, there was much canvassing to be one of them; and without my having asked anyone, Lord Lansdowne proposed me and got me elected. If I am right in my supposition, it was a queer concatenation of events that my father not eating cheese half a century before in Holland led to my election as a member of the Athenaeum.

Early in life my father occasionally wrote down a short account of some curious event and conversation, which are enclosed in a separate envelope.

The sharpness of his observation led him to predict with remarkable skill the course of any illness, and he suggested endless small details of relief. I was told that a young doctor in Shrewsbury, who disliked my father, used to say that he was wholly unscientific, but owned that his power of predicting the end of an illness was unparalleled. Formerly, when he thought that I should be a doctor, he talked much to me about his patients. In the old days the practice of bleeding largely was universal, but my father maintained that far more evil was thus caused than good done, and he advised me if ever I was myself ill not to allow any doctor to take from me more than an extremely small quantity of blood. Long before typhoid fever was recognised as distinct, my father told me that two utterly distinct kinds of illness were confounded under the name of typhus fever. He was vehement against drinking and was convinced of both the direct and inherited evil effects of alcohol when habitually taken, even in moderate quantity, in a very large majority of cases. But he admitted and advanced instances of certain persons who could drink largely during their whole lives without apparently suffering any evil effects; and he believed that he could often beforehand tell who would thus not suffer. He himself never drank a drop of any alcoholic fluid. This remark reminds me of a case showing how a witness under the most favourable circumstances may be wholly mistaken. A gentleman-farmer was strongly urged by my father not to drink and was encouraged by being told that he himself never touched any spirituous liquor. Whereupon the gentleman said, "Come, come, Doctor, that won't do—though it is very kind of you to say so for my sake—for I know that you take a very large glass of hot gin and water every evening after your dinner." [6] So my father asked him how he knew this. The man answered, "My cook was your kitchen-maid for two or three years, and she saw the butler every day prepare and take to you the gin and water." The explanation was that my father had the odd habit of drinking hot water in a very tall and large glass after his dinner; and the butler used first to put some cold water in the glass, which the girl mistook for gin, and then filled it up with boiling water from the kitchen boiler.

My father used to tell me many little things which he had found useful in his medical practice. Thus ladies often cried much while telling him their troubles and thus caused much loss of his precious time. He soon found that begging them to command and restrain themselves always made them weep the more, so that afterwards he always encouraged them to go on crying, saying that this would relieve them more than anything else, with the invariable result that they soon ceased to cry and he could hear what they had to say and give his advice. When patients who were very ill craved for some strange and unnatural food my father asked them what had put such an idea into their heads; if they answered that they did not know, he would allow

[6] This belief still survives and was mentioned to my brother in 1884 by an old inhabitant of Shrewsbury.

them to try the food, and often with success, as he trusted to their having a kind of instinctive desire; but if they answered that they had heard that the food in question had done good to someone else, he firmly refused his assent.

He gave one day an odd little specimen of human nature. When a very young man, he was called in to consult with the family physician in the case of a gentleman of much distinction in Shropshire. The old doctor told the wife that the illness was of such a nature that it must end fatally. My father took a different view and maintained that the gentleman would recover; he was proved quite wrong in all respects (I think by autopsy), and he owned his error. He was then convinced that he should never again be consulted by this family; but after a few months the widow sent for him, having dismissed the old family doctor. My father was so much surprised at this that he asked a friend of the widow to find out why he was again consulted. The widow answered her friend that "she would never again see that odious old doctor who said from the first that her husband would die, while Dr. Darwin always maintained that he would recover!" In another case my father told a lady that her husband would certainly die. Some months afterwards he saw the widow, who was a very sensible woman, and she said, "You are a very young man, and allow me to advise you always to give, as long as you possibly can, hope to any near relation nursing a patient. You made me despair, and from that moment I lost strength." My father said that he had often since seen the paramount importance, for the sake of the patient, of keeping up the hope and with it the strength of the nurse in charge. This he sometimes found it difficult to do compatibly with truth. One old gentleman, however, Mr. Pemberton, caused him no such perplexity. He was sent for by Mr. Pemberton, who said, "From all that I have seen and heard of you I believe you are the sort of man who will speak the truth, and if I ask you will tell me when I am dying. Now I much desire that you should attend me, if you will promise, whatever I may say, always to declare that I am not going to die." My father acquiesced on this understanding that his words should in fact have no meaning.

My father possessed an extraordinary memory, especially for dates, so that he knew when he was very old the day of the birth, marriage, and death of a multitude of persons in Shropshire;

and he once told me that this power annoyed him, for if he once
heard a date he could not forget it and thus the deaths of many
friends were often recalled to his mind. Owing to his strong
memory he knew an extraordinary number of curious stories,
which he liked to tell, as he was a great talker. He was generally
in high spirits and laughed and joked with everyone—often with
his servants—with the utmost freedom; yet he had the art of
making everyone obey him to the letter. Many persons were
much afraid of him. I remember my father telling us one day
with a laugh that several persons had asked him whether Miss
Piggott (a grand old lady in Shropshire) had called on him, so
that at last he enquired why they asked him and was told that
Miss Piggott, whom my father had somehow mortally offended,
was telling everybody that she would call and tell "that fat old
doctor very plainly what she thought of him." She had already
called, but her courage had failed, and no one could have been
more courteous and friendly. As a boy I went to stay at the
house of Major B——, whose wife was insane; and the poor
creature, as soon as she saw me, was in the most abject state of
terror that I ever saw, weeping bitterly and asking me over and
over again, "Is your father coming?," but was soon pacified.
On my return home, I asked my father why she was so frightened, and he answered he [was] very glad to hear it, as he had
frightened her on purpose, feeling sure that she could be kept in
safety, and much happier without any restraint, if her husband
could influence her whenever she became at all violent by proposing to send for Dr. Darwin; and these words succeeded perfectly during the rest of her long life.

My father was very sensitive, so that many small events annoyed or pained him much. I once asked him, when he was old
and could not walk, why he did not drive out for exercise; and
he answered, "Every road out of Shrewsbury is associated in my
mind with some painful event." Yet he was generally in high
spirits. He was easily made very angry, but as his kindness was
unbounded he was widely and deeply loved.

He was a cautious and good man of business, so that he
hardly ever lost money by an investment and left to his children
a very large property. I remember a story showing how easily
utterly false beliefs originate and spread. Mr. E——, a squire of
one of the oldest families in Shropshire and head partner in a

bank, committed suicide. My father was sent for as a matter of form and found him dead. I may mention by the way, to show how matters were managed in those old days, that because Mr. E—— was a rather great man and universally respected no inquest was held over his body. My father, in returning home, thought it proper to call at the bank (where he had an account) to tell the managing partner of the event, as it was not improbable it would cause a run on the bank. Well, the story was spread far and wide that my father went into the bank, drew out all his money, left the bank, came back again, and said, "I may just tell you that Mr. E—— has killed himself," and then departed. It seems that it was then a common belief that money withdrawn from a bank was not safe until the person had passed out through the door of the bank. My father did not hear this story till some little time afterwards, when the managing partner said that he had departed from his invariable rule of never allowing anyone to see the account of another man by having shown the ledger with my father's account to several persons, as this proved that my father had not drawn out a penny on that day. It would have been dishonourable in my father to have used his professional knowledge for his private advantage. Nevertheless the supposed act was greatly admired by some persons; and many years afterwards a gentleman remarked, "Ah, Doctor, what a splendid man of business you were in so cleverly getting all your money safe out of that bank."

My father's mind was not scientific, and he did not try to generalise his knowledge under general laws; yet he formed a theory for almost everything which occurred. I do not think that I gained much from him intellectually; but his example ought to have been of much moral service to all his children. One of his golden rules (a hard one to follow) was, "Never become the friend of anyone whom you cannot respect."

With respect to my father's father, the author of the *Botanic Garden,* etc., I have put together all the facts which I could collect in his published *Life.*

Having said this much about my father, I will add a few words about my brother and sisters.

My brother Erasmus possessed a remarkably clear mind, with extensive and diversified tastes and knowledge in literature, art, and even in science. For a short time he collected and dried plants and during a somewhat longer time experimented in chemistry. He was extremely agreeable, and his wit often reminded me of that in the letters and works of Charles Lamb. He was very kind-hearted; but his health from his boyhood had been weak, and as a consequence he failed in energy. His spirits were not high, sometimes low, more especially during early and middle manhood. He read much, even whilst a boy, and at school encouraged me to read, lending me books. Our minds and tastes were, however, so different that I do not think that I owe much to him intellectually—nor to my four sisters, who possessed very different characters, and some of them had strongly marked characters. All were extremely kind and affectionate towards me during their whole lives. I am inclined to agree with Francis Galton in believing that education and environment produce only a small effect on the mind of anyone and that most of our qualities are innate.

The above sketch of my brother's character was written before that which was published in Carlyle's Remembrances and which appears to me to have little truth and no merit.

Looking back as well as I can at my character during my school life, the only qualities which at this period promised well for the future were that I had strong and diversified tastes, much zeal for whatever interested me, and a keen pleasure in understanding any complex subject or thing. I was taught Euclid by a private tutor, and I distinctly remember the intense satisfaction which the clear geometrical proofs gave me. I remember with equal distinctness the delight which my uncle gave me (the father of Francis Galton) by explaining the principle of the vernier of a barometer. With respect to diversified tastes, independently of science, I was fond of reading various books, and I used to sit for hours reading the historical plays of Shakespeare, generally in an old window in the thick walls of the school. I read also other poetry, such as the recently published poems of Byron, Scott, and Thomson's *Seasons*. I mention this because later in life I wholly lost, to my great regret, all pleasure from poetry of any kind, including Shakespeare. In connection with pleasure from poetry I may add that in 1822 a vivid delight in scenery was first awakened in my mind, during a riding tour on the borders of Wales, and which has lasted longer than any other aesthetic pleasure.

Early in my school-days a boy had a copy of the *Wonders of the World,* which I often read and disputed with other boys about the veracity of some of the statements; and I believe this book first gave me a wish to travel in remote countries, which was ultimately fulfilled by the voyage of the *Beagle.* In the latter part of my school life I became passionately fond of shooting, and I do not believe that anyone could have shown more zeal for the most holy cause than I did for shooting birds. How well I remember killing my first snipe, and my excitement was so great that I had much difficulty in reloading my gun from the trembling of my hands. This taste long continued, and I became a very good shot. When at Cambridge I used to practise throwing up my gun to my shoulder before a looking-glass to see that I threw it up straight. Another and better plan was to get a friend to wave about a lighted candle and then to fire at it with a cap on the nipple, and if the aim was accurate the little puff of air would blow out the candle. The explosion of the cap caused a sharp crack, and I was told that the Tutor of the College remarked, "What an extraordinary thing it is, Mr. Darwin seems to spend hours in cracking a horse-whip in his room, for I often hear the crack when I pass under his windows."

I had many friends amongst the school-boys whom I loved dearly, and I think that my disposition was then very affectionate. Some of these boys were rather clever, but I may add on the principle of "noscitur a socio" that not one of them ever became in the least distinguished.

With respect to science, I continued collecting minerals with much zeal, but quite unscientifically—all that I cared for was a new *named* mineral, and I hardly attempted to classify them. I must have observed insects with some little care, for when ten years old (1819) I went for three weeks to Plas Edwards on the sea-coast in Wales, I was very much interested and surprised at seeing a large black and scarlet Hemipterous insect, many moths (Zygaena) and a Cicindela, which are not found in Shropshire. I almost made up my mind to begin collecting all the insects which I could find dead, for on consulting my sister I concluded that it was not right to kill insects for the sake of making a collection. From reading White's *Selborne* I took much pleasure in watching the habits of birds and even made notes on the subject. In my simplicity I remember wondering why every gentleman did not become an ornithologist.

Towards the close of my school life, my brother worked hard at chemistry and made a fair laboratory with proper apparatus in the tool-house in the garden, and I was allowed to aid him as a servant in most of his experiments. He made all the gases and many compounds, and I read with care several books on chemistry, such as Henry and Parkes' *Chemical Catechism*. The subject interested me greatly, and we often used to go on working till rather late at night. This was the best part of my education at school, for it showed me practically the meaning of experimental science. The fact that we worked at chemistry somehow got known at school, and as it was an unprecedented fact I was nicknamed "Gas." I was also once publicly rebuked by the head-master, Dr. Butler, for thus wasting my time over such useless subjects; and he called me very unjustly a "poco curante," and as I did not understand what he meant it seemed to me a fearful reproach.

As I was doing no good at school, my father wisely took me away at a rather earlier age than usual and sent me (October 1825) to Edinburgh University [7] with my brother, where I stayed for two years or sessions. My brother was completing his medical studies, though I do not believe he ever really intended to practise, and I was sent there to commence them. But soon after this period I became convinced from various small circumstances that my father would leave me property enough to subsist on with some comfort, though I never imagined that I should be so rich a man as I am; but my belief was sufficient to check any strenuous effort to learn medicine.

The instruction at Edinburgh was altogether by lectures, and these were intolerably dull, with the exception of those on chemistry by Hope; but to my mind there are no advantages and many disadvantages in lectures compared with reading. Dr.

[7] He lodged at Mrs. Mackay's, 11 Lothian Street. What little the records of Edinburgh University can reveal has been published in the *Edinburgh Weekly Dispatch,* May 22, 1888, and in the *St. James's Gazette,* Feb. 16, 1888. From the latter journal it appears that he and his brother Erasmus made more use of the library than was usual among the students of their time.

Duncan's lectures on Materia Medica at eight o'clock on a winter's morning are something fearful to remember. Dr. Munro made his lectures on human anatomy as dull as he was himself, and the subject disgusted me. It has proved one of the greatest evils in my life that I was not urged to practise dissection, for I should soon have got over my disgust, and the practice would have been invaluable for all my future work. This has been an irremediable evil, as well as my incapacity to draw. I also attended regularly the clinical wards in the hospital. Some of the cases distressed me a good deal, and I still have vivid pictures before me of some of them; but I was not so foolish as to allow this to lessen my attendance. I cannot understand why this part of my medical course did not interest me in a greater degree, for during the summer before coming to Edinburgh I began attending some of the poor people, chiefly children and women in Shrewsbury. I wrote down as full an account as I could of the cases with all the symptoms and read them aloud to my father, who suggested further inquiries and advised me what medicines to give, which I made up myself. At one time I had at least a dozen patients, and I felt a keen interest in the work.[8] My father, who was by far the best judge of character whom I ever knew, declared that I should make a successful physician—meaning by this one who got many patients. He maintained that the chief element of success was exciting confidence; but what he saw in me which convinced him that I should create confidence I know not. I also attended on two occasions the operating theatre in the hospital at Edinburgh and saw two very bad operations, one on a child, but I rushed away before they were completed. Nor did I ever attend again, for hardly any inducement would have been strong enough to make me do so, this being long before the blessed days of chloroform. The two cases fairly haunted me for many a long year.

My brother stayed only one year at the University, so that during the second year I was left to my own resources; and this was an advantage, for I became well acquainted with several young men fond of natural science. One of these was Ainsworth, who afterwards published his travels in Assyria; he was a Wer-

---

[8] I have heard him call to mind the pride he felt at the results of the successful treatment of a whole family with tartar emetic.

nerian geologist and knew a little about many subjects but was superficial and very glib with his tongue. Dr. Coldstream was a very different young man, prim, formal, highly religious, and most kind-hearted; he afterwards published some good zoological articles. A third young man was Hardie, who would, I think, have made a good botanist but died early in India. Lastly, Dr. Grant, my senior by several years; but how I became acquainted with him I cannot remember. He published some first-rate zoological papers, but after coming to London as Professor in University College he did nothing more in science—a fact which has always been inexplicable to me. I knew him well; he was dry and formal in manner but with much enthusiasm beneath this outer crust. He one day, when we were walking together, burst forth in high admiration of Lamarck and his views on evolution. I listened in silent astonishment and, as far as I can judge, without any effect on my mind. I had previously read the *Zoönomia* of my grandfather, in which similar views are maintained, but without producing any effect on me. Nevertheless it is probable that the hearing rather early in life such views maintained and praised may have favoured my upholding them under a different form in my *Origin of Species*. At this time I admired greatly the *Zoönomia;* but on reading it a second time after an interval of ten or fifteen years I was much disappointed, the proportion of speculation being so large to the facts given.

Drs. Grant and Coldstream attended much to marine zoology, and I often accompanied the former to collect animals in the tidal pools which I dissected as well as I could. I also became friends with some of the Newhaven fishermen and sometimes accompanied them when they trawled for oysters, and thus got many specimens. But from not having had any regular practice in dissection and from possessing only a wretched microscope my attempts were very poor. Nevertheless I made one interesting little discovery and read, about the beginning of the year 1826, a short paper on the subject before the Plinian Socy. This was that the so-called ova of Flustra had the power of independent movement by means of cilia and were in fact larvae. In another short paper I showed that little globular bodies which had been supposed to be the young state of *Fucus loreus* were the egg-cases of the worm-like *Pontobdella muricata*.

The Plinian Society [9] was encouraged and, I believe, founded by Professor Jameson. It consisted of students and met in an underground room in the University for the sake of reading papers on natural science and discussing them. I used regularly to attend, and the meetings had a good effect on me in stimulating my zeal and giving me new congenial acquaintances. One evening a poor young man got up, and after stammering for a prodigious length of time, blushing crimson, he at last slowly got out the words, "Mr. President, I have forgotten what I was going to say." The poor fellow looked quite overwhelmed, and all the members were so surprised that no one could think of a word to say to cover his confusion. The papers which were read to our little society were not printed, so that I had not the satisfaction of seeing my paper in print; but I believe Dr. Grant noticed my small discovery in his excellent memoir on Flustra.

I was also a member of the Royal Medical Society and attended pretty regularly, but as the subjects were exclusively medical I did not much care about them. Much rubbish was talked there, but there were some good speakers, of whom the best was the present Sir J. Kay-Shuttleworth. Dr. Grant took me occasionally to the meetings of the Wernerian Society, where various papers on natural history were read, discussed, and afterwards published in the Transactions. I heard Audubon deliver there some interesting discourses on the habits of N. American birds, sneering somewhat unjustly at Waterton. By the way, a Negro lived in Edinburgh who had travelled with Waterton and gained his livelihood by stuffing birds, which he did excellently; he gave me lessons for payment, and I used often to sit with him, for he was a very pleasant and intelligent man.

Mr. Leonard Horner also took me once to a meeting of the Royal Society of Edinburgh, where I saw Sir Walter Scott in the chair as President, and he apologised to the meeting as not feeling fitted for such a position. I looked at him and at the whole scene with some awe and reverence; and I think it was owing to this visit during my youth and to my having attended the Royal Medical Society that I felt the honour of being elected a few years ago an honorary member of both these Societies more than any other similar honour. If I had been told at that time that

[9] The society was founded in 1823 and expired about 1848 (*Edinburgh Weekly Dispatch,* May 22, 1888).

I should one day have been thus honoured I declare that I should have thought it as ridiculous and improbable as if I had been told that I should be elected King of England.

During my second year in Edinburgh I attended Jameson's lectures on geology and zoology, but they were incredibly dull. The sole effect they produced on me was the determination never as long as I lived to read a book on geology or in any way to study the science. Yet I feel sure that I was prepared for a philosophical treatment of the subject, for an old Mr. Cotton in Shropshire who knew a good deal about rocks had pointed out to me, two or three years previously, a well-known large erratic boulder in the town of Shrewsbury called the bell-stone. He told me that there was no rock of the same kind nearer than Cumberland or Scotland, and he solemnly assured me that the world would come to an end before anyone would be able to explain how this stone came where it now lay. This produced a deep impression on me, and I meditated over this wonderful stone. So that I felt the keenest delight when I first read of the action of icebergs in transporting boulders, and I gloried in the progress of geology. Equally striking is the fact that I, though now only sixty-seven years old, heard Professor Jameson in a field lecture at Salisbury Craigs discoursing on a trap-dyke, with amygdaloidal margins and the strata indurated on each side, with volcanic rocks all around us, and say that it was a fissure filled with sediment from above, adding with a sneer that there were men who maintained that it had been injected from beneath in a molten condition. When I think of this lecture, I do not wonder that I determined never to attend to geology.

From attending Jameson's lectures I became acquainted with the curator of the museum, Mr. Macgillivray, who afterwards published a large and excellent book on the birds of Scotland. He had not much the appearance or manners of the gentleman. I had much interesting natural-history talk with him, and he was very kind to me. He gave me some rare shells, for I at that time collected marine mollusca, but with no great zeal.

My summer vacations during these two years were wholly given up to amusements, though I always had some book in hand, which I read with interest. During the summer of 1826 I took a long walking tour with two friends with knapsacks on our backs through North Wales. We walked thirty miles most days, including one day the ascent of Snowdon. I also went with my sister Caroline a riding tour in North Wales, a servant with saddle-bags carrying our clothes. The autumns were devoted to shooting, chiefly at Mr. Owen's at Woodhouse and at my Uncle Jos's [10] at Maer. My zeal was so great that I used to place my shooting boots open by my bed-side when I went to bed, so as not to lose half a minute in putting them on in the morning. And on one occasion I reached a distant part of the Maer estate on the twentieth of August for black-game shooting before I could see; I then toiled on with the gamekeeper the whole day through thick heath and young Scotch firs. I kept an exact record of every bird which I shot throughout the whole season. One day when shooting at Woodhouse with Captain Owen, the eldest son and Major Hill, his cousin, afterwards Lord Berwick, both of whom I liked very much, I thought myself shamefully used, for every time after I had fired and thought that I had killed a bird one of the two acted as if loading his gun and cried out, "You must not count that bird, for I fired at the same time," and the game-keeper, perceiving the joke, backed them up. After some hours they told me the joke, but it was no joke to me, for I had shot a large number of birds but did not know how many and could not add them to my list, which I used to do by making a knot in a piece of string tied to a button-hole. This my wicked friends had perceived.

How I did enjoy shooting, but I think that I must have been half-consciously ashamed of my zeal, for I tried to persuade myself that shooting was almost an intellectual employment, it required so much skill to judge where to find most game and to hunt the dogs well.

One of my autumnal visits to Maer in 1827 was memorable from meeting there Sir J. Mackintosh, who was the best conversor I ever listened to. I heard afterwards with a glow of pride that he had said, "There is something in that young man that interests me." This must have been chiefly due to his perceiving that I listened with much interest to everything which he said, for I was as ignorant as a pig about his subjects of history, politics, and moral philosophy. To hear of praise from an eminent person, though no doubt apt or certain to excite vanity, is, I think, good for a young man, as it helps to keep him in the right course.

[10] Josiah Wedgwood, the son of the founder of the Etruria Works.

My visits to Maer during these two and the three succeeding years were quite delightful, independently of the autumnal shooting. Life there was perfectly free; the country was very pleasant for walking or riding; and in the evening there was much very agreeable conversation, not so personal as it generally is in large family parties, together with music. In the summer the whole family used often to sit on the steps of the old portico, with the flower-garden in front and with the steep wooded bank, opposite the house, reflected in the lake, with here and there a fish rising or a water-bird paddling about. Nothing has left a more vivid picture on my mind than these evenings at Maer. I was also attached to and greatly revered my Uncle Jos. He was silent and reserved so as to be a rather awful man, but he sometimes talked openly with me. He was the very type of an upright man with the clearest judgment. I do not believe that any power on earth could have made him swerve an inch from what he considered the right course. I used to apply to him in my mind the well-known ode of Horace, now forgotten by me, in which the words "nec vultus tyranni, &c." [11] come in.

Cambridge, 1828–1831

After having spent two sessions in Edinburgh, my father perceived or he heard from my sisters that I did not like the thought of being a physician, so he proposed that I should become a clergyman. He was very properly vehement against my turning an idle sporting man, which then seemed my probable destination. I asked for some time to consider, as from what little I had heard and thought on the subject I had scruples about declaring my belief in all the dogmas of the Church of England, though otherwise I liked the thought of being a country clergyman. Accordingly I read with care *Pearson on the Creed* and a few other books on divinity; and as I did not then in the least doubt the strict and literal truth of every word in the Bible, I soon per-

[11] Justum et tenacem propositi virum
Non civium ardor prava jubentium,
Non vultus instantis tyranni
Mente quatit solida.

suaded myself that our Creed must be fully accepted. It never struck me how illogical it was to say that I believed in what I could not understand and what is in fact unintelligible. I might have said with entire truth that I had no wish to dispute any dogma; but I never was such a fool as to feel and say "credo quia incredibile."

Considering how fiercely I have been attacked by the orthodox it seems ludicrous that I once intended to be a clergyman. Nor was this intention and my father's wish ever formally given up, but died a natural death when on leaving Cambridge I joined the *Beagle* as naturalist. If the phrenologists are to be trusted I was well fitted in one respect to be a clergyman. A few years ago the secretaries of a German psychological society asked me earnestly by letter for a photograph of myself; and some time afterwards I received the proceedings of one of the meetings in which it seemed that the shape of my head had been the subject of a public discussion, and one of the speakers declared that I had the bump of Reverence developed enough for ten priests.

As it was decided that I should be a clergyman, it was necessary that I should go to one of the English universities and take a degree; but as I had never opened a classical book since leaving school I found to my dismay that in the two intervening years I had actually forgotten, incredible as it may appear, almost everything which I had learnt even to some few of the Greek letters. I did not therefore proceed to Cambridge at the usual time in October but worked with a private tutor in Shrewsbury and went to Cambridge after the Christmas vacation, early in 1828. I soon recovered my school standard of knowledge and could translate easy Greek books, such as Homer and the Greek Testament, with moderate facility.

During the three years which I spent at Cambridge my time was wasted, as far as the academical studies were concerned, as completely as at Edinburgh and at school. I attempted mathematics and even went during the summer of 1828 with a private tutor (a very dull man) to Barmouth, but I got on very slowly. The work was repugnant to me, chiefly from my not being able to see any meaning in the early steps in algebra. This impatience was very foolish, and in after years I have deeply regretted that I did not proceed far enough at least to understand something of the great leading principles of mathematics, for men thus endowed seem to have an extra sense. But I do not believe that I should ever have succeeded beyond a very low grade. With respect to Classics I did nothing except attend a few compulsory college lectures, and the attendance was almost nominal. In my second year I had to work for a month or two to pass the Little Go, which I did easily. Again in my last year I worked with some earnestness for my final degree of B.A. and brushed up my Classics together with a little algebra and Euclid, which latter gave me much pleasure, as it did whilst at school. In order to pass the B.A. examination it was also necessary to get up Paley's *Evidences of Christianity* and his *Moral Philosophy*. This was done in a thorough manner, and I am convinced that I could have written out the whole of the *Evidences* with perfect correctness, but not of course in the clear language of Paley. The logic of this book and as I may add of his *Natural Theology* gave me as much delight as did Euclid. The careful study of these works, without attempting to learn any part by rote, was the only part of the academical course which, as I then felt and as I still believe, was of the least use to me in the education of my mind. I did not at that time trouble myself about Paley's premises; and taking these on trust I was charmed and convinced by the long line of argumentation. By answering well the examination questions in Paley, by doing Euclid well, and by not failing miserably in Classics I gained a good place among the *hoi polloi,* or crowd of men who do not go in for honours. Oddly enough I cannot remember how high I stood, and my memory fluctuates between the fifth, tenth, or twelfth name on the list.[1]

Public lectures on several branches were given in the University, attendance being quite voluntary; but I was so sickened with lectures at Edinburgh that I did not even attend Sedgwick's eloquent and interesting lectures. Had I done so I should probably have become a geologist earlier than I did. I attended, however, Henslow's lectures on botany and liked them much for their extreme clearness and the admirable illustrations; but I did not study botany. Henslow used to take his pupils, including several of the older members of the University, field excursions, on foot or in coaches, to distant places, or in a barge down the river, and lectured on the rarer plants or animals which were observed. These excursions were delightful.

[1] Tenth in the list of January 1831.

Although as we shall presently see there were some redeeming features in my life at Cambridge, my time was sadly wasted there and worse than wasted. From my passion for shooting and for hunting and, when this failed, for riding across country I got into a sporting set, including some dissipated low-minded young men. We used often to dine together in the evening, though these dinners often included men of a higher stamp, and we sometimes drank too much, with jolly singing and playing at cards afterwards. I know that I ought to feel ashamed of days and evenings thus spent, but as some of my friends were very pleasant and we were all in the highest spirits I cannot help looking back to these times with much pleasure.[2]

But I am glad to think that I had many other friends of a widely different nature. I was very intimate with Whitley,[3] who was afterwards Senior Wrangler, and we used continually to take long walks together. He inoculated me with a taste for pictures and good engravings, of which I bought some. I frequently went to the Fitzwilliam Gallery, and my taste must have been fairly good, for I certainly admired the best pictures, which I discussed with the old curator. I read also with much interest Sir J. Reynolds' book. This taste, though not natural to me, lasted for several years, and many of the pictures in the National Gallery in London gave me much pleasure—that of Sebastian del Piombo exciting in me a sense of sublimity.

I also got into a musical set, I believe by means of my warmhearted friend Herbert,[4] who took a high wrangler's degree. From associating with these men and hearing them play, I acquired a strong taste for music and used very often to time my walks so as to hear on week days the anthem in King's College Chapel. This gave me intense pleasure, so that my backbone would sometimes shiver. I am sure that there was no affectation or mere imitation in this taste, for I used generally to go by myself to King's College, and I sometimes hired the chorister boys to sing in my rooms. Nevertheless I am so utterly destitute of an ear that I cannot perceive a discord or keep time

[2] I gather from some of my father's contemporaries that he has exaggerated the Bacchanalian nature of these parties.

[3] Rev. C. Whitley, Hon. Canon of Durham, formerly Reader in Natural Philosophy in Durham University.

[4] John Maurice Herbert, County Court Judge of Cardiff and the Monmouth Circuit.

and hum a tune correctly; and it is a mystery how I could possibly have derived pleasure from music.

My musical friends soon perceived my state and sometimes amused themselves by making me pass an examination, which consisted in ascertaining how many tunes I could recognise when they were played rather more quickly or slowly than usual. "God save the King" when thus played was a sore puzzle. There was another man with almost as bad an ear as I had, and strange to say he played a little on the flute. Once I had the triumph of beating him in one of our musical examinations.

But no pursuit at Cambridge was followed with nearly so much eagerness or gave me so much pleasure as collecting beetles. It was the mere passion for collecting, for I did not dissect them and rarely compared their external characters with published descriptions, but got them named anyhow. I will give a proof of my zeal: One day, on tearing off some old bark, I saw two rare beetles and seized one in each hand; then I saw a third and new kind, which I could not bear to lose, so that I popped the one which I held in my right hand into my mouth. Alas, it ejected some intensely acrid fluid which burnt my tongue, so that I was forced to spit the beetle out, which was lost, as well as the third one.

I was very successful in collecting and invented two new methods: I employed a labourer to scrape, during the winter, moss off old trees and place [it] in a large bag, and likewise to collect the rubbish at the bottom of the barges in which reeds are brought from the fens, and thus I got some very rare species. No poet ever felt more delight at seeing his first poem published than I did at seeing in Stephen's *Illustrations of British Insects* the magic words "captured by C. Darwin, Esq." I was introduced to entomology by my second cousin, W. Darwin Fox, a clever and most pleasant man who was then at Christ's College and with whom I became extremely intimate. Afterwards I became well acquainted, and went out collecting, with Albert Way of Trinity, who in after years became a well-known archaeologist; also with H. Thompson of the same College, afterwards a leading agriculturist, chairman of a great railway, and Member of Parliament. It seems therefore that a taste for collecting beetles is some indication of future success in life!

I am surprised what an indelible impression many of the

beetles which I caught at Cambridge have left on my mind. I can remember the exact appearance of certain posts, old trees, and banks where I made a good capture. The pretty *Panagaeus crux-major* was a treasure in those days, and here at Down I saw a beetle running across a walk and on picking it up instantly perceived that it differed slightly from *P. crux-major,* and it turned out to be *P. quadripunctatus,* which is only a variety or closely allied species, differing from it very slightly in outline. I had never seen in those old days Licinus alive, which to an uneducated eye hardly differs from many other black Carabidous beetles; but my sons found here a specimen, and I instantly recognised that it was new to me; yet I had not looked at a British beetle for the last twenty years.

I have not as yet mentioned a circumstance which influenced my whole career more than any other. This was my friendship with Prof. Henslow. Before coming up to Cambridge I had heard of him from my brother as a man who knew every branch of science, and I was accordingly prepared to reverence him. He kept open house once every week,[5] where all undergraduates and several older members of the University, who were attached to science, used to meet in the evening. I soon got, through Fox, an invitation, and went there regularly. Before long I became well acquainted with Henslow and during the latter half of my time at Cambridge took long walks with him on most days; so that I was called by some of the dons "the man who walks with Henslow"; and in the evening I was very often asked to join his family dinner. His knowledge was great in botany, entomology, chemistry, mineralogy, and geology. His strongest taste was to draw conclusions from long-continued minute observations. His judgment was excellent, and his whole mind well-balanced; but I do not suppose that anyone would say that he possessed much original genius.

He was deeply religious and so orthodox that he told me one day he should be grieved if a single word of the Thirty-nine Articles were altered. His moral qualities were in every way admirable. He was free from every tinge of vanity or other

[5] The *Cambridge Ray Club,* which in 1887 attained its fiftieth anniversary, is the direct descendant of these meetings, having been founded to fill the blank caused by the discontinuance, in 1836, of Henslow's Friday evenings. See Professor Babington's pamphlet, *The Cambridge Ray Club* (1887).

petty feeling; and I never saw a man who thought so little about himself or his own concerns. His temper was imperturbably good, with the most winning and courteous manners; yet, as I have seen, he could be roused by any bad action to the warmest indignation and prompt action. I once saw in his company in the streets of Cambridge almost as horrid a scene as could have been witnessed during the French Revolution. Two body-snatchers had been arrested and whilst being taken to prison had been torn from the constable by a crowd of the roughest men, who dragged them by their legs along the muddy and stony road. They were covered from head to foot with mud, and their faces were bleeding either from having been kicked or from the stones; they looked like corpses, but the crowd was so dense that I got only a few momentary glimpses of the wretched creatures. Never in my life have I seen such wrath painted on a man's face as was shown by Henslow at this horrid scene. He tried repeatedly to penetrate the mob; but it was simply impossible. He then rushed away to the mayor, telling me not to follow him, to get more policemen. I forget the issue, except that the two were got into the prison before being killed.

Henslow's benevolence was unbounded, as he proved by his many excellent schemes for his poor parishioners, when in after years he held the living of Hitcham. My intimacy with such a man ought to have been and I hope was an inestimable benefit. I cannot resist mentioning a trifling incident which showed his kind consideration. Whilst examining some pollen-grains on a damp surface I saw the tubes exserted and instantly rushed off to communicate my surprising discovery to him. Now I do not suppose any other Professor of Botany could have helped laughing at my coming in such a hurry to make such a communication. But he agreed how interesting the phenomenon was and explained its meaning, but made me clearly understand how well it was known; so I left him not in the least mortified, but well pleased at having discovered for myself so remarkable a fact, but determined not to be in such a hurry again to communicate my discoveries.

Dr. Whewell was one of the older and distinguished men who sometimes visited Henslow, and on several occasions I walked home with him at night. Next to Sir J. Mackintosh he was the best converser on grave subjects to whom I ever listened.

Leonard Jenyns (grandson of the famous Soames Jenyns), who afterwards published some good essays in natural history, often stayed with Henslow, who was his brother-in-law. At first I disliked him from his somewhat grim and sarcastic expression, and it is not often that a first impression is lost; but I was completely mistaken and found him very kindhearted, pleasant, and with a good stock of humour. I visited him at his parsonage on the borders of the Fens [Swaffham Bulbeck] and had many a good walk and talk with him about natural history. I became also acquainted with several other men older than me who did not care much about science but were friends of Henslow. One was a Scotchman, brother of Sir Alexander Ramsay and tutor of Jesus College; he was a delightful man but did not live for many years. Another was Mr. Dawes, afterwards Dean of Hereford and famous for his success in the education of the poor. These men and others of the same standing, together with Henslow, used sometimes to take distant excursions into the country which I was allowed to join, and they were most agreeable.

Looking back, I infer that there must have been something in me a little superior to the common run of youths, otherwise the above-mentioned men, so much older than me and higher in academical position, would never have allowed me to associate with them. Certainly I was not aware of any such superiority, and I remember one of my sporting friends, Turner, who saw me at work on my beetles, saying that I should some day be a Fellow of the Royal Society, and the notion seemed to me preposterous.

During my last year at Cambridge I read with care and profound interest Humboldt's *Personal Narrative*. This work and Sir J. Herschel's *Introduction to the Study of Natural Philosophy* stirred up in me a burning zeal to add even the most humble contribution to the noble structure of natural science. No one or a dozen other books influenced me nearly so much as these two. I copied out from Humboldt long passages about Teneriffe and read them aloud on one of the above-mentioned excursions to (I think) Henslow, Ramsay, and Dawes; for on a previous occasion I had talked about the glories of Teneriffe, and some of the party declared they would endeavour to go there, but I think that they were only half in earnest. I was, however, quite in earnest and got an introduction to a merchant in London to inquire about ships; but the scheme was of course knocked on the head by the voyage of the *Beagle*.

My summer vacations were given up to collecting beetles, to some reading and short tours. In the autumn my whole time was devoted to shooting, chiefly at Woodhouse and Maer, and sometimes with young Eyton of Eyton. Upon the whole the three years which I spent at Cambridge were the most joyful in my happy life; for I was then in excellent health and almost always in high spirits.

As I had at first come up to Cambridge at Christmas, I was forced to keep two terms after passing my final examination, at the commencement of 1831; and Henslow then persuaded me to begin the study of geology. Therefore on my return to Shropshire I examined sections and coloured a map of parts round Shrewsbury. Professor Sedgwick intended to visit N. Wales in the beginning of August to pursue his famous geological investigation amongst the older rocks, and Henslow asked him to allow me to accompany him.[6] Accordingly he came and slept at my father's house.

A short conversation with him during this evening produced a strong impression on my mind. Whilst examining an old gravel-pit near Shrewsbury a labourer told me that he had found in it a large worn tropical Volute shell, such as may be seen on the chimney-pieces of cottages; and as he would not sell the shell I was convinced that he had really found it in the pit. I told Sedgwick of the fact, and he at once said (no doubt truly) that it must have been thrown away by someone into the pit; but then added, if really embedded there it would be the greatest misfortune to geology, as it would overthrow all that we know about the superficial deposits of the midland counties. These gravel-beds belonged in fact to the glacial

[6] In connection with this tour my father used to tell a story about Sedgwick: They had started from their inn one morning and had walked a mile or two when Sedgwick suddenly stopped and vowed that he would return, being certain "that damned scoundrel" (the waiter) had not given the chambermaid the sixpence entrusted to him for the purpose. He was ultimately persuaded to give up the project, seeing that there was no reason for suspecting the waiter of perfidy.

period, and in after years I found in them broken arctic shells. But I was then utterly astonished at Sedgwick not being delighted at so wonderful a fact as a tropical shell being found near the surface in the middle of England. Nothing before had ever made me thoroughly realise, though I had read various scientific books, that science consists in grouping facts so that general laws or conclusions may be drawn from them.

Next morning we started for Llangollen, Conway, Bangor, and Capel Curig. This tour was of decided use in teaching me a little how to make out the geology of a country. Sedgwick often sent me on a line parallel to his, telling me to bring back specimens of the rocks and to mark the stratification on a map. I have little doubt that he did this for my good, as I was too ignorant to have aided him. On this tour I had a striking instance how easy it is to overlook phenomena, however conspicuous, before they have been observed by anyone. We spent many hours in Cwm Idwal, examining all the rocks with extreme care, as Sedgwick was anxious to find fossils in them, but neither of us saw a trace of the wonderful glacial phenomena all around us; we did not notice the plainly scored rocks, the perched boulders, the lateral and terminal moraines. Yet these phenomena are so conspicuous that, as I declared in a paper published many years afterwards in the *Philosophical Magazine*,[7] a house burnt down by fire did not tell its story more plainly than did this valley. If it had still been filled by a glacier the phenomena would have been less distinct than they now are.

At Capel Curig I left Sedgwick and went in a straight line by compass and map across the mountains to Barmouth, never following any track unless it coincided with my course. I thus came on some strange wild places and enjoyed much this manner of travelling. I visited Barmouth to see some Cambridge friends who were reading there, and thence returned to Shrewsbury and to Maer for shooting; for at that time I should have thought myself mad to give up the first days of partridge-shooting for geology or any other science.

[7] *Philosophical Magazine* (1842).

Voyage of the Beagle: From
Dec. 27, 1831, to Oct. 2, 1836

On returning home from my short geological tour in N. Wales, I found a letter from Henslow informing me that Captain Fitz-Roy was willing to give up part of his own cabin to any young man who would volunteer to go with him without pay as naturalist to the Voyage of the *Beagle*. I have given as I believe in my M.S. Journal on account of all the circumstances which then occurred; I will here only say that I was instantly eager to accept the offer, but my father strongly objected, adding the words fortunate for me: "If you can find any man of common sense who advises you to go, I will give my consent." So I wrote that evening and refused the offer. On the next morning I went to Maer to be ready for September 1, and whilst out shooting, my uncle [1] sent for me, offering to drive me over to Shrewsbury and talk with my father. As my uncle thought it would be wise in me to accept the offer, and as my father always maintained that he was one of the most sensible men in the world, he at once consented in the kindest manner. I had been rather extravagant at Cambridge and to console my father said "that I should be deuced clever to spend more than my allowance whilst on board the *Beagle*"; but he answered with a smile, "But they all tell me you are very clever."

Next day I started for Cambridge to see Henslow, and thence to London to see Fitz-Roy, and all was soon arranged. Afterwards, on becoming very intimate with Fitz-Roy, I heard that I had run a very narrow risk of being rejected on account of the shape of my nose! He was an ardent disciple of Lavater and was convinced that he could judge a man's character by the outline of his features; and he doubted whether anyone with my nose could possess sufficient energy and determination for the voyage. But I think he was afterwards well-satisfied that my nose had spoken falsely.

Fitz-Roy's character was a singular one, with many very noble features: He was devoted to his duty, generous to a fault, bold, determined, indomitably energetic, and an ardent friend to all under his sway. He would undertake any sort of trouble

[1] Josiah Wedgwood.

to assist those whom he thought deserved assistance. He was a handsome man, strikingly like a gentleman, with highly courteous manners, which resembled those of his maternal uncle, the famous Lord Castlereagh, as I was told by the Minister at Rio. Nevertheless he must have inherited much in his appearance from Charles II, for Dr. Wallich gave me a collection of photographs which he had made, and I was struck with the resemblance of one to Fitz-Roy; on looking at the name, I found it Ch. E. Sobieski Stuart, Count d'Albanie,[2] illegitimate descendant of the same monarch.

Fitz-Roy's temper was a most unfortunate one. This was shown not only by passion but by fits of long-continued moroseness against those who had offended him. His temper was usually worst in the early morning, and with his eagle eye he could generally detect something amiss about the ship and was then unsparing in his blame. The junior officers when they relieved each other in the forenoon used to ask "whether much hot coffee had been served out this morning?," which meant, how was the Captain's temper? He was also somewhat suspicious and occasionally in very low spirits, on one occasion bordering on insanity. He seemed to me often to fail in sound judgment or common sense. He was extremely kind to me but was a man very difficult to live with on the intimate terms which necessarily followed from our messing by ourselves in the same cabin. We had several quarrels, for when out of temper he was utterly unreasonable. For instance, early in the voyage at Bahia in Brazil he defended and praised slavery, which I abominated, and told me that he had just visited a great slave-owner, who had called up many of his slaves and asked them whether they were happy, and whether they wished to be free, and all answered "No." I then asked him, perhaps with a sneer, whether he thought that the answers of slaves in the presence of their master was worth anything. This made him excessively angry, and he said that as I doubted his word, we could not live any longer together. I thought that I should have been compelled to leave the ship; but as soon as the news spread, which it did

---

[2] The Count d'Albanie's claim to royal descent has been shown to be based on a myth. See the *Quarterly Review* (1847), Vol. LXXXI, p. 83; also Hayward's *Biographical and Critical Essays* (1873), Vol. II, p. 201.

quickly, as the captain sent for the first lieutenant to assuage his anger by abusing me, I was deeply gratified by receiving an invitation from all the gun-room officers to mess with them. But after a few hours Fitz-Roy showed his usual magnanimity by sending an officer to me with an apology and a request that I would continue to live with him. I remember another instance of his candour. At Plymouth before we sailed, he was extremely angry with a dealer in crockery who refused to exchange some article purchased in his shop. The Captain asked the man the price of a very expensive set of china and said, "I should have purchased this if you had not been so disobliging." As I knew that the cabin was amply stocked with crockery, I doubted whether he had any such intention; and I must have shown my doubts in my face, for I said not a word. After leaving the shop he looked at me, saying, "You do not believe what I have said," and I was forced to own that it was so. He was silent for a few minutes and then said, "You are right, and I acted wrongly in my anger at the blackguard."

At Conception in Chile, poor Fitz-Roy was sadly overworked and in very low spirits; he complained bitterly to me that he must give a great party to all the inhabitants of the place. I remonstrated and said that I could see no such necessity on his part under the circumstances. He then burst out into a fury, declaring that I was the sort of man who would receive any favours and make no return. I got up and left the cabin without saying a word and returned to Conception, where I was then lodging. After a few days I came back to the ship and was received by the Captain as cordially as ever, for the storm had by that time quite blown over. The first Lieutenant, however, said to me: "Confound you, philosopher, I wish you would not quarrel with the skipper; the day you left the ship I was dead-tired [the ship was refitting], and he kept me walking the deck till midnight abusing you all the time." The difficulty of living on good terms with a Captain of a Man-of-War is much increased by its being almost mutinous to answer him as one would answer anyone else; and by the awe in which he is held —or was held in my time, by all on board. I remember hearing a curious instance of this in the case of the Purser of the *Adventure*—the ship which sailed with the *Beagle* during the first voyage. The Purser was in a store in Rio de Janeiro, purchasing rum from the ship's company, and a little gentleman in plain clothes walked in. The Purser said to him, "Now, sir, be so kind as to taste this rum and give me your opinion of it." The gentleman did as he was asked, and soon left the store. The store-keeper then asked the Purser whether he knew that he had been speaking to the Captain of a Line of Battleships which had just come into the harbour. The poor Purser was struck dumb with horror; he let the glass of spirit drop from his hand onto the floor and immediately went on board, and no persuasion, as an officer on the *Adventure* assured me, could make him go on shore again for fear of meeting the Captain after his dreadful act of familiarity.

I saw Fitz-Roy only occasionally after our return home, for I was always afraid of unintentionally offending him, and did so once, almost beyond mutual reconciliation. He was afterwards very indignant with me for having published so unorthodox a book (for he became very religious) as the *Origin of Species*. Towards the close of his life he was, as I fear, much impoverished, and this was largely due to his generosity. Anyhow, after his death a subscription was raised to pay his debts. His end was a melancholy one, namely suicide, exactly like that of his uncle Ld. Castlereagh, whom he resembled closely in manner and appearance.

His character was in several respects one of the most noble which I have ever known, though tarnished by grave blemishes.

The voyage of the *Beagle* has been by far the most important event in my life and has determined my whole career; yet it depended on so small a circumstance as my uncle offering to drive me thirty miles to Shrewsbury, which few uncles would have done, and on such a trifle as the shape of my nose. I have always felt that I owe to the voyage the first real training or education of my mind. I was led to attend closely to several branches of natural history, and thus my powers of observation were improved, though they were already fairly developed.

The investigation of the geology of all the places visited was far more important, as reasoning here comes into play. On first examining a new district nothing can appear more hopeless than the chaos of rocks; but by recording the stratification and nature of the rocks and fossils at many points, always reasoning and predicting what will be found elsewhere, light soon begins to dawn on the district, and the structure of the whole becomes more or less intelligible. I had brought with me the first volume of Lyell's *Principles of Geology,* which I studied attentively; and this book was of the highest service to me in many ways. The very first place which I examined, namely St. Jago in the Cape Verde islands, showed me clearly the wonderful superiority of Lyell's manner of treating geology, compared with that of any other author whose works I had with me or ever afterwards read.

Another of my occupations was collecting animals of all classes, briefly describing and roughly dissecting many of the marine ones; but from not being able to draw and from not having sufficient anatomical knowledge a great pile of MS. which I made during the voyage has proved almost useless. I thus lost much time, with the exception of that spent in acquiring some knowledge of the Crustaceans, as this was of service when in after years I undertook a monograph of the Cirripedia.

During some part of the day I wrote my Journal and took much pains in describing carefully and vividly all that I had seen; and this was good practice. My Journal served, also, in part as letters to my home, and portions were sent to England whenever there was an opportunity.

The above various special studies were, however, of no importance compared with the habit of energetic industry and of concentrated attention to whatever I was engaged in which I then acquired. Everything about which I thought or read was made to bear directly on what I had seen and was likely to see; and this habit of mind was continued during the five years of the voyage. I feel sure that it was this training which has enabled me to do whatever I have done in science.

Looking backwards, I can now perceive how my love for science gradually preponderated over every other taste. During the first two years my old passion for shooting survived in nearly full force, and I shot myself all the birds and animals for my collection; but gradually I gave up my gun more and more, and finally altogether to my servant, as shooting interfered with my work, more especially with making out the geological structure of a country. I discovered, though unconsciously and insensibly, that the pleasure of observing and reasoning was a much higher one than that of skill and sport. The primeval instincts of the barbarian slowly yielded to the acquired tastes of the civilized man. That my mind became developed through my pursuits during the voyage is rendered probable by a remark made by my father, who was the most acute observer whom I ever saw, of a sceptical disposition, and far from being a believer in phrenology; for on first seeing me after the voyage, he turned round to my sisters and exclaimed, "Why, the shape of his head is quite altered."

To return to the voyage. On September 11 (1831) I paid a flying visit with Fitz-Roy to the *Beagle* at Plymouth. Thence to Shrewsbury to wish my father and sisters a long farewell. On October 24 I took up my residence at Plymouth and remained there until December 27, when the *Beagle* finally left the shores of England for her circumnavigation of the world. We made two earlier attempts to sail but were driven back each time by heavy gales. These two months at Plymouth were the most miserable which I ever spent, though I exerted myself in various ways. I was out of spirits at the thought of leaving all my family and friends for so long a time, and the weather seemed to me inexpressibly gloomy. I was also troubled with palpitations and pain about the heart, and like many a young ignorant man, especially one with a smattering of medical knowledge, was convinced that I had heart-disease. I did not consult any doctor, as I fully expected to hear the verdict that I was not fit for the voyage, and I was resolved to go at all hazards.

I need not here refer to the events of the voyage—where we went and what we did—as I have given a sufficiently full account in my published Journal. The glories of the vegetation of the Tropics rise before my mind at the present time more vividly than anything else. Though the sense of sublimity which the great deserts of Patagonia and the forest-clad mountains of Tierra del Fuego excited in me has left an indelible impression on my mind. The sight of a naked savage in his native land is an event which can never be forgotten. Many of my excursions on horseback through wild countries, or in the boats, some of which lasted several weeks, were deeply interesting; their discomfort and some degree of danger were at that time hardly a drawback and none at all afterwards. I also reflect with high satisfaction on some of my scientific work, such as solving the problem of coral-islands, and making out the geological structure of certain islands, for instance St. Helena. Nor must I pass over the discovery of the singular relations of the animals and plants inhabiting the several islands of the Galapagos archipelago, and of all of them to the inhabitants of South America.

As far as I can judge of myself I worked to the utmost during the voyage from the mere pleasure of investigation, and from my strong desire to add a few facts to the great mass of facts in natural science. But I was also ambitious to take a fair place among scientific men—whether more ambitious or less so than most of my fellow-workers I can form no opinion.

The geology of St. Jago is very striking yet simple: A stream of lava formerly flowed over the bed of the sea, formed of triturated recent shells and corals, which it has baked into a hard white rock. Since then the whole island has been upheaved. But the line of white rock revealed to me a new and important fact, namely that there had been afterwards subsidence round the craters, which had since been in action and had poured forth lava. It then first dawned on me that I might perhaps write a book on the geology of the various countries visited, and this made me thrill with delight. That was a memorable hour to me, and how distinctly I can call to mind the low cliff of lava beneath which I rested, with the sun glaring hot, a few strange desert plants growing near, and with living corals in the tidal pools at my feet. Later in the voyage Fitz-Roy asked to read some of my Journal and declared it would be worth publishing; so here was a second book in prospect!

Towards the close of our voyage I received a letter whilst at Ascension, in which my sisters told me that Sedgwick had called on my father and said that I should take a place among the leading scientific men. I could not at the time understand how he could have learnt anything of my proceedings, but I heard (I believe afterwards) that Henslow had read some of the letters which I wrote to him before the Philosophical Soc. of Cambridge [3] and had printed them for private distribution. My collection of fossil bones, which had been sent to Henslow, also excited considerable attention amongst palaeontologists. After reading this letter I clambered over the mountains of Ascension

[3] Read at the meeting held Nov. 16, 1835, and printed in a pamphlet of 31 pp. for distribution among the members of the Society.

with a bounding step and made the volcanic rocks resound under my geological hammer! All this shows how ambitious I was; but I think that I can say with truth that in after years, though I cared in the highest degree for the approbation of such men as Lyell and Hooker, who were my friends, I did not care much about the general public. I do not mean to say that a favourable review or a large sale of my books did not please me greatly; but the pleasure was a fleeting one, and I am sure that I have never turned one inch out of my course to gain fame.

From My Return to England Oct. 2, 1836, to My Marriage Jan. 29, 1839

These two years and three months were the most active ones which I ever spent, though I was occasionally unwell and so lost some time. After going backwards and forwards several times between Shrewsbury, Maer, Cambridge, and London, I settled in lodgings at Cambridge [1] on December 13, where all my collections were under the care of Henslow. I stayed here three months and got my minerals and rocks examined by the aid to Prof. Miller.

I began preparing my Journal of travels, which was not hard work, as my MS. Journal had been written with care, and my chief labour was making an abstract of my more interesting scientific results. I sent also, at the request of Lyell, a short account of my observations on the elevation of the coast of Chile to the Geological Society.

On March 7, 1837, I took lodgings in Great Marlborough Street in London and remained there for nearly two years, until I was married. During these two years I finished my Journal, read several papers before the Geological Society, began preparing the MS. for my *Geological Observations,* and arranged for the publication of the *Zoology of the Voyage of the Beagle.* In July I opened my first note-book for facts in rela-

[1] In Fitzwilliam Street.

tion to the *Origin of Species,* about which I had long reflected, and never ceased working on for the next twenty years.

During these two years I also went a little into society and acted as one of the hon. secretaries of the Geological Society. I saw a great deal of Lyell. One of his chief characteristics was his sympathy with the work of others; and I was as much astonished as delighted at the interest which he showed when on my return to England I explained to him my views on coral reefs. This encouraged me greatly, and his advice and example had much influence on me. During this time I saw also a good deal of Robert Brown "facile princeps botanicorum." I used often to call and sit with him during his breakfast on Sunday mornings, and he poured forth a rich treasure of curious observations and acute remarks, but they almost always related to minute points, and he never with me discussed large and general questions in science.

During these two years I took several short excursions as a relaxation, and one longer one to the parallel roads of Glen Roy, an account of which was published in the *Philosophical Transactions.* This paper was a great failure, and I am ashamed of it. Having been deeply impressed with what I had seen of the elevation of the land in S. America, I attributed the parallel lines to the action of the sea; but I had to give up this view when Agassiz propounded his glacier-lake theory. Because no other explanation was possible under our then state of knowledge, I argued in favour of sea-action; and my error has been a good lesson to me never to trust in science to the principle of exclusion.

As I was not able to work all day at science I read a good deal during these two years on various subjects, including some metaphysical books, but I was not at [all] well fitted for such studies. About this time I took much delight in Wordsworth's and Coleridge's poetry, and can boast that I read the *Excursion* twice through. Formerly Milton's *Paradise Lost* had been my chief favourite, and in my excursions during the voyage of the *Beagle,* when I could take only a single small volume, I always chose Milton.

Religious Belief

During these two years [1] I was led to think much about religion. Whilst on board the *Beagle* I was quite orthodox, and I remember being heartily laughed at by several of the officers (though themselves orthodox) for quoting the Bible as an unanswerable authority on some point of morality. I suppose it was the novelty of the argument that amused them. But I had gradually come, by this time, to see that the Old Testament from its manifestly false history of the world, with the Tower of Babel, the rainbow as a sign, etc., etc., and from its attributing to God the feelings of a revengeful tyrant, was no more to be trusted than the sacred books of the Hindoos, or the beliefs of any barbarian. The question then continually rose before my mind and would not be banished—is it credible that if God were now to make a revelation to the Hindoos, would he permit it to be connected with the belief in Vishnu, Siva, &c., as Christianity is connected with the Old Testament? This appeared to me utterly incredible.

By further reflecting that the clearest evidence would be requisite to make any sane man believe in the miracles by which Christianity is supported, that the more we know of the fixed laws of nature the more incredible do miracles become, that the men at that time were ignorant and credulous to a degree almost incomprehensible by us, that the Gospels cannot be proved to have been written simultaneously with the events, that they differ in many important details, far too important as it seemed to me to be admitted as the usual inaccuracies of eyewitnesses—by such reflections as these, which I give not as having the least novelty or value, but as they influenced me, I gradually came to disbelieve in Christianity as a divine revelation. The fact that many false religions have spread over large portions of the earth like wild-fire had some weight with me. Beautiful as is the morality of the New Testament, it can hardly be denied that its perfection depends in part on the interpretation which we now put on metaphors and allegories.

But I was very unwilling to give up my belief; I feel sure of this, for I can well remember often and often inventing day-

---

[1] October 1836 to January 1839.

366

dreams of old letters between distinguished Romans and manuscripts being discovered at Pompeii or elsewhere which confirmed in the most striking manner all that was written in the Gospels. But I found it more and more difficult, with free scope given to my imagination, to invent evidence which would suffice to convince me. Thus disbelief crept over me at a very slow rate but was at last complete. The rate was so slow that I felt no distress and have never since doubted even for a single second that my conclusion was correct. I can indeed hardly see how anyone ought to wish Christianity to be true; for if so the plain language of the text seems to show that the men who do not believe, and this would include my father, brother, and almost all my best friends, will be everlastingly punished.

And this is a damnable doctrine.

Although I did not think much about the existence of a personal God until a considerably later period of my life, I will here give the vague conclusions to which I have been driven. The old argument of design in nature, as given by Paley, which formerly seemed to me so conclusive, fails, now that the law of natural selection has been discovered. We can no longer argue that, for instance, the beautiful hinge of a bivalve shell must have been made by an intelligent being, like the hinge of a door by man. There seems to be no more design in the variability of organic beings and in the action of natural selection than in the course which the wind blows. Everything in nature is the result of fixed laws. But I have discussed this subject at the end of my book on the *Variation of Domestic Animals and Plants*,[2] and the argument there given has never, as far as I can see, been answered.

But passing over the endless beautiful adaptations which we everywhere meet with, it may be asked how can the generally beneficent arrangement of the world be accounted for? Some

[2] My father asks whether we are to believe that the forms are preordained of the broken fragments of rock which are fitted together by man to build his houses. If not, why should we believe that the variations of domestic animals or plants are preordained for the sake of the breeder? "But if we give up the principal in one case . . . no shadow of reason can be assigned for the belief that variations alike in nature and the result of the same general laws, which have been the groundwork through natural selection of the formation of the most perfectly adapted animals in the world, man included, were intentionally and specially guided " —*Variations of Animals and Plants,* 1st Edit., Vol. II, p. 431.

writers indeed are so much impressed with the amount of suffering in the world that they doubt, if we look to all sentient
beings, whether there is more of misery or of happiness—
whether the world as a whole is a good or a bad one. According to my judgment happiness decidedly prevails, though this
would be very difficult to prove. If the truth of this conclusion
be granted, it harmonises well with the effects which we might
expect from natural selection. If all the individuals of any
species were habitually to suffer to an extreme degree they
would neglect to propagate their kind; but we have no reason
to believe that this has ever or at least often occurred. Some
other considerations, moreover, lead to the belief that all
sentient beings have been formed so as to enjoy, as a general
rule, happiness.

Everyone who believes, as I do, that all the corporeal and
mental organs (excepting those which are neither advantageous
or disadvantageous to the possessor) of all beings have been
developed through natural selection, or the survival of the
fittest, together with use or habit, will admit that these organs
have been formed so that their possessors may compete suc-
cessfully with other beings and thus increase in number. Now
an animal may be led to pursue that course of action which is
the most beneficial to the species by suffering, such as pain,
hunger, thirst, and fear—or by pleasure, as in eating and
drinking and in the propagation of the species, &c., or by both
means combined, as in the search for food. But pain or suffering of any kind, if long continued, causes depression and lessens the power of action, yet is well adapted to make a creature
guard itself against any great or sudden evil. Pleasurable sensations, on the other hand, may be long continued without any
depressing effect; on the contrary they stimulate the whole system to increased action. Hence it has come to pass that most
or all sentient beings have been developed in such a manner
through natural selection that pleasurable sensations serve as
their habitual guides. We see this in the pleasure from exertion,
even occasionally from great exertion of the body or mind—
in the pleasure of our daily meals, and especially in the pleasure
derived from sociability and from loving our families. The sum
of such pleasures as these, which are habitual or frequently recurrent, give, as I can hardly doubt, to most sentient beings an

excess of happiness over misery, although many occasionally suffer much. Such suffering is quite compatible with the belief in natural selection, which is not perfect in its action but tends only to render each species as successful as possible in the battle for life with other species, in wonderfully complex and changing circumstances.

That there is much suffering in the world no one disputes. Some have attempted to explain this in reference to man by imagining that it serves for his moral improvement. But the number of men in the world is as nothing compared with that of all other sentient beings, and these often suffer greatly without any moral improvement. A being so powerful and so full of knowledge as a God who could create the universe is to our finite minds omnipotent and omniscient, and it revolts our understanding to suppose that his benevolence is not unbounded, for what advantage can there be in the sufferings of millions of the lower animals throughout almost endless time? This very old argument from the existence of suffering against the existence of an intelligent first cause seems to me a strong one; whereas, as just remarked, the presence of much suffering agrees well with the view that all organic beings have been developed through variation and natural selection.

At the present day the most usual argument for the existence of an intelligent God is drawn from the deep inward conviction and feelings which are experienced by most persons. But it cannot be doubted that Hindoos, Mahomadans, and others might argue in the same manner and with equal force in favour of the existence of one God, or of many Gods, or—as with the Buddhists—of no God. There are also many barbarian tribes who cannot be said with any truth to believe in what we call God: They believe indeed in spirits or ghosts, and it can be explained, as Tyler and Herbert Spencer have shown, how such a belief would be likely to arise.

Formerly I was led by feelings such as those just referred to (although I do not think that the religious sentiment was ever strongly developed in me) to the firm conviction of the existence of God, and of the immortality of the soul. In my Journal I wrote that whilst standing in the midst of the grandeur of a Brazilian forest "it is not possible to give an adequate idea

※
see
change
in tone
bet.
early &
late
man.

of the higher feelings of wonder, admiration, and devotion which fill and elevate the mind." I well remember my conviction that there is more in man than the mere breath of his body. But now the grandest scenes would not cause any such convictions and feelings to rise in my mind. It may be truly said that I am like a man who has become colour-blind, and the universal belief by men of the existence of redness makes my present loss of perception of not the least value as evidence. This argument would be a valid one if all men of all races had the same inward conviction of the existence of one God; but we know that this is very far from being the case. Therefore I cannot see that such inward convictions and feelings are of any weight as evidence of what really exists. The state of mind which grand scenes formerly excited in me, and which was intimately connected with a belief in God, did not essentially differ from that which is often called the sense of sublimity; and however difficult it may be to explain the genesis of this sense, it can hardly be advanced as an argument for the existence of God, any more than the powerful though vague and similar feelings excited by music.

With respect to immortality, nothing shows me how strong and almost instinctive a belief it is as the consideration of the view now held by most physicists, namely that the sun with all the planets will in time grow too cold for life, unless indeed some great body dashes into the sun and thus gives it fresh life. Believing as I do that man in the distant future will be a far more perfect creature than he now is, it is an intolerable thought that he and all other sentient beings are doomed to complete annihilation after such long-continued slow progress. To those who fully admit the immortality of the human soul, the destruction of our world will not appear so dreadful.

Another source of conviction in the existence of God, connected with the reason and not with the feelings, impresses me as having much more weight. This follows from the extreme difficulty or rather impossibility of conceiving this immense and wonderful universe, including man with his capacity of looking far backwards and far into futurity, as the result of blind chance or necessity. When thus reflecting I feel compelled to look to a First Cause having an intelligent mind in some degree analogous to that of man; and I deserve to be called a theist.

This conclusion was strong in my mind about the time, as far as I can remember, when I wrote the *Origin of Species;* and it is since that time that it has very gradually with many fluctuations become weaker. But then arises the doubt—can the mind of man, which has, as I fully believe, been developed from a mind as low as that possessed by the lowest animal, be trusted when it draws such grand conclusions? May not these be the result of the connection between cause and effect which strikes us as a necessary one, but probably depends merely on inherited experience? Nor must we overlook the probability of the constant inculcation in a belief in God on the minds of children producing so strong and perhaps an inherited effect on their brains not yet fully developed that it would be as difficult for them to throw off their belief in God as for a monkey to throw off its instinctive fear and hatred of a snake.

I cannot pretend to throw the least light on such abstruse problems. The mystery of the beginning of all things is insoluble by us; and I for one must be content to remain an agnostic.

A man who has no assured and ever present belief in the existence of a personal God, or of a future existence with retribution and reward, can have for his rule of life, as far as I can see, only to follow those impulses and instincts which are the strongest or which seem to him the best ones. A dog acts in this manner, but he does so blindly. A man, on the other hand, looks forwards and backwards, and compares his various feelings, desires, and recollections. He then finds, in accordance with the verdict of all the wisest men, that the highest satisfaction is derived from following certain impulses, namely the social instincts. If he acts for the good of others, he will receive the approbation of his fellow men and gain the love of those with whom he lives; and this latter gain undoubtedly is the highest pleasure on this earth. By degrees it will become intolerable to him to obey his sensuous passions rather than his higher impulses, which when rendered habitual may be almost called instincts. His reason may occasionally tell him to act in opposition to the opinion of others, whose approbation he will then not receive; but he will still have the solid satisfaction of knowing that he has followed his innermost guide or conscience. As for myself, I believe that I have acted rightly in steadily following and devoting my life to science. I feel no remorse from having committed any great sin, but have often and often regretted that I have not done more direct good to my fellow creatures. My sole and poor excuse is much ill-health and my mental constitution, which makes it extremely difficult for me to turn from one subject or occupation to another. I can imagine with high satisfaction giving up my whole time to philanthropy, but not a portion of it; though this would have been a far better line of conduct.

Nothing is more remarkable than the spread of scepticism or rationalism during the latter half of my life. Before I was engaged to be married, my father advised me to conceal carefully my doubts, for he said that he had known extreme misery thus caused with married persons. Things went on pretty well until the wife or husband became out of health, and then some women suffered miserably by doubting about the salvation of their husbands, thus making them likewise to suffer. My father added that he had known during his whole long life only three women who were sceptics; and it should be remembered that he knew well a multitude of persons and possessed extraordinary power of winning confidence. When I asked him who the three women were, he had to own with respect to one of them, his sister-in-law Kitty Wedgwood, that he had no good evidence, only the vaguest hints, aided by the conviction that so clear-sighted a woman could not be a believer. At the present time, with my small acquaintance, I know (or have known) several married ladies who believe very little more than their husbands. My father used to quote an unanswerable argument, by which an old lady, a Mrs. Barlow, who suspected him of unorthodoxy, hoped to convert him: "Doctor, I know that sugar is sweet in my mouth, and I know that my Redeemer liveth."

From My Marriage, Jan. 29, 1839, and Residence in Upper Gower Street to Our Leaving London and Settling at Down, Sept. 14, 1842

You all know well your mother, and what a good mother she has ever been to all of you. She has been my greatest blessing, and I can declare that in my whole life I have never heard her utter one word which I had rather have been unsaid. She has never failed in the kindest sympathy towards me, and has borne with the utmost patience my frequent complaints from ill-health and discomfort. I do not believe she has ever missed an opportunity of doing a kind action to anyone near her. I marvel at my good fortune that she, so infinitely my superior in every single moral quality, consented to be my wife. She has been my wise adviser and cheerful comforter throughout life, which without her would have been during a very long period a miserable one from ill health. She has earned the love and admiration of every soul near her.

(Mem.: her beautiful letter to myself preserved, shortly after our marriage.)

I have indeed been most happy in my family, and I must say to you my children that not one of you has ever given me one minute's anxiety, except on the score of health. There are, I suspect, very few fathers of five sons who could say this with entire truth. When you were very young it was my delight to play with you all, and I think with a sigh that such days can never return. From your earliest days to now that you are grown up, you have all, sons and daughters, ever been most pleasant, sympathetic, and affectionate to us and to one another. When all or most of you are at home (as, thank Heavens, happens pretty frequently) no party can be, according to my taste, more agreeable, and I wish for no other society. We have suffered only one very severe grief in the death of Annie at Malvern on April 24, 1851, when she was just over ten years old. She was a most sweet and affectionate child, and I feel sure would have grown into a delightful woman. But I need say nothing here of her character, as I wrote a short sketch of

maudlin
sentamentality?
or genuine love

it shortly after her death. Tears still sometimes come into my eyes when I think of her sweet ways.

During the three years and eight months whilst we resided in London I did less scientific work, though I worked as hard as I possibly could, then during any other equal length of time in my life. This was owing to frequently recurring unwellness and to one long and serious illness. The greater part of my time, when I could do anything, was devoted to my work on *Coral Reefs,* which I had begun before my marriage, and of which the last proof-sheet was corrected on May 6, 1842. This book, though a small one, cost me twenty months of hard work, as I had to read every work on the islands of the Pacific and to consult many charts. It was thought highly of by scientific men, and the theory therein given is, I think, now well-established.

No other work of mine was begun in so deductive a spirit as this, for the whole theory was thought out on the west coast of S. America before I had seen a true coral reef. I had therefore only to verify and extend my views by a careful examination of living reefs. But it should be observed that I had during the two previous years been incessantly attending to the effects on the shores of S. America of the intermittent elevation of the land, together with denudation and the deposition of sediment. This necessarily led me to reflect much on the effects of subsidence, and it was easy to replace in imagination the continued deposition of sediment by the upward growth of coral. To do this was to form my theory of the formation of barrier-reefs and atolls.

Besides my work on coral reefs, during my residence in London I read before the Geological Society papers on the Erratic Boulders of S. America,[1] on Earthquakes,[2] and on the Formation by the Agency of Earth-worms of Mould.[3] I also continued to superintend the publication of the *Zoology of the Voyage of the Beagle.* Nor did I ever intermit collecting facts bearing on the origin of species; and I could sometimes do this when I could do nothing else from illness.

In the summer of 1842 I was stronger than I had been for

[1] *Geolog. Soc. Proc.* III (1842).
[2] *Geolog. Trans.* V (1840).
[3] *Geolog. Soc. Proc.* II (1838).

some time and took a little tour by myself in N. Wales, for the sake of observing the effects of the old glaciers which formerly filled all the larger valleys. I published a short account of what I saw in the *Philosophical Magazine.*[4] This excursion interested me greatly, and it was the last time I was ever strong enough to climb mountains or to take long walks, such as are necessary for geological work.

During the early part of our life in London I was strong enough to go into general society and saw a good deal of several scientific men and other more or less distinguished men. I will give my impressions with respect to some of them, though I have little to say worth saying.

I saw more of Lyell than of any other man both before and after my marriage. His mind was characterised, as it appeared to me, by clearness, caution, sound judgment, and a good deal of originality. When I made any remark to him on geology, he never rested until he saw the whole case clearly and often made me see it more clearly than I had done before. He would advance all possible objections to my suggestion, and even after these were exhausted would long remain dubious. A second characteristic was his hearty sympathy with the work of other scientific men.

On my return from the voyage of the *Beagle,* I explained to him my views on coral reefs, which differed from his, and I was greatly surprised and encouraged by the vivid interest which he showed. On such occasions, while absorbed in thought, he would throw himself into the strangest attitudes, often resting his head on the seat of a chair, while standing up. His delight in science was ardent, and he felt the keenest interest in the future progress of mankind. He was very kind-hearted, and thoroughly liberal in his religious beliefs or rather disbeliefs; but he was a strong theist. His candour was highly remarkable. He exhibited this by becoming a convert to the Descent-theory, though he had gained much fame by opposing Lamarck's views, and this after he had grown old. He reminded me that I had many years before said to him, when discussing the opposition of the old school of geologists to his new views, "What a good thing it would be, if every scientific man was to die when sixty years old, as afterwards he would be sure to oppose all

[4] *Philosophical Magazine* (1842).

new doctrines." But he hoped that now he might be allowed to live. He had a strong sense of humour and often told amusing anecdotes. He was very fond of society, especially of eminent men and of persons high in rank; and this over-estimation of a man's position in the world seemed to me his chief foible. He used to discuss with Lady Lyell as a most serious question whether or not they should accept some particular invitation. But as he would not dine out more than three times a week on account of the loss of time he was justified in weighing his invitations with some care. He looked forward to going out oftener in the evening with advancing years, as to a great reward; but the good time never came, as his strength failed.

The science of geology is enormously indebted to Lyell—more so, as I believe, than to any other man who ever lived. When I was starting on the voyage of the *Beagle,* the sagacious Henslow, who, like all other geologists, believed at that time in successive cataclysms, advised me to get and study the first volume of the *Principles,* which had then just been published, but on no account to accept the views therein advocated. How differently would anyone now speak of the *Principles!* I am proud to remember that the first place, namely St. Jago, in the Cape Verde Archipelago, which I geologised, convinced me of the infinite superiority of Lyell's views over those advocated in any other work known to me.

The powerful effects of Lyell's works could formerly be plainly seen in the different progress of the science in France and England. The present total oblivion of Elie de Beaumont's wild hypotheses, such as his *Craters of Elevation* and *Lines of Elevation* (which latter hypothesis I heard Sedgwick at the Geolog. Soc. lauding to the skies), may be largely attributed to Lyell.

All the leading geologists were more or less known by me at the time when geology was advancing with triumphant steps. I liked most of them, with the exception of Buckland, who though very good-humoured and good-natured seemed to me a vulgar and almost coarse man. He was incited more by a craving for notoriety, which sometimes made him act like a buffoon, than by a love of science. He was not, however, selfish in his desire for notoriety; for Lyell, when a very young man, consulted him about communicating a poor paper to the

Geolog. Soc. which had been sent him by a stranger, and Buckland answered, "You had better do so, for it will be headed, 'Communicated by Charles Lyell,' and thus your name will be brought before the public."

The services rendered to geology by Murchison by his classification of the older formations cannot be overestimated; but he was very far from possessing a philosophical mind. He was very kind-hearted and would exert himself to the utmost to oblige anyone. The degree to which he valued rank was ludicrous, and he displayed this feeling and his vanity with the simplicity of a child. He related with the utmost glee to a large circle, including many mere acquaintances, in the rooms of the Geolog. Soc. how the Czar Nicholas, when in London, had patted him on the shoulder and had said, alluding to his geological work, "Mon ami, Russia is grateful to you," and then Murchison added, rubbing his hands together, "The best of it was that Prince Albert heard it all." He announced one day to the Council of the Geolog. Soc. that his great work on the Silurian system was at last published; and he then looked at all who were present and said, "You will every one of you find your name in the Index," as if this was the height of glory.

I saw a good deal of Robert Brown, "facile Princeps Botanicorum," as he was called by Humboldt; and before I was married I used to go and sit with him almost every Sunday morning. He seemed to me to be chiefly remarkable for the minuteness of his observations and their perfect accuracy. He never propounded to me any large scientific views in biology. His knowledge was extraordinarily great, and much died with him, owing to his excessive fear of ever making a mistake. He poured out his knowledge to me in the most unreserved manner, yet was strangely jealous on some points. I called on him two or three times before the voyage of the *Beagle,* and on one occasion he asked me to look through a microscope and describe what I saw. This I did and believe now that it was the marvellous currents of protoplasm in some vegetable cell. I then asked him what I had seen; but he answered me, who was then hardly more than a boy and on the point of leaving England for five years, "That is my little secret." I suppose that he was afraid that I might steal his discovery. Hooker told me that he was a complete miser, and knew himself to be

a miser, about his dried plants; and he would not lend specimens to Hooker, who was describing the plants of Tierra del
Fuego, although well knowing that he himself would never
make any use of the collections from this country. On the other
hand he was capable of the most generous actions. When old,
much out of health and quite unfit for any exertion, he daily
visited (as Hooker told me) an old man-servant, who lived at
a distance and whom he supported, and read aloud to him. This
is enough to make up for any degree of scientific penuriousness
or jealousy. He was rather given to sneering at anyone who
wrote about what he did not fully understand: I remember
praising Whewell's *History of the Inductive Sciences* to him,
and he answered, "Yes, I suppose that he has read the prefaces
of very many books."

I often saw Owen, whilst living in London, and admired him
greatly, but was never able to understand his character and
never became intimate with him. After the publication of the
Origin of Species he became my bitter enemy, not owing to
any quarrel between us but as far as I could judge out of jealousy at its success. Poor dear Falconer, who was a charming
man, had a very bad opinion of him, being convinced that he
was not only ambitious, very envious, and arrogant, but un-
truthful and dishonest. His power of hatred was certainly un-
surpassed. When in former days I used to defend Owen, Falconer often said, "You will find him out some day," and so it
has proved.

At a somewhat later period I became very intimate with
Hooker, who has been one of my best friends throughout life.
He is a delightfully pleasant companion and most kind-hearted.
One can see at once that he is honourable to the back-bone.
His intellect is very acute, and he has great power of generalisation. He is the most untirable worker that I have ever seen,
and will sit the whole day working with the microscope and be
in the evening as fresh and pleasant as ever. He is in all ways
very impulsive and somewhat peppery in temper; but the clouds
pass away almost immediately. He once sent me an almost
savage letter from a cause which will appear ludicrously small
to an outsider, viz. because I maintained for a time the silly
notion that our coal-plants had lived in shallow water in the
sea. His indignation was all the greater because he could not

pretend that he should ever have suspected that the mangrove (and a few other marine plants which I named) had lived in the sea if they had been found only in a fossil state. On another occasion he was almost equally indignant because I rejected with scorn the notion that a continent had formerly extended between Australia and S. America. I have known hardly any man more lovable than Hooker.

A little later I became intimate with Huxley. His mind is as quick as a flash of lightning and as sharp as a razor. He is the best talker whom I have known. He never writes and never says anything flat. From his conversation no one would suppose that he could cut up his opponents in so trenchant a manner as he can do and does do. He has been a most kind friend to me and would always take any trouble for me. He has been the mainstay in England of the principle of the gradual evolution of organic beings. Much splendid work as he has done in zoology, he would have done far more if his time had not been so largely consumed by official and literary work, and by his efforts to improve the education of the country. He would allow me to say anything to him: Many years ago I thought that it was a pity that he attacked so many scientific men, although I believe that he was right in each particular case, and I said so to him. He denied the charge indignantly, and I answered that I was very glad to hear that I was mistaken. We had been talking about his well-deserved attacks on Owen, so I said after a time, "How well you have exposed Ehrenberg's blunders"; he agreed and added that it was necessary for science that such mistakes should be exposed. Again after a time, I added: "Poor Agassiz has fared ill under your hands." Again I added another name, and now his bright eyes flashed on me, and he burst out laughing, anathematising me in some manner. He is a splendid man and has worked well for the good of mankind.

I may here mention a few other eminent men whom I have occasionally seen, but I have little to say about them worth saying. I felt a high reverence for Sir J. Herschel and was delighted to dine with him at his charming house at the C. of Good Hope and afterwards at his London house. I saw him, also, on a few other occasions. He never talked much, but every word which he uttered was worth listening to. He was very shy and he often had a distressed expression. Lady Caroline Bell, at whose house I dined at the C. of Good Hope, admired Herschel much, but said that he always came into a room as if he knew that his hands were dirty, and that he knew that his wife knew that they were dirty.

I once met at breakfast at Sir R. Murchison's house the illustrious Humboldt, who honoured me by expressing a wish to see me. I was a little disappointed with the great man, but my anticipations probably were too high. I can remember nothing distinctly about our interview, except that Humboldt was very cheerful and talked much.

I used to call pretty often on Babbage and regularly attended his famous evening parties. He was always worth listening to, but he was a disappointed and discontented man; and his expression was often or generally morose. I do not believe that he was half as sullen as he pretended to be. One day he told me that he had invented a plan by which all fires could be effectively stopped, but added, "I shan't publish it—damn them all, let all their houses be burnt." The "all" were the inhabitants of London. Another day he told me that he had seen a pump on a road-side in Italy, with a pious inscription on it to the effect that the owner had erected the pump for the love of God and his country, that the tired wayfarer might drink. This led Babbage to examine the pump closely, and he soon discovered that every time that a wayfarer pumped some water for himself, he pumped a larger quantity into the owner's house. Babbage then added, "There is only one thing which I hate more than piety, and that is patriotism." But I believe that his bark was much worse than his bite.

Herbert Spencer's conversation seemed to me very interesting, but I did not like him particularly and did not feel that I could easily have become intimate with him. I think that he was extremely egotistical. After reading any of his books, I generally feel enthusiastic admiration for his transcendent talents, and have often wondered whether in the distant future he would rank with such great men as Descartes, Leibnitz, etc., about whom, however, I know very little. Nevertheless I am not conscious of having profited in my own work by Spencer's writings. His deductive manner of treating every subject is wholly opposed to my frame of mind. His conclusions never convince me; and over and over again I have said to myself, after reading one of his discussions, "Here would be a fine subject for half a dozen years' work." His fundamental generalisations (which have been compared in importance by some persons with Newton's laws!)—which I daresay may be very valuable under a philosophical point of view—are of such a nature that they do not seem to me to be of any strictly scientific use. They partake more of the nature of definitions than of laws of nature. They do not aid one in predicting what will happen in any particular case. Anyhow they have not been of any use to me.

Speaking of H. Spencer reminds me of Buckle,[5] whom I once met at Hensleigh Wedgwood's. I was very glad to learn from him his system of collecting facts. He told me that he bought all the books which he read and made a full index to each of the facts which he thought might prove serviceable to him, and that he could always remember in what book he had read anything, for his memory was wonderful. I then asked him how at first he could judge what facts would be serviceable, and he answered that he did not know, but that a sort of instinct guided him. From this habit of making indices, he was enabled to give the astonishing number of references on all sorts of subjects which may be found in his *History of Civilisation*. This book I thought most interesting and read it twice; but I doubt whether his generalisations are worth anything. H. Spencer told me that he had never read a line of it! Buckle was a great talker, and I listened to him without saying hardly a word, nor indeed could I have done so, for he left no gaps. When Effie began to sing, I jumped up and said that I must listen to her. This I suppose offended him, for after I had moved away, he turned round to a friend and said (as was overheard by my brother), "Well, Mr. Darwin's books are much better than his conversation." What he really meant was that I did not properly appreciate his conversation.

Of other great literary men, I once met Sydney Smith at Dean Milman's house. There was something inexplicably amusing in every word which he uttered. Perhaps this was partly due to the expectation of being amused. He was talking about Lady Cork, who was then extremely old. This was the lady,

[5] Henry Thomas Buckle, 1821–1862. Self-educated historian.

who, as he said, was once so much affected by one of his charity sermons that she *borrowed* a guinea from a friend to put into the plate. He now said, "It is generally believed that my dear old friend Lady Cork has been overlooked"; and he said this in such a manner that no one could for a moment doubt that he meant that his dear old friend had been overlooked by the devil. How he managed to express this I know not.

I likewise once met Macaulay at Lord Stanhope's (the historian's) house, and as there was only one other man at dinner I had a grand opportunity of hearing him converse, and he was very agreeable. He did not talk at all too much; nor indeed could such a man talk too much, as long as he allowed others to turn the stream of his conversation, and this he did allow.

Lord Stanhope once gave me a curious little proof of the accuracy and fulness of Macaulay's memory: Many historians used often to meet at Lord Stanhope's house, and, in discussing various subjects, they would sometimes differ from Macaulay, and formerly they often referred to some book to see who was right; but latterly, as Lord Stanhope noticed, no historian ever took this trouble, and whatever Macaulay said was final.

On another occasion I met at Ld. Stanhope's house one of his parties of historians and other literary men, and amongst them were Motley and Grote. After luncheon I walked about Chevening Park for nearly an hour with Grote and was much interested by his conversation and pleased by the simplicity and absence of all pretension in his manners.

I met another set of great men at breakfast at Ld. Stanhope's house in London. After breakfast was quite over, Monckton Milnes (Ld. Houghton now) walked in and, after looking round, exclaimed (justifying Sidney Smith's nickname of "the cool of the evening"), "Well, I declare, you are all very premature."

Long ago I dined occasionally with the old Earl Stanhope, the father of the historian. I have heard that his father, the democratic earl, well-known at the time of the French Revolution, had his son educated as a blacksmith, as he declared that every man ought to know some trade. The old earl whom I knew was a strange man, but what little I saw of him I liked much. He was frank, genial, and pleasant. He had strongly marked features, with a brown complexion, and his clothes, when I saw him, were all brown. He seemed to believe in everything which was to others utterly incredible. He said one day to me, "Why don't you give up your fiddle-faddle of geology and zoology, and turn to the occult sciences?" The historian (then Ld. Mahon) seemed shocked at such a speech to me, and his charming wife much amused.

The last man whom I will mention is Carlyle, seen by me several times at my brother's house and two or three times at my own house. His talk was very racy and interesting, just like his writings, but he sometimes went on too long on the same subject. I remember a funny dinner at my brother's, where, amongst a few others, were Babbage and Lyell, both of whom liked to talk. Carlyle, however, silenced everyone by haranguing during the whole dinner on the advantages of silence. After dinner, Babbage, in his grimmest manner, thanked Carlyle for his very interesting Lecture on Silence.

Carlyle sneered at almost everyone. One day in my house he called Grote's *History* "a fetid quagmire, with nothing spiritual about it." I always thought, until his *Reminiscences* appeared, that his sneers were partly jokes, but this now seems rather doubtful. His expression was that of a depressed, almost despondent, yet benevolent man; and it is notorious how heartily he laughed. I believe that his benevolence was real, though stained by not a little jealousy. No one can doubt about his extraordinary power of drawing vivid pictures of things and men—far more vivid, as it appears to me, than any drawn by Macaulay. Whether his pictures of men were true ones is another question.

He has been all-powerful in impressing some grand moral truths on the minds of men. On the other hand, his views about slavery were revolting. In his eyes might was right. His mind seemed to me a very narrow one, even if all branches of science, which he despised, are excluded. It is astonishing to me that Kingsley should have spoken of him as a man well fitted to advance science. He laughed to scorn the idea that a mathematician, such as Whewell, could judge, as I maintained he could, of Goethe's views on light. He thought it a most ridiculous thing that anyone should care whether a glacier moved a little quicker or a little slower, or moved at all. As far as I could judge, I never met a man with a mind so ill adapted for scientific research.

Whilst living in London, I attended as regularly as I could the meetings of several scientific societies and acted as secretary to the Geological Society. But such attendance, and ordinary society, suited my health so badly that we resolved to live in the country, which we both preferred and have never repented of.

Residence at Down, From Sept. 14, 1842, to the Present Time, 1876

After several fruitless searches in Surrey and elsewhere, we found this house and purchased it. I was pleased with the diversified appearance of the vegetation proper to a chalk district, and so unlike what I had been accustomed to in the Midland counties, and still more pleased with the extreme quietness and rusticity of the place. It is not, however, quite so retired a place as a writer in a German periodical makes it, who says that my house can be approached only by a mule-track! Our fixing ourselves here has answered admirably in one way which we did not anticipate, namely, by being very convenient for frequent visits from our children, who never miss an opportunity of doing so when they can.

Few persons can have lived a more retired life than we have done. Besides short visits to the houses of relations, and occasionally to the seaside or elsewhere, we have gone nowhere. During the first part of our residence we went a little into society and received a few friends here; but my health almost always suffered from the excitement, violent shivering and vomiting attacks being thus brought on. I have therefore been compelled for many years to give up all dinner-parties; and this has been somewhat of a deprivation to me, as such parties always put me into high spirits. From the same cause I have been able to invite here very few scientific acquaintances. Whilst I was young and strong I was capable of very warm attachments, but of late years, though I still have very friendly feelings towards many persons, I have lost the power of becoming deeply attached to anyone, not even so deeply to my good and dear friends Hooker and Huxley, as I should formerly have been. As far as I can judge, this grievous loss of feeling has gradually crept over me, from the expectation of much distress afterwards from exhaustion having become firmly associated in my mind with seeing and talking with anyone for an hour, except my wife and children.

My chief enjoyment and sole employment throughout life has been scientific work; and the excitement from such work makes me for the time forget, or drives quite away, my daily discomfort. I have therefore nothing to record during the rest of my life, except the publication of my several books. Perhaps a few details how they arose may be worth giving.

My Several Publications

In the early part of 1844 my observations on the Volcanic Islands visited during the voyage of the *Beagle* were published. In 1845 I took much pains in correcting a new edition of my *Journal of Researches,* which was originally published in 1839 as part of Fitz-Roy's work. The success of this my first literary child always tickles my vanity more than that of any of my other books. Even to this day it sells steadily in England and the United States, and has been translated for the second time into German, and into French and other languages. This success of a book of travels, especially of a scientific one, so many years after its first publication is surprising. Ten thousand copies have now been sold in England of the second edition. In 1846 my *Geological Observations on South America* were published. I record in a little diary, which I have always kept, that my three geological books (*Coral Reefs* included) consumed four and a half years' steady work; "and now it is ten years since my return to England. How much time have I lost by illness?" I have nothing to say about these three books except that to my surprise new editions have lately been called for.[1] In October 1846 I began to work on Cirripedia. When on the coast of Chile, I found a most curious form, which burrowed into the shells of Concholepas and which differed so much from all other Cirripedes that I had to form a new suborder for its sole reception. Lately an allied burrowing genus has been found on the shores of Portugal. To understand the structure of my new Cirripede I had to examine and dissect many of the common forms, and this gradually led me on to take up the whole group. I worked steadily on the subject for the next eight years, and ultimately published two thick volumes describing all the known living species, and two thin quartos on the extinct species. I do not doubt that Sir E. Lytton Bulwer had me in his mind when he introduces in one of his novels a Professor Long, who had written two huge volumes on limpets.

Although I was employed during eight years on this work, yet I record in my diary that about two years out of this time was lost by illness. On this account I went in 1848 for some months to Malvern for hydropathic treatment, which did me much good, so that on my return home I was able to resume work. So much was I out of health that when my dear father died on November 13, 1847, I was unable to attend his funeral or to act as one of his executors.

My work on the Cirripedia possesses, I think, considerable value, as besides describing several new and remarkable forms I made out the homologies of the various parts—I discovered the cementing apparatus, though I blundered dreadfully about the cement glands—and lastly I proved the existence in certain genera of minute males complemental to and parasitic on the hermaphrodites. This latter discovery has at last been fully confirmed, though at one time a German writer was pleased to attribute the whole account to my fertile imagination. The Cirripedes form a highly varying and difficult group of species to class, and my work was of considerable use to me when I had to discuss in the *Origin of Species* the principles of a natural classification. Nevertheless, I doubt whether the work was worth the consumption of so much time.

From September 1854 onwards I devoted all my time to

[1] *Geological Observations* (1876), 2nd Edit. *Coral Reefs* (1874), 2nd Edit.

arranging my huge pile of notes, to observing, and experimenting, in relation to the transmutation of species. During the voyage of the *Beagle* I had been deeply impressed by discovering in the Pampean formation great fossil animals covered with armour like that on the existing armadillos; secondly, by the manner in which closely allied animals replace one another in proceeding southwards over the Continent; and, thirdly, by the South American character of most of the productions of the Galapagos archipelago, and more especially by the manner in which they differ slightly on each island of the group—none of these islands appearing to be very ancient in a geological sense.

It was evident that such facts as these, as well as many others, could be explained on the supposition that species gradually become modified; and the subject haunted me. But it was equally evident that neither the action of the surrounding conditions nor the will of the organisms (especially in the case of plants) could account for the innumerable cases in which organisms of every kind are beautifully adapted to their habits of life—for instance, a woodpecker or tree-frog to climb trees, or a seed for dispersal by hooks or plumes. I had always been much struck by such adaptations, and until these could be explained it seemed to me almost useless to endeavour to prove by indirect evidence that species have been modified.

After my return to England it appeared to me that by following the example of Lyell in geology, and by collecting all facts which bore in any way on the variation of animals and plants under domestication and nature, some light might perhaps be thrown on the whole subject. My first note-book was opened in July 1837. I worked on true Baconian principles and without any theory collected facts on a wholesale scale, more especially with respect to domesticated productions, by printed enquiries, by conversation with skilful breeders and gardeners, and by extensive reading. When I see the list of books of all kinds which I read and abstracted, including whole series of journals and transactions, I am surprised at my industry. I soon perceived that selection was the keystone of man's success in making useful races of animals and plants. But how selection could be applied to organisms living in a state of nature remained for some time a mystery to me.

In October 1838, that is, fifteen months after I had begun my systematic inquiry, I happened to read for amusement Malthus on *Population* and, being well prepared to appreciate the struggle for existence which everywhere goes on from long-continued observation of the habits of animals and plants, it at once struck me that under these circumstances favourable variations would tend to be preserved and unfavourable ones to be destroyed. The result of this would be the formation of new species. Here, then, I had at last got a theory by which to work; but I was so anxious to avoid prejudice that I determined not for some time to write even the briefest sketch of it. In June 1842 I first allowed myself the satisfaction of writing a very brief abstract of my theory in pencil in thirty-five pages; and this was enlarged during the summer of 1844 into one of two hundred and thirty pages, which I had fairly copied out and still possess.

But at that time I overlooked one problem of great importance; and it is astonishing to me, except on the principle of Columbus and his egg, how I could have overlooked it and its solution. This problem is the tendency in organic beings descended from the same stock to diverge in character as they become modified. That they have diverged greatly is obvious from the manner in which species of all kinds can be classed under genera, genera under families, families under sub-orders, and so forth; and I can remember the very spot in the road, whilst in my carriage, when to my joy the solution occurred to me; and this was long after I had come to Down. The solution, as I believe, is that the modified offspring of all dominant and increasing forms tend to become adapted to many and highly diversified places in the economy of nature.

Early in 1856 Lyell advised me to write out my views pretty fully, and I began at once to do so on a scale three or four times as extensive as that which was afterwards followed in my *Origin of Species;* yet it was only an abstract of the materials which I had collected, and I got through about half the work on this scale. But my plans were overthrown, for early in the summer of 1858 Mr. Wallace, who was then in the Malay archipelago, sent me an essay *On the Tendency of Varieties to depart indefinitely from the Original Type;* and this essay contained exactly the same theory as mine. Mr. Wallace expressed the wish that if I thought well of his essay I should send it to Lyell for perusal.

The circumstances under which I consented at the request of Lyell and Hooker to allow of an extract from my MS., together with a letter to Asa Gray, dated September 5, 1857, to be published at the same time with Wallace's Essay are given in the *Journal of the Proceedings of the Linnaean Society,* 1858, p. 45. I was at first very unwilling to consent, as I thought Mr. Wallace might consider my doing so unjustifiable, for I did not then know how generous and noble was his disposition. The extract from my MS. and the letter to Asa Gray had neither been intended for publication and were badly written. Mr. Wallace's essay, on the other hand, was admirably expressed and quite clear. Nevertheless, our joint productions excited very little attention, and the only published notice of them which I can remember was by Professor Haughton of Dublin, whose verdict was that all that was new in them was false, and what was true was old. This shows how necessary it is that any new view should be explained at considerable length in order to arouse public attention.

In September 1858 I set to work by the strong advice of Lyell and Hooker to prepare a volume on the transmutation of species, but was often interrupted by ill health and short visits to Dr. Lane's delightful hydropathic establishment at Moor Park. I abstracted the MS. begun on a much larger scale in 1856, and completed the volume on the same reduced scale. It cost me thirteen months and ten days' hard labour. It was published under the title of the *Origin of Species* in November 1859. Though considerably added to and corrected in the later editions, it has remained substantially the same book.

It is no doubt the chief work of my life. It was from the first highly successful. The first small edition of 1250 copies was sold on the day of publication, and a second edition of 3000 copies soon afterwards. Sixteen thousand copies have now (1876) been sold in England, and considering how stiff a book it is, this is a large sale. It has been translated into almost every European tongue, even into such languages as Spanish, Bohemian, Polish, and Russian. It has also, according to

Miss Bird, been translated into Japanese, and is there much
studied.[2] Even an essay in Hebrew has appeared on it, showing
that the theory is contained in the Old Testament! The reviews
were very numerous; for a time I collected all that appeared
on the *Origin* and on my related books, and these amount (excluding newspaper reviews) to two hundred and sixty-five; but
after a time I gave up the attempt in despair. Many separate
essays and books on the subject have appeared; and in Germany a catalogue or bibliography on "Darwinismus" has appeared every year or two.

The success of the *Origin* may, I think, be attributed in
large part to my having long before written two condensed
sketches and to my having finally abstracted a much larger
manuscript, which was itself an abstract. By this means I was
enabled to select the more striking facts and conclusions. I
had also, during many years, followed a golden rule, namely,
that whenever a published fact, a new observation, or thought
came across me which was opposed to my general results, to
make a memorandum of it without fail and at once; for I had
found by experience that such facts and thoughts were far more
apt to escape from the memory than favourable ones. Owing
to this habit, very few objections were raised against my views
which I had not at least noticed and attempted to answer.

It has sometimes been said that the success of the *Origin*
proved "that the subject was in the air," or "that men's minds
were prepared for it." I do not think that this is strictly true,
for I occasionally sounded not a few naturalists, and never happened to come across a single one who seemed to doubt about
the permanence of species. Even Lyell and Hooker, though
they would listen with interest to me, never seemed to agree. I
tried once or twice to explain to able men what I meant by
natural selection, but signally failed. What I believe was strictly
true is that innumerable well-observed facts were stored in the
minds of naturalists, ready to take their proper places as soon
as any theory which would receive them was sufficiently explained. Another element in the success of the book was its
moderate size; and this I owe to the appearance of Mr. Wallace's essay; had I published on the scale in which I began to
write in 1856, the book would have been four or five times as

[2] Miss Bird is mistaken, as I learn from Professor Mitsukuri.

large as the *Origin,* and very few would have had the patience
to read it.

I gained much by my delay in publishing from about 1839,
when the theory was clearly conceived, to 1859; and I lost
nothing by it, for I cared very little whether men attributed
most originality to me or Wallace; and his essay no doubt aided
in the reception of the theory. I was forestalled in only one
important point, which my vanity has always made me regret,
namely, the explanation by means of the Glacial period of the
presence of the same species of plants and of some few animals on distant mountain summits and in the arctic regions.
This view pleased me so much that I wrote it out *in extenso,*
and it was read by Hooker some years before E. Forbes published his celebrated memoir on the subject.[3] In the very few
points in which we differed, I still think that I was in the right.
I have never, of course, alluded in print to my having independently worked out this view.

Hardly any point gave me so much satisfaction when I was
at work on the *Origin* as the explanation of the wide difference
in many classes between the embryo and the adult animal,
and of the close resemblance of the embryos within the same
class. No notice of this point was taken, as far as I remember,
in the early reviews of the *Origin,* and I recollect expressing my
surprise on this head in a letter to Asa Gray. Within late years
several reviewers have given the whole credit of the idea to
Fritz Müller and Häckel, who undoubtedly have worked it out
much more fully, and in some respects more correctly, than
I did. I had materials for a whole chapter on the subject, and
I ought to have made the discussion longer, for it is clear that
I failed to impress my readers; and he who succeeds in doing so
deserves, in my opinion, all the credit.

This leads me to remark that I have almost always been
treated honestly by my reviewers, passing over those without
scientific knowledge as not worthy of notice. My views have
often been grossly misrepresented, bitterly opposed, and ridi-
culed, but this has been generally done, as I believe, in good
faith. I must, however, except Mr. Mivart, who as an American expressed it in a letter, has acted towards me "like a pet-
tifogger," or as Huxley has said, "like an Old Bailey lawyer."

[3] *Geol. Survey Mem.* (1846).

On the whole I do not doubt that my works have been over and over again greatly overpraised. I rejoice that I have avoided controversies, and this I owe to Lyell, who many years ago, in reference to my geological works, strongly advised me never to get entangled in a controversy, as it rarely did any good and caused a miserable loss of time and temper.

Whenever I have found out that I have blundered, or that my work has been imperfect, and when I have been contemptuously criticised, and even when I have been overpraised, so that I have felt mortified, it has been my greatest comfort to say hundreds of times to myself that "I have worked as hard and as well as I could, and no man can do more than this." I remember when in Good Success Bay, in Tierra del Fuego, thinking (and I believe that I wrote home to the effect) that I could not employ my life better than in adding a little to natural science. This I have done to the best of my abilities, and critics may say what they like, but they cannot destroy this conviction.

During the two last months of the year 1859 I was fully occupied in preparing a second edition of the *Origin,* and by an enormous correspondence. On January 7, 1860, I began arranging my notes for my work on the *Variation of Animals and Plants under Domestication;* but it was not published until the beginning of 1868, the delay having been caused partly by frequent illnesses, one of which lasted seven months, and partly by having been tempted to publish on other subjects which at the time interested me more.

On May 15, 1862, my little book on the *Fertilisation of Orchids,* which cost me ten months' work, was published; most of the facts had been slowly accumulated during several previous years. During the summer of 1839 and, I believe, during the previous summer, I was led to attend to the cross-fertilisation of flowers by the aid of insects, from having come to the conclusion in my speculations on the origin of species that crossing played an important part in keeping specific forms constant. I attended to the subject more or less during every subsequent summer; and my interest in it was greatly enhanced by having procured and read in November 1841, through the advice of Robert Brown, a copy of C. K. Sprengel's wonderful book, *Das entdeckte Geheimnis der Natur.* For some years before 1862 I had specially attended to the fertilisation of our British orchids; and it seemed to me the best plan to prepare as complete a treatise on this group of plants as well as I could, rather than to utilise the great mass of matter which I had slowly collected with respect to other plants.

My resolve proved a wise one, for since the appearance of my book a surprising number of papers and separate works on the fertilisation of all kinds of flowers have appeared; and these are far better done than I could possibly have effected. The merits of poor old Sprengel, so long overlooked, are now fully recognised many years after his death.

During this same year I published in the *Journal of the Linnean Society* a paper *On the Two Forms, or Dimorphic Condition of Primula* and during the next five years five other papers on dimorphic and trimorphic plants. I do not think anything in my scientific life has given me so much satisfaction as making out the meaning of the structure of these plants. I had noticed in 1838 or 1839 the dimorphism of *Linum flavum,* and had at first thought that it was merely a case of unmeaning variability. But on examining the common species of Primula, I found that the two forms were much too regular and constant to be thus viewed. I therefore became almost convinced that the common cowslip and primrose were on the high-road to become dioecious—that the short pistil in the one form and the short stamens in the other form were tending towards abortion. The plants were therefore subjected under this point of view to trial; but as soon as the flowers with short pistils fertilised with pollen from the short stamens were found to yield more seeds than any other of the four possible unions, the abortion theory was knocked on the head. After some additional experiment, it became evident that the two forms, though both were perfect hermaphrodites, bore almost the same relation to one another as do the two sexes of an ordinary animal. With Lythrum we have the still more wonderful case of three forms standing in a similar relation to one another. I afterwards found that the offspring from the union of two plants belonging to the same forms presented a close and curious analogy with hybrids from the union of two distinct species.

In the autumn of 1864 I finished a long paper on *Climbing Plants* and sent it to the Linnaean Society. The writing of this paper cost me four months, but I was so unwell when I received the proof-sheets that I was forced to leave them very badly and often obscurely expressed. The paper was little noticed, but when in 1875 it was corrected and published as a separate book it sold well. I was led to take up this subject by reading a short paper by Asa Gray, published in 1858, on the movements of the tendrils of a Cucurbitacean plant. He sent me seeds, and on raising some plants I was so much fascinated and perplexed by the revolving movements of the tendrils and stems, which movements are really very simple though appearing at first very complex, that I procured various other kinds of climbing plants and studied the whole subject. I was all the more attracted to it from not being at all satisfied with the explanation which Henslow gave us in his lectures about twining plants, namely, that they had a natural tendency to grow up in a spire. This explanation proved quite erroneous. Some of the adaptations displayed by climbing plants are as beautiful as those by orchids for ensuring cross-fertilisation.

My *Variation of Animals and Plants under Domestication* was begun, as already stated, in the beginning of 1860 but was not published until the beginning of 1868. It is a big book, and cost me four years and two months' hard labour. It gives all my observations and an immense number of facts collected from various sources about our domestic productions. In the second volume the causes and laws of variation, inheritance, &c., are discussed, as far as our present state of knowledge permits. Towards the end of the work I give my well-abused hypothesis of Pangenesis. An unverified hypothesis is of little or no value. But if anyone should hereafter be led to make observations by which some such hypothesis could be established, I shall have done good service, as an astonishing number of isolated facts can thus be connected together and rendered intelligible. In 1875 a second and largely corrected edition, which cost me a good deal of labour, was brought out.

My *Descent of Man* was published in Feb. 1871. As soon as I had become, in the year 1837 or 1838, convinced that species were mutable productions, I could not avoid the belief that man must come under the same law. Accordingly I collected notes on the subject for my own satisfaction, and not for a long time with any intention of publishing. Although in the *Origin of Species* the derivation of any particular species is never discussed, yet I thought it best, in order that no honourable man should accuse me of concealing my views, to add that by the work in question "light would be thrown on the origin of man and his history." It would have been useless and injurious to the success of the book to have paraded without giving any evidence my conviction with respect to his origin.

But when I found that many naturalists fully accepted the doctrine of the evolution of species, it seemed to me advisable to work up such notes as I possessed and to publish a special treatise on the origin of man. I was the more glad to do so, as it gave me an opportunity of fully discussing sexual selection —a subject which had always greatly interested me. This subject, and that of the variation of our domestic productions, together with the causes and laws of variation, inheritance, etc., and the intercrossing of plants, are the sole subjects which I have been able to write about in full, so as to use all the materials which I had collected. The *Descent of Man* took me three years to write, but then as usual some of this time was lost by ill health, and some was consumed by preparing new editions and other minor works. A second and largely corrected edition of the *Descent* appeared in 1874.

My book on the *Expression of the Emotions in Men and Animals* was published in the autumn of 1872. I had intended to give only a chapter on the subject in the *Descent of Man,* but as soon as I began to put my notes together I saw that it would require a separate treatise.

My first child was born on December 27, 1839, and I at once commenced to make notes on the first dawn of the various expressions which he exhibited, for I felt convinced, even at this early period, that the most complex and fine shades of expression must all have had a gradual and natural origin. During the summer of the following year, 1840, I read Sir C. Bell's admirable work on expression, and this greatly increased the interest which I felt in the subject, though I could not at all agree with his belief that various muscles had been specially created for the sake of expression. From this time forward I occasionally attended to the subject, both with respect to man and our domesticated animals. My book sold largely, 5267 copies having been disposed of on the day of publication.

In the summer of 1860 I was idling and resting near Hartfield, where two species of Drosera abound; and I noticed that numerous insects had been entrapped by the leaves. I carried home some plants, and on giving them insects saw the movements of the tentacles, and this made me think it probable that the insects were caught for some special purpose. Fortunately a crucial test occurred to me, that of placing a large number of leaves in various nitrogenous and non-nitrogenous fluids of equal density; and as soon as I found that the former alone excited energetic movements, it was obvious that here was a fine new field for investigation.

During subsequent years, whenever I had leisure, I pursued my experiments, and my book on *Insectivorous Plants* was published July 1875—that is, sixteen years after my first observations. The delay in this case, as with all my other books, has been a great advantage to me; for a man after a long interval can criticise his own work almost as well as if it were that of another person. The fact that a plant should secrete, when properly excited, a fluid containing an acid and ferment, closely analogous to the digestive fluid of an animal, was certainly a remarkable discovery.

During this autumn of 1876 I shall publish on the *Effects of Cross- and Self-Fertilisation in the Vegetable Kingdom.* This book will form a complement to that on the *Fertilisation of Orchids,* in which I showed how perfect were the means for cross-fertilisation, and here I shall show how important are the results. I was led to make, during eleven years, the numerous experiments recorded in this volume by a mere accidental observation; and indeed it required the accident to be repeated before my attention was thoroughly aroused to the remarkable fact that seedlings of self-fertilised parentage are inferior, even in the first generation, in height and vigour to seedlings of cross-fertilised parentage. I hope also to republish a revised edition of my book on orchids and hereafter my papers on dimorphic and trimorphic plants, together with some additional observations on allied points which I never have had time to arrange.

My strength will then probably be exhausted, and I shall be ready to exclaim "Nunc dimittis."

The Effects of Cross- and Self-Fertilisation was published in the autumn of 1876; and the results there arrived at explain, as I believe, the endless and wonderful contrivances for the transportal of pollen from one plant to another of the same species. I now believe, however, chiefly from the observations of Hermann Müller, that I ought to have insisted more strongly than I did on the many adaptations for self-fertilisation, though I was well aware of many such adaptations. A much enlarged edition of my *Fertilisation of Orchids* was published in 1877.

In this same year *The Different Forms of Flowers,* etc., appeared, and in 1880 a second edition. This book consists chiefly of the several papers on heterostyled flowers originally published by the Linnaean Society, corrected, with much new matter added, together with observations on some other cases in which the same plant bears two kinds of flowers. As before remarked, no little discovery of mine ever gave me so much pleasure as the making out the meaning of heterostyled flowers. The results of crossing such flowers in an illegitimate manner I believe to be very important as bearing on the sterility of hybrids, although these results have been noticed by only a few persons.

In 1879 I had a translation of Dr. Ernst Krause's *Life of Erasmus Darwin* published, and I added a sketch of his character and habits from materials in my possession. Many persons have been much interested by this little life, and I am surprised that only 800 or 900 copies were sold. Owing to my having accidentally omitted to mention that Dr. Krause had enlarged and corrected his article in German before it was translated, Mr. Samuel Butler abused me with almost insane virulence. How I offended him so bitterly I have never been able to understand. The subject gave rise to some controversy in the Athenaeum newspaper and *Nature*. I laid all the documents before some good judges, viz. Huxley, Leslie Stephen, Litchfield, etc., and they were all unanimous that the attack was so baseless that it did not deserve any public answer, for I had already expressed privately my regret to Mr. Butler for my accidental omission. Huxley consoled me by quoting some

German lines from Goethe, who had been attacked by someone, to the effect "that every Whale has its Louse."

In 1880 I published, with Frank's assistance, our *Power of Movement in Plants*. This was a tough piece of work. The book bears somewhat the same relation to my little book on *Climbing Plants* which *Cross-Fertilisation* did to the *Fertilisation of Orchids;* for in accordance with the principles of evolution it was impossible to account for climbing plants having been developed in so many widely different groups unless all kinds of plants possess some slight power of movement of an analogous kind. This I proved to be the case, and I was further led to a rather wide generalisation, viz., that the great and important classes of movements, excited by light, the attraction of gravity, &c., are all modified forms of the fundamental movement of circumnutation. It has always pleased me to exalt plants in the scale of organised beings; and I therefore felt an especial pleasure in showing how many and what admirably well adapted movements the tip of a root possesses.

I have now (May 1, 1881) sent to the printers the MS. of a little book on *The Formation of Vegetable Mould through the Action of Worms*. This is a subject of but small importance; and I know not whether it will interest any readers,[4] but it has interested me. It is the completion of a short paper read before the Geological Society more than forty years ago, and has revived old geological thoughts.

I have now mentioned all the books which I have published, and these have been the milestones in my life, so that little remains to be said. I am not conscious of any change in my mind during the last thirty years, excepting in one point presently to be mentioned; nor indeed could any change have been expected unless one of general deterioration. But my father lived to his eighty-third year with his mind as lively as ever it was, and all his faculties undimmed; and I hope that I may die before my mind fails to a sensible extent. I think that I have become a little more skilful in guessing right explanations and in devising experimental tests; but this may probably be the result of mere practice and of a larger store of knowledge. I have as much difficulty as ever in expressing myself clearly and concisely,

[4] Between November 1881 and February 1884, 8500 copies were sold.

and this difficulty has caused me a very great loss of time; but it has had the compensating advantage of forcing me to think long and intently about every sentence, and thus I have been often led to see errors in reasoning and in my own observations or those of others.

There seems to be a sort of fatality in my mind leading me to put at first my statement and proposition in a wrong or awkward form. Formerly I used to think about my sentences before writing them down; but for several years I have found that it saves time to scribble in a vile hand whole pages as quickly as I possibly can, contracting half the words, and then correct deliberately. Sentences thus scribbled down are often better ones than I could have written deliberately.

Having said this much about my manner of writing, I will add that with my larger books I spend a good deal of time over the general arrangement of the matter. I first make the rudest outline in two or three pages, and then a larger one in several pages, a few words or one word standing for a whole discussion or series of facts. Each of these headings is again enlarged and often transformed before I begin to write *in extenso*. As in several of my books facts observed by others have been very extensively used, and as I have always had several quite distinct subjects in hand at the same time, I may mention that I keep from thirty to forty large portfolios, in cabinets with labelled shelves, into which I can at once put a detached reference or memorandum. I have bought many books and at their ends I make an index of all the facts that concern my work, or, if the book is not my own, write out a separate abstract, and of such abstracts I have a large drawer full. Before beginning on any subject I look to all the short indices and make a general and classified index, and by taking the one or more proper portfolios I have all the information collected during my life ready for use.

I have said that in one respect my mind has changed during the last twenty or thirty years. Up to the age of thirty, or beyond it, poetry of many kinds, such as the works of Milton, Gray, Byron, Wordsworth, Coleridge, and Shelley, gave me great pleasure, and even as a school-boy I took intense delight in Shakespeare, especially in the historical plays. I have also said that formerly pictures gave me considerable, and music very great, delight. But now for many years I cannot endure to read a line of poetry: I have tried lately to read Shakespeare, and found it so intolerably dull that it nauseated me. I have also almost lost any taste for pictures or music. Music generally sets me thinking too energetically on what I have been at work on, instead of giving me pleasure. I retain some taste for fine scenery, but it does not cause me the exquisite delight which it formerly did. On the other hand, novels which are works of the imagination, though not of a very high order, have been for years a wonderful relief and pleasure to me, and I often bless all novelists. A surprising number have been read aloud to me, and I like all if moderately good, and if they do not end unhappily—against which a law ought to be passed. A novel, according to my taste, does not come into the first class unless it contains some person whom one can thoroughly love, and if it be a pretty woman all the better.

This curious and lamentable loss of the higher aesthetic tastes is all the odder, as books on history, biographies and travels (independently of any scientific facts which they may contain), and essays on all sorts of subjects interest me as much as ever they did. My mind seems to have become a kind of machine for grinding general laws out of large collections of facts, but why this should have caused the atrophy of that part of the brain alone, on which the higher tastes depend, I cannot conceive. A man with a mind more highly organised or better constituted than mine would not I suppose have thus suffered; and if I had to live my life again I would have made a rule to read some poetry and listen to some music at least once every week; for perhaps the parts of my brain now atrophied could thus have been kept active through use. The loss of these tastes is a loss of happiness, and may possibly be injurious to the intellect, and more probably to the moral character, by enfeebling the emotional part of our nature.

My books have sold largely in England, have been translated into many languages, and passed through several editions in foreign countries. I have heard it said that the success of a work abroad is the best test of its enduring value. I doubt whether this is at all trustworthy; but judged by this standard my name ought to last for a few years. Therefore it may be worth while for me to try to analyse the mental qualities and the conditions on which my success has depended, though I am aware that no man can do this correctly.

I have no great quickness of apprehension or wit which is so remarkable in some clever men, for instance Huxley. I am therefore a poor critic: A paper or book, when first read, generally excites my admiration, and it is only after considerable reflection that I perceive the weak points. My power to follow a long and purely abstract train of thought is very limited; I should, moreover, never have succeeded with metaphysics or mathematics. My memory is extensive, yet hazy; it suffices to make me cautious by vaguely telling me that I have observed or read something opposed to the conclusion which I am drawing, or on the other hand in favour of it, and after a time I can generally recollect where to search for my authority. So poor in one sense is my memory that I have never been able to remember for more than a few days a single date or a line of poetry.

Some of my critics have said, "Oh, he is a good observer, but has no power of reasoning." I do not think that this can be true, for the *Origin of Species* is one long argument from the beginning to the end, and it has convinced not a few able men. No one could have written it without having some power of reasoning. I have a fair share of invention and of common sense or judgment, such as every fairly successful lawyer or doctor must have, but not, I believe, in any higher degree.

On the favourable side of the balance, I think that I am superior to the common run of men in noticing things which easily escape attention, and in observing them carefully. My industry has been nearly as great as it could have been in the observation and collection of facts. What is far more important, my love of natural science has been steady and ardent. This pure love has, however, been much aided by the ambition to be esteemed by my fellow naturalists. From my early youth I have had the strongest desire to understand or explain whatever I observed—that is, to group all facts under some general laws. These causes combined have given me the patience to reflect or ponder for any number of years over any unexplained problem. As far as I can judge, I am not apt to follow blindly the lead of other men. I have steadily endeavoured to keep my mind free, so as to give up any hypothesis, however much beloved (and I cannot resist forming one on every subject), as soon as facts are shown to be opposed to it. Indeed I have had no choice but to act in this manner, for, with the exception of the coral reefs, I cannot remember a single first-formed hypothesis which had not after a time to be given up or greatly modified. This has naturally led me to distrust greatly deductive reasoning in the mixed sciences. On the other hand, I am not very sceptical, a frame of mind which I believe to be injurious to the progress of science; a good deal of .scepticism in a scientific man is advisable to avoid much loss of time; for I have met with not a few men, who I feel sure have often thus been deterred from experiment or observations, which would have proved directly or indirectly serviceable.

In illustration, I will give the oddest case which I have known. A gentleman (who, as I afterwards heard, was a good local botanist) wrote to me from the Eastern counties that the seeds or beans of the common field-bean had this year everywhere grown on the wrong side of the pod. I wrote back, asking for further information, as I did not understand what was meant; but I did not receive any answer for a long time. I then saw in two newspapers, one published in Kent and the other in Yorkshire, paragraphs stating that it was a most remarkable fact that "the beans this year had all grown on the wrong side." So I thought that there must be some foundation for so general a statement. Accordingly, I went to my gardener, an old Kentish man, and asked him whether he had heard anything about it; and he answered, "Oh, no, sir, it must be a mistake, for the beans grow on the wrong side only on Leap-year, and this is not Leap-year." I then asked him how they grew on common years and how on Leap-years, but soon found out that he knew absolutely nothing of how they grew at any time; but he stuck to his belief.

After a time I heard from my first informant, who, with many apologies, said that he should not have written to me had he not heard the statement from several intelligent farmers; but that he had since spoken again to every one of them, and not one knew in the least what he had himself meant. So that here a belief—if indeed a statement with no definite idea attached to it can be called a belief—had spread over almost the whole of England without any vestige of evidence. I have known in the course of my life only three intentionally falsified statements, and one of these may have been a hoax (and there have been several scientific hoaxes) which, however, took in an American agricultural journal. It related to the formation in Holland of a new breed of oxen by the crossing of distinct species of Bos (some of which I happen to know are sterile together), and the author had the impudence to state that he had corresponded with me, and that I had been deeply impressed with the importance of his results. The article was sent to me by the editor of an English agricult. journal, asking for my opinion before republishing it.

A second case was an account of several varieties raised by the author from several species of Primula, which had spontaneously yielded a full complement of seed, although the parent plants had been carefully protected from the access of insects. This account was published before I had discovered the meaning of heterostylism, and the whole statement must have been fraudulent, or there was neglect in excluding insects so gross as to be scarcely credible.

The third case was more curious: Mr. Huth published in his book on consanguineous marriage some long extracts from a Belgian author, who stated that he had interbred rabbits in the closest manner for very many generations without the least injurious effects. The account was published in a most respectable journal, that of the Royal Medical Soc. of Belgium; but I could not avoid feeling doubts—I hardly know why, except that there were no accidents of any kind, and my experience in breeding animals made me think this improbable.

So with much hesitation I wrote to Prof. Van Beneden asking him whether the author was a trustworthy man. I soon heard in answer that the Society had been greatly shocked by discovering that the whole account was a fraud. The writer had been publicly challenged in the *Journal* to say where he had resided and kept his large stock of rabbits while carrying on his experiments, which must have consumed several years, and no answer could be extracted from him. I informed poor Mr. Huth that the account which formed the cornerstone of his argument was fraudulent; and he in the most honourable manner immediately had a slip printed to this effect to be inserted in all future copies of his book which might be sold.

My habits are methodical, and this has been of not a little use for my particular line of work. Lastly, I have had ample leisure from not having to earn my own bread. Even ill health, though it has annihilated several years of my life, has saved me from the distractions of society and amusement.

Therefore, my success as a man of science, whatever this may have amounted to, has been determined, as far as I can judge, by complex and diversified mental qualities and conditions. Of these the most important have been the love of science, unbounded patience in long reflecting over any subject, industry in observing and collecting facts, and a fair share of invention as well as of common-sense. With such moderate abilities as I possess, it is truly surprising that thus I should have influenced to a considerable extent the beliefs of scientific men on some important points.

August 3, 1876

This sketch of my life was begun about May 28 at Hopedene, and since then I have written for nearly an hour on most afternoons

*The Formation of Vegetable Mould
through the Action of Worms*

CHAPTER I

Habits of Worms

Earth-worms are distributed throughout the world under the form of a few genera, which externally are closely similar to one another. The British species of Lumbricus have never been carefully monographed; but we may judge of their probable number from those inhabiting neighbouring countries. In Scandinavia there are eight species, according to Eisen; [1] but two of these rarely burrow in the ground, and one inhabits very wet places or even lives under the water. We are here concerned only with the kinds which bring up earth to the surface in the form of castings. Hoffmeister says that the species in Germany are not well known, but gives the same number as Eisen, together with some strongly marked varieties. [2]

Earth-worms abound in England in many different stations. Their castings may be seen in extraordinary numbers on commons and chalk-downs, so as almost to cover the whole surface, where the soil is poor and the grass short and thin. But they are almost or quite as numerous in some of the London parks, where the grass grows well and the soil appears rich. Even on the same field worms are much more frequent in some places than in others, without any visible difference in the nature of the soil. They abound in paved court-yards close to houses; and an instance will be given in which they had burrowed through the floor of a very damp cellar. I have seen worms in black peat in a boggy field; but they are extremely rare, or quite absent in the drier, brown, fibrous peat, which is so much valued by gardeners. On dry, sandy or gravelly tracks, where heath with some gorse, ferns, coarse grass, moss, and lichens alone grow, hardly any worms can be found. But in many parts of England, wherever a path crosses a heath, its

[1] *Bidrag till Skandinaviens Oligochætfauna* (1871).
[2] *Die bis jetzt bekannten Arten aus der Familie der Regenwürmer* (1845).

surface becomes covered with a fine short sward. Whether this change of vegetation is due to the taller plants being killed by the occasional trampling of man and animals, or to the soil being occasionally manured by the droppings from animals, I do not know.[3] On such grassy paths worm-castings may often be seen. On a heath in Surrey, which was carefully examined, there were only a few castings on these paths, where they were much inclined; but on the more level parts, where a bed of fine earth had been washed down from the steeper parts and had accumulated to a thickness of a few inches, worm-castings abounded. These spots seemed to be overstocked with worms, so that they had been compelled to spread to a distance of a few feet from the grassy paths, and here their castings had been thrown up among the heath; but beyond this limit, not a single casting could be found. A layer, though a thin one, of fine earth, which probably long retains some moisture, is in all cases, as I believe, necessary for their existence; and the mere compression of the soil appears to be in some degree favourable to them, for they often abound in old gravel walks, and in foot-paths across fields.

Beneath large trees few castings can be found during certain seasons of the year, and this is apparently due to the moisture having been sucked out of the ground by the innumerable roots of the trees; for such places may be seen covered with castings after the heavy autumnal rains. Although most coppices and woods support many worms, yet in a forest of tall and ancient beech-trees in Knole Park, where the ground beneath was bare of all vegetation, not a single casting could be found over wide spaces, even during the autumn. Nevertheless, castings were abundant on some grass-covered glades and indentations which penetrated this forest. On the mountains of North Wales and on the Alps, worms, as I have been informed, are in most places rare; and this may perhaps be due to the close proximity of the subjacent rocks, into which worms cannot burrow dur-

---

[3] There is even some reason to believe that pressure is actually favourable to the growth of grasses, for Professor Buckman, who made many observations on their growth in the experimental gardens of the Royal Agricultural College, remarks (*Gardeners' Chronicle* [1854], p. 619): "Another circumstance in the cultivation of grasses in the separate form or small patches is the impossibility of rolling or treading them firmly, without which no pasture can continue good."

ing the winter so as to escape being frozen. Dr. McIntosh, however, found worm-castings at a height of fifteen hundred feet on Schiehallion in Scotland. They are numerous on some hills near Turin at from two thousand to three thousand feet above the sea, and at a great altitude on the Nilgiri Mountains in South India and on the Himalaya.

Earth-worms must be considered as terrestrial animals, though they are still in one sense semi-aquatic, like the other members of the great class of annelids to which they belong. M. Perrier found that their exposure to the dry air of a room for only a single night was fatal to them. On the other hand he kept several large worms alive for nearly four months, completely submerged in water.[4] During the summer when the ground is dry, they penetrate to a considerable depth and cease to work, as they do during the winter when the ground is frozen. Worms are nocturnal in their habits, and at night may be seen crawling about in large numbers, but usually with their tails still inserted in their burrows. By the expansion of this part of their bodies, and with the help of the short, slightly reflexed bristles, with which their bodies are armed, they hold so fast that they can seldom be dragged out of the ground without being torn into pieces.[5] During the day they remain in their burrows, except at the pairing season, when those which inhabit adjoining burrows expose the greater part of their bodies for an hour or two in the early morning. Sick individuals, which are generally affected by the parasitic larvae of a fly, must also be accepted, as they wander about during the day and die on the surface. After heavy rain succeeding dry weather, an astonishing number of dead worms may sometimes be seen lying on the ground. Mr. Galton informs me that on one such occasion (March 1881), the dead worms averaged one for every two and a half paces in length on a walk in Hyde Park, four paces in width. He counted no less that forty-five dead worms in one place in a length of sixteen paces. From

[4] I shall have occasion often to refer to M. Perrier's admirable memoir, "Organisation des Lombriciens terrestres" in *Archives de Zoologie expérimentale* (1874), Tom III, p. 372. C. F. Morren (*De Lumbrici terrestris* [1829], p. 14) found that worms endured immersion for fifteen to twenty days in summer, but that in winter they died when thus treated.

[5] Morren, *De Lumbrici terrestris*, &c. (1829), p. 67.

the facts above given, it is not probable that these worms could have been drowned, and if they had been drowned they would have perished in their burrows. I believe that they were already sick, and that their deaths were merely hastened by the ground being flooded.

It has often been said that under ordinary circumstances healthy worms never, or very rarely, completely leave their burrows at night; but this is an error, as White of Selborne long ago knew. In the morning, after there has been heavy rain, the film of mud or of very fine sand over gravel-walks is often plainly marked with their tracks. I have noticed this from August to May, both months included, and it probably occurs during the two remaining months of the year when they are wet. On these occasions, very few dead worms could anywhere be seen. On January 31, 1881, after a long-continued and unusually severe frost with much snow, as soon as a thaw set in, the walks were marked with innumerable tracks. On one occasion, five tracks were counted crossing a space of only an inch square. They could sometimes be traced either to or from the mouths of the burrows in the gravel-walks, for distances between two or three up to fifteen yards. I have never seen two tracks leading to the same burrow; nor is it likely, from what we shall presently see of their sense-organs, that a worm could find its way back to its burrow after having once left it. They apparently leave their burrows on a voyage of discovery, and thus they find new sites to inhabit.

Morren states [6] that worms often lie for hours almost motionless close beneath the mouths of their burrows. I have occasionally noticed the same fact with worms kept in pots in the house; so that by looking down into their burrows, their heads could just be seen. If the ejected earth or rubbish over the burrows be suddenly removed, the end of the worm's body may very often be seen rapidly retreating. This habit of lying near the surface leads to their destruction to an immense extent. Every morning during certain seasons of the year, the thrushes and blackbirds on all the lawns throughout the country draw out of their holes an astonishing number of worms; and this they could not do, unless they lay close to the surface. It is not probable that worms behave in this manner for

[6] *De Lumbrici terrestris,* &c., p. 14.

the sake of breathing fresh air, for we have seen that they can live for a long time under water. I believe that they lie near the surface for the sake of warmth, especially in the morning; and we shall hereafter find that they often coat the mouths of their burrows with leaves, apparently to prevent their bodies from coming into close contact with the cold damp earth. It is said that they completely close their burrows during the winter.

Structure. A few remarks must be made on this subject. The body of a large worm consists of from one hundred to two hundred almost cylindrical rings or segments, each furnished with minute bristles. The muscular system is well developed. Worms can crawl backwards as well as forwards, and by the aid of their affixed tails can retreat with extraordinary rapidity into their burrows. The mouth is situated at the anterior end of the body, and is provided with a little projection (lobe or lip, as it has been variously called) which is used for prehension. Internally, behind the mouth, there is a strong pharynx, shown in the accompanying diagram (fig. 1) * which is pushed forwards when the animal eats, and this part corresponds, according to Perrier, with the protrudable trunk or proboscis of other annelids. The pharynx leads into the oesophagus, on each side of which in the lower part there are three pairs of large glands, which secrete a surprising amount of carbonate of lime. These calciferous glands are highly remarkable, for nothing like them is known in any other animal. Their use will be discussed when we treat of the digestive process. In most of the species, the oesophagus is enlarged into a crop in front of the gizzard. This latter organ is lined with a smooth thick chitinous membrane, and is surrounded by weak longitudinal, but by powerful transverse muscles. Perrier saw these muscles in energetic action; and, as he remarks, the trituration of the food must be chiefly effected by this organ, for worms possess no jaws or teeth of any kind. Grains of sand and small stones, from the one twentieth to a little more than the one tenth inch in diameter, may generally be found in their gizzards and intestines. As it is certain that worms swallow many little stones, independently of those swallowed while excavating their burrows, it is probable that they serve, like mill-stones,

* In the original edition.—S.E.H.

to triturate their food. The gizzard opens into the intestine, which runs in a straight course to the vent at the posterior end of the body. The intestine presents a remarkable structure, the typhosolis, or, as the old anatomists called it, an intestine within an intestine; and Claparède [7] has shown that this consists of a deep longitudinal involution of the walls of the intestine, by which means an extensive absorbent surface is gained.

The circulatory system is well developed. Worms breathe by their skin, as they do not possess any special respiratory organs. The two sexes are united in the same individual, but two individuals pair together. The nervous system is fairly well developed; and the two almost confluent cerebral ganglia are situated very near to the anterior end of the body.

Senses. Worms are destitute of eyes, and at first I thought that they were quite insensible to light; for those kept in confinement were repeatedly observed by the aid of a candle, and others out of doors by the aid of a lantern, yet they were rarely alarmed, although extremely timid animals. Other persons have found no difficulty in observing worms at night by the same means.[8]

Hoffmeister, however, states [9] that worms, with the exception of a few individuals, are extremely sensitive to light; but he admits that in most cases a certain time is requisite for its action. These statements led me to watch on many successive nights worms kept in pots, which were protected from currents of air by means of glass plates. The pots were approached very gently, in order that no vibration of the floor should be caused. When under these circumstances worms were illuminated by a bull's-eye lantern having slides of dark red and blue glass, which intercepted so much light that they could be seen only with some difficulty, they were not at all affected by this amount of light, however long they were exposed to it. The light, as far as I could judge, was brighter than that from the full moon. Its colour apparently made no difference in the result. When they were illuminated by a candle, or even

---

[7] *Histolog. Untersuchungen über die Regenwürmer. "Zeitschrift für wissenschaft. Zoologie"* (1869), B. XIX, p. 611.

[8] For instance, Mr. Bridgman and Mr. Newman (*The Zoologist* [1849], Vol. VII, p. 2576), and some friends who observed worms for me.

[9] *Familie der Regenwürmer* (1845), p. 18.

by a bright paraffin lamp, they were not usually affected at first. Nor were they when the light was alternately admitted and shut off. Sometimes, however, they behaved very differently, for as soon as the light fell on them, they withdrew into their burrows with almost instantaneous rapidity. This occurred perhaps once out of a dozen times. When they did not withdraw instantly, they often raised the anterior tapering ends of their bodies from the ground, as if their attention was aroused or as if surprise was felt; or they moved their bodies from side to side as if feeling for some object. They appeared distressed by the light; but I doubt whether this was really the case, for on two occasions after withdrawing slowly, they remained for a long time with their anterior extremities protruding a little from the mouths of their burrows, in which position they were ready for instant and complete withdrawal.

When the light from a candle was concentrated by means of a large lens on the anterior extremity, they generally withdrew instantly; but this concentrated light failed to act perhaps once out of half a dozen trials. The light was on one occasion concentrated on a worm lying beneath water in a saucer, and it instantly withdrew into its burrow. In all cases the duration of the light, unless extremely feeble, made a great difference in the result; for worms left exposed before a paraffin lamp or a candle invariably retreated into their burrows within from five to fifteen minutes; and if in the evening the pots were illuminated before the worms had come out of their burrows, they failed to appear.

From the foregoing facts it is evident that light affects worms by its intensity and by its duration. It is only the anterior extremity of the body, where the cerebral ganglia lie, which is affected by light, as Hoffmeister asserts, and as I observed on many occasions. If this part is shaded, other parts of the body may be fully illuminated, and no effect will be produced. As these animals have no eyes, we must suppose that the light passes through their skins, and in some manner excites their cerebral ganglia. It appeared at first probable that the different manner in which they were affected on different occasions might be explained, either by the degree of extension of their skin and its consequent transparency, or by some particular incidence of the light; but I could discover no such relation. One thing was manifest, namely, that when worms were employed in dragging leaves into their burrows or in eating them, and even during the short intervals whilst they rested from their work, they either did not perceive the light or were regardless of it; and this occurred even when the light was concentrated on them through a large lens. So, again, whilst they are paired, they will remain for an hour or two out of their burrows, fully exposed to the morning light; but it appears from what Hoff-meister says that a light will occasionally cause paired individuals to separate.

When a worm is suddenly illuminated and dashes like a rabbit into its burrow—to use the expression employed by a friend—we are at first led to look at the action as a reflex one. The irritation of the cerebral ganglia appears to cause certain muscles to contract in an inevitable manner, independently of the will or consciousness of the animal, as if it were an automaton. But the different effect which a light produced on different occasions, and especially the fact that a worm when in any way employed and in the intervals of such employment, whatever set of muscles and ganglia may then have been brought into play, is often regardless of light, are opposed to the view of the sudden withdrawal being a simple reflex action. With the higher animals, when close attention to some object leads to the disregard of the impressions which other objects must be producing on them, we attribute this to their attention being then absorbed; and attention implies the presence of a mind. Every sportsman knows that he can approach animals whilst they are grazing, fighting, or courting much more easily than at other times. The state, also, of the nervous system of the higher animals differs much at different times, for instance, a horse is much more readily startled at one time than at another. The comparison here implied between the actions of one of the higher animals and of one so low in the scale as an earthworm may appear far-fetched; for we thus attribute to the worm attention and some mental power, nevertheless I can see no reason to doubt the justice of the comparison.

Although worms cannot be said to possess the power of vision, their sensitiveness to light enables them to distinguish between day and night; and they thus escape extreme danger from the many diurnal animals which prey on them. Their withdrawal into their burrows during the day appears, however, to have become an habitual action; for worms kept in pots covered by glass-plates, over which sheets of black paper were spread, and placed before a north-east window, remained during the day-time in their burrows and came out every night; and they continued thus to act for a week. No doubt a little light may have entered between the sheets of glass and the blackened paper; but we know from the trials with coloured glass, that worms are indifferent to a small amount of light.

Worms appear to be less sensitive to moderate radiant heat than to a bright light. I judge of this from having held at different times a poker heated to dull redness near some worms, at a distance which caused a very sensible degree of warmth in my hand. One of them took no notice; a second withdrew into its burrow, but not quickly; the third and fourth much more quickly, and the fifth as quickly as possible. The light from a candle, concentrated by a lens and passing through a sheet of glass which would intercept most of the heat-rays, generally caused a much more rapid retreat than did the heated poker. Worms are sensitive to a low temperature, as may be inferred from their not coming out of their burrows during a frost.

Worms do not possess any sense of hearing. They took not the least notice of the shrill notes from a metal whistle, which was repeatedly sounded near them; nor did they of the deepest and loudest tones of a bassoon. They were indifferent to shouts, if care was taken that the breath did not strike them. When placed on a table close to the keys of a piano, which was played as loudly as possible, they remained perfectly quiet.

Although they are indifferent to undulations in the air audible by us, they are extremely sensitive to vibrations in any solid object. When the pots containing two worms which had remained quite indifferent to the sound of the piano were placed on this instrument, and the note C in the bass clef was struck, both instantly retreated into their burrows. After a time they emerged, and when G above the line in the treble clef was struck they again retreated. Under similar circumstances on another night one worm dashed into its burrow on a very high note being struck only once, and the other worm when C in the treble clef was struck. On these occasions the worms were not touching the sides of the pots, which stood in saucers; so that the vibrations, before reaching their bodies, had to pass from the sounding board of the piano, through the saucer, the bottom of the pot, and the damp, not very compact earth on which they lay with their tails in their burrows. They often showed their sensitiveness when the pot in which they lived, or the table on which the pot stood, was accidentally and lightly struck; but they appeared less sensitive to such jars than to the vibrations of the piano; and their sensitiveness to jars varied much at different times. It has often been said that if the ground is beaten or otherwise made to tremble worms believe that they are pursued by a mole and leave their burrows. I beat the ground in many places where worms abounded, but not one emerged. When, however, the ground is dug with a fork and is violently disturbed beneath a worm, it will often crawl quickly out of its burrow.

The whole body of a worm is sensitive to contact. A slight puff of air from the mouth causes an instant retreat. The glass plates placed over the pots did not fit closely, and blowing through the very narrow chinks thus left often sufficed to cause a rapid retreat. They sometimes perceived the eddies in the air caused by quickly removing the glass plates. When a worm first comes out of its burrow, it generally moves the much extended anterior extremity of its body from side to side in all directions, apparently as an organ of touch; and there is some reason to believe, as we shall see in the next chapter, that they are thus enabled to gain a general notion of the form of an object. Of all their senses that of touch, including in this term the perception of a vibration, seems much the most highly developed.

In worms the sense of smell apparently is confined to the perception of certain odours, and is feeble. They were quite indifferent to my breath, as long as I breathed on them very gently. This was tried, because it appeared possible that they might thus be warned of the approach of an enemy. They exhibited the same indifference to my breath whilst I chewed some tobacco, and while a pellet of cotton-wool with a few drops of mille-fleurs perfume or of acetic acid was kept in my mouth. Pellets of cotton-wool soaked in tobacco juice, and in mille-fleurs perfume, and in paraffin, were held with pincers and were waved about within two or three inches of several worms, but they took no notice. On one or two occasions, however, when acetic acid had been placed on the pellets, the worms appeared a little uneasy, and this was probably due to the irritation of their skins. The perception of such unnatural odours would be of no service to worms; and as such timid creatures would almost certainly exhibit some signs of any new impression, we may conclude that they did not perceive these odours.

The result was different when cabbage-leaves and pieces of onion were employed, both of which are devoured with much relish by worms. Small square pieces of fresh and half-decayed cabbage-leaves and of onion bulbs were on nine occasions buried in my pots, beneath about a quarter of an inch of common garden soil; and they were always discovered by the worms. One bit of cabbage was discovered and removed in the course of two hours; three were removed by the next morning, that is, after a single night; two others after two nights; and the seventh bit after three nights. Two pieces of onion were discovered and removed after three nights. Bits of fresh raw meat, of which worms are very fond, were buried, and were not discovered within forty-eight hours, during which time they had not become putrid. The earth above the various buried objects was generally pressed down only slightly, so as not to prevent the emission of any odour. On two occasions, however, the surface was well watered, and was thus rendered somewhat compact. After the bits of cabbage and onion had been removed, I looked beneath them to see whether the worms had accidentally come up from below, but there was no sign of a burrow; and twice the buried objects were laid on pieces of tin-foil which were not in the least displaced. It is of course possible that the worms whilst moving about on the surface of the ground, with their tails affixed within their burrows, may have poked their heads into the places where the above objects were buried; but I have never seen worms acting in this manner. Some pieces of cabbage-leaf and of onion were twice buried beneath very fine ferruginous sand, which was slightly pressed down and well watered, so as to be rendered very compact, and these pieces were never discovered. On a third occasion the same kind of sand was neither pressed down nor watered, and the pieces of cabbage were discovered and removed after the second night. These several facts indicate that worms possess some power of smell; and that they discover by this means odoriferous and much-coveted kinds of food.

It may be presumed that all animals which feed on various substances possess the sense of taste, and this is certainly the case with worms. Cabbage-leaves are much liked by worms; and it appears that they can distinguish between different varieties; but this may perhaps be owing to differences in their texture. On eleven occasions pieces of the fresh leaves of a common green variety and of the red variety used for pickling were given them, and they preferred the green, the red being either wholly neglected or much less gnawed. On two other occasions, however, they seemed to prefer the red. Half-decayed leaves of the red variety and fresh leaves of the green were attacked about equally. When leaves of the cabbage, horse-radish (a favourite food) and of the onion were given together, the latter were always and manifestly preferred. Leaves of the cabbage, lime-tree, Ampelopis, parsnip (Pastinaca), and celery (Apium) were likewise given together; and those of the celery were first eaten. But when leaves of cabbage, turnip, beet, celery, wild cherry, and carrots were given together, the two latter kinds, especially those of the carrot, were preferred to all the others, including those of celery. It was also manifest after many trials that wild cherry leaves were greatly preferred to those of the lime-tree and hazel (Corylus). According to Mr. Bridgman the half-decayed leaves of *Phlox verna* are particularly liked by worms.[10]

Pieces of the leaves of cabbage, turnip, horse-radish, and onion were left on the pots during twenty-two days, and were all attacked and had to be renewed; but during the whole of this time leaves of an Artemisia and of the culinary sage, thyme, and mint, mingled with the above leaves, were quite neglected excepting those of the mint, which were occasionally and very slightly nibbled. These latter four kinds of leaves do not differ in texture in a manner which could make them disagreeable to worms; they all have a strong taste, but so have the four first mentioned kinds of leaves; and the wide difference in the result must be attributed to a preference by the worms for one taste over another.

[10] *The Zoologist* (1849), Vol. VII, p. 2576.

Mental Qualities. There is little to be said on this head. We have seen that worms are timid. It may be doubted whether they suffer as much pain when injured, as they seem to express by their contortions. Judging by their eagerness for certain kinds of food, they must enjoy the pleasure of eating. Their sexual passion is strong enough to overcome for a time their dread of light. They perhaps have a trace of social feeling, for they are not disturbed by crawling over each other's bodies, and they sometimes lie in contact. According to Hoffmeister they pass the winter either singly or rolled up with others into a ball at the bottom of their burrows.[11] Although worms are so remarkably deficient in the several sense-organs, this does not necessarily preclude intelligence, as we know from such cases as those of Laura Bridgman; and we have seen that when their attention is engaged, they neglect impressions to which they would otherwise have attended; and attention indicates the presence of a mind of some kind. They are also much more easily excited at certain times than at others. They perform a few actions instinctively, that is, all the individuals, including the young, perform such actions in nearly the same fashion. This is shown by the manner in which the species of Perichaeta eject their castings, so as to construct towers; also by the manner in which the burrows of the common earth-worm are smoothly lined with fine earth and often with little stones, and the mouths of their burrows with leaves. One of their strongest instincts is the plugging up the mouths of their burrows with various objects; and very young worms act in this manner. But some degree of intelligence appears, as we shall see in the next chapter, to be exhibited in this work—a result which has surprised me more than anything else in regard to worms.

Food and Digestion. Worms are omnivorous. They swallow an enormous quantity of earth, out of which they extract any digestible matter which it may contain; but to this subject I must recur. They also consume a large number of half-decayed leaves of all kinds, excepting a few which have an unpleasant taste or are too tough for them; likewise petioles, peduncles and decayed flowers. But they will also consume fresh leaves, as I have found by repeated trials. According to Morren [12]

[11] *Familie der Regenwürmer*, p. 13.
[12] *De Lumbrici terrestris*, p. 19.

they will eat particles of sugar and liquorice; and the worms which I kept drew many bits of dry starch into their burrows, and a large bit had its angles well rounded by the fluid poured out of their mouths. But as they often drag particles of soft stone, such as of chalk, into their burrows, I feel some doubt whether the starch was used as food. Pieces of raw and roasted meat were fixed several times by long pins to the surface of the soil in my pots, and night after night the worms could be seen tugging at them, with the edges of the pieces engulfed in their mouths, so that much was consumed. Raw fat seems to be preferred even to raw meat or to any other substance which was given them, and much was consumed. They are cannibals, for the two halves of a dead worm placed in two of the pots were dragged into the burrows and gnawed; but as far as I could judge, they prefer fresh to putrid meat, and in so far I differ from Hoffmeister.

Léon Frédéricq states [13] that the digestive fluid of worms is of the same nature as the pancreatic secretion of the higher animals; and this conclusion agrees perfectly with the kinds of food which worms consume. Pancreatic juice emulsifies fat, and we have just seen how greedily worms devour fat; it dissolves fibrin, and worms eat raw meat; it converts starch into grape-sugar with wonderful rapidity, and we shall presently show that the digestive fluid of worms acts on starch.[14] But they live chiefly on half-decayed leaves; and these would be useless to them unless they could digest the cellulose forming the cell-walls; for it is well known that all other nutritious substances are almost completely withdrawn from leaves, shortly before they fall off. It has, however, now been ascertained that cellulose, though very little or not at all attacked by the gastric secretion of the higher animals, is acted on by that from the pancreas.[15]

The half-decayed or fresh leaves which worms intend to devour, are dragged into the mouths of their burrows to a depth of from one to three inches, and are then moistened with a secreted fluid. It has been assumed that this fluid serves to

[13] *Archives de Zoologie expérimentale* (1878), Tom. VII, p. 394.

[14] On the action of the pancreatic ferment, see *A Text-Book of Physiology* (1878) by Michael Foster, 2nd edit., pp. 198–203.

[15] Schmulewitsch, "Action des Sucs digestifs sur la Cellulose," *Bull. de l'Acad. Imp. de St. Pétersbourg* (1879), Tom. XXV, p. 549.

hasten their decay; but a large number of leaves were twice
pulled out of the burrows of worms and kept for many weeks
in a very moist atmosphere under a bell-glass in my study; and
the parts which had been moistened by the worms did not decay more quickly in any plain manner than the other parts.
When fresh leaves were given in the evening to worms kept in
confinement and examined early on the next morning, therefore not many hours after they had been dragged into the burrows, the fluid with which they were moistened, when tested
with neutral litmus paper, showed an alkaline reaction. This
was repeatedly found to be the case with celery, cabbage, and
turnip leaves. Parts of the same leaves which had not been
moistened by the worms were pounded with a few drops of
distilled water, and the juice thus extracted was not alkaline.
Some leaves, however, which had been drawn into burrows out
of doors, at an unknown antecedent period, were tried, and
though still moist, they rarely exhibited even a trace of alkaline
reaction.

The fluid with which the leaves are bathed acts on them
whilst they are fresh or nearly fresh, and in a remarkable manner; for it quickly kills and discolours them. Thus the ends of
a fresh carrot-leaf, which had been dragged into a burrow, were
found after twelve hours of a dark brown tint. Leaves of celery,
turnip, maple, elm, lime, thin leaves of ivy, and occasionally
those of the cabbage were similarly acted on. The end of a leaf
of *Triticum repens,* still attached to a growing plant, had been
drawn into a burrow, and this part was dark brown and dead,
whilst the rest of the leaf was fresh and green. Several leaves of
lime and elm removed from burrows out of doors were found
affected in different degrees. The first change appears to be that
the veins become of a dull reddish-orange. The cells with
chlorophyll next lose more or less completely their green colour,
and their contents finally become brown. The parts thus affected often appeared almost black by reflected light; but when
viewed as a transparent object under the microscope, minute
specks of light were transmitted, and this was not the case with
the unaffected parts of the same leaves. These effects, however,
merely show that the secreted fluid is highly injurious or poisonous to leaves; for nearly the same effects were produced in
from one to two days on various kinds of young leaves, not

only by artificial pancreatic fluid, prepared with or without thymol, but quickly by a solution of thymol by itself. On one occasion leaves of Corylus were much discoloured by being kept for eighteen hours in pancreatic fluid, without any thymol. With young and tender leaves immersion in human saliva during rather warm weather acted in the same manner as the pancreatic fluid, but not so quickly. The leaves in all these cases often became infiltrated with the fluid.

Large leaves from an ivy plant growing on a wall were so tough that they could not be gnawed by worms, but after four days they were affected in a peculiar manner by the secretion poured out of their mouths. The upper surfaces of the leaves, over which the worms had crawled, as was shown by the dirt left on them, were marked in sinuous lines, by either a continuous or broken chain of whitish and often star-shaped dots, about 2 mm. in diameter. The appearance thus presented was curiously like that of a leaf, into which the larva of some minute insect had burrowed. But my son Francis, after making and examining sections, could nowhere find that the cell-walls had been broken down or that the epidermis had been penetrated. When the section passed through the whitish dots, the grains of chlorophyll were seen to be more or less discoloured, and some of the palisade and mesophyll cells contained nothing but broken down granular matter. These effects must be attributed to the transudation of the secretion through the epidermis into the cells.

The secretion with which worms moisten leaves likewise acts on the starch granules within the cells. My son examined some leaves of the ash and many of the lime, which had fallen off the trees and had been partly dragged into worm-burrows. It is known that with fallen leaves the starch-grains are preserved in the guard-cells of the stomata. Now in several cases the starch had partially or wholly disappeared from these cells, in the parts which had been moistened by the secretion; while they were still well preserved in the other parts of the same leaves. Sometimes the starch was dissolved out of only one of the two guard-cells. The nucleus in one case had disappeared, together with the starch-granules. The mere burying of lime-leaves in damp earth for nine days did not cause the destruction of the starch-granules. On the other hand, the immersion of fresh lime and cherry leaves for eighteen hours in artificial pancreatic fluid led to the dissolution of the starch-granules in the guard-cells as well as in the other cells.

From the secretion with which the leaves are moistened being alkaline, and from its acting both on the starch-granules and on the protoplasmic contents of the cells, we may infer that it resembles in nature not saliva,[16] but pancreatic secretion; and we know from Frédéricq that a secretion of this kind is found in the intestines of worms. As the leaves which are dragged into the burrows are often dry and shrivelled, it is indispensable for their disintegration by the unarmed mouths of worms that they should first be moistened and softened; and fresh leaves, however soft and tender they may be, are similarly treated, probably from habit. The result is that they are partially digested before they are taken into the alimentary canal. I am not aware of any other case of extra-stomachal digestion having been recorded. The boa-constrictor bathes its prey with saliva, but this is solely for lubricating it. Perhaps the nearest analogy may be found in such plants as Drosera and Dionaea; for here animal matter is digested and converted into peptone not within a stomach, but on the surfaces of the leaves.

Calciferous Glands. These glands (see fig. 1), judging from their size and from their rich supply of blood-vessels, must be of much importance to the animal. But almost as many theories have been advanced on their use as there have been observers. They consist of three pairs, which in the common earth-worm debouch into the alimentary canal in advance of the gizzard, but posteriorly to it in Urochtaea and some other genera.[17] The two posterior pairs are formed by lamellae, which according to Claparède, are diverticula from the oesophagus.[18] These lamellae are coated with a pulpy cellular layer, with the outer cells lying free in infinite numbers. If one of these glands is punctured and squeezed, a quantity of white pulpy matter exudes, consisting of these free cells. They are minute, and vary in diameter from two to six μ. They contain in their centres a little excessively fine granular matter; but they look so like oil

---

[16] Claparède doubts whether saliva is secreted by worms; see *Zeitschrift für wissenschaft. Zoologie* (1869), B. xix, p. 601.

[17] Perrier, *Archives de Zoolog. expér.*, July, 1874, pp. 416, 419.

[18] *Zeitschrift für wissenschaft. Zoologie* (1869), B. xix, pp. 603–606.

globules that Claparède and others at first treated them with
ether. This produces no effect; but they are quickly dissolved
with effervescence in acetic acid, and when oxalate of ammonia
is added to the solution a white precipitate is thrown down. We
may therefore conclude that they contain carbonate of lime. If
the cells are immersed in a very little acid, they become more
transparent, look like ghosts, and are soon lost to view; but if
much acid is added, they disappear instantly. After a very large
number have been dissolved, a flocculent residue is left, which
apparently consists of the delicate ruptured cell-walls. In the
two posterior pairs of glands the carbonate of lime contained in
the cells occasionally aggregates into small rhombic crystals or
into concretions, which lie between the lamellae; but I have
seen only one, and Claparède only a very few such cases.

The two anterior glands differ a little in shape from the four
posterior ones by being more oval. They differ also conspicu-
ously in generally containing several small, or two or three
larger, or a single very large concretion of carbonate of lime,
as much as 1.5 mm. in diameter. When a gland includes only
a few very small concretions, or, as sometimes happens, none
at all, it is easily overlooked. The large concretions are round
or oval, and exteriorly almost smooth. One was found which
filled up not only the whole gland, as is often the case, but its
neck; so that it resembled an olive-oil flask in shape. These concretions when broken are seen to be more or less crystalline in
structure. How they escape from the gland is a marvel; but that
they do escape is certain, for they are often found in the gizzard, intestines, and in the castings of worms, both with those
kept in confinement and those in a state of nature.

Claparède says very little about the structure of the two
anterior glands, and he supposes that the calcareous matter of
which the concretions are formed is derived from the four
posterior glands. But if an anterior gland which contains only
small concretions is placed in acetic acid and afterwards dissected, or if sections are made of such a gland without being
treated with acid, lamellae like those in the posterior glands and
coated with cellular matter could be plainly seen, together with
a multitude of free calciferous cells readily soluble in acetic
acid. When a gland is completely filled with a single large concretion, there are no free cells, as these have been all consumed

in forming the concretion. But if such a concretion, or one of only moderately large size, is dissolved in acid much membranous matter is left, which appears to consist of the remains of the formerly active lamellae. After the formation and expulsion of a large concretion, new lamellae must be developed in some manner. In one section made by my son, the process had apparently commenced, although the gland contained two rather large concretions, for near the walls several cylindrical and oval pipes were intersected, which were lined with cellular matter and were quite filled with free calciferous cells. A great enlargement in one direction of several oval pipes would give rise to the lamellae.

Besides the free calciferous cells in which no nucleus was visible, other and rather larger free cells were seen on three occasions; and these contained a distinct nucleus and nucleolus. They were only so far acted on by acetic acid that the nucleus was thus rendered more distinct. A very small concretion was removed from between two of the lamellae within an anterior gland. It was embedded in pulpy cellular matter, with many free calciferous cells, together with a multitude of the larger, free, nucleated cells, and these latter cells were not acted on by acetic acid, while the former were dissolved. From this and other such cases I am led to suspect that the calciferous cells are developed from the larger nucleated ones; but how this is effected was not ascertained.

When an anterior gland contains several minute concretions, some of these are generally angular or crystalline in outline, while the greater number are rounded with an irregular mulberry-like surface. Calciferous cells adhered to many parts of these mulberry-like masses, and their gradual disappearance could be traced while they still remained attached. It was thus evident that the concretions are formed from the lime contained within the free calciferous cells. As the smaller concretions increase in size, they come into contact and unite, thus enclosing the now functionless lamellae; and by such steps the formation of the largest concretions could be followed. Why the process regularly takes place in the two anterior glands, and only rarely in the four posterior glands is quite unknown. Morren says that these glands disappear during the winter; and I have seen some instances of this fact, and others in which either the anterior or posterior glands were at this season so shrunk and empty that they could be distinguished only with much difficulty.

With respect to the function of the calciferous glands, it is probable that they primarily serve as organs of excretion, and secondarily as an aid to digestion. Worms consume many fallen leaves; and it is known that lime goes on accumulating in leaves until they drop off the parent-plant, instead of being reabsorbed into the stem or roots, like various other organic and inorganic substances.[19] The ashes of a leaf of an acacia have been known to contain as much as seventy-two per cent of lime. Worms therefore would be liable to become charged with this earth, unless there were some special means for its excretion; and the calciferous glands are well adapted for this purpose. The worms which live in mould close over the chalk often have their intestines filled with this substance, and their castings are almost white. Here it is evident that the supply of calcareous matter must be superabundant. Nevertheless with several worms collected on such a site, the calciferous glands contained as many free calciferous cells, and fully as many and large concretions, as did the glands of worms which lived where there was little or no lime; and this indicates that the lime is an excretion, and not a secretion poured into the alimentary canal for some special purpose.

On the other hand, the following considerations render it highly probable that the carbonate of lime, which is excreted by the glands, aids the digestive process under ordinary circumstances. Leaves during their decay generate an abundance of various kinds of acids, which have been grouped together under the term of humus acids. We shall have to recur to this subject in our fifth chapter, and I need here only say that these acids act strongly on carbonate of lime. The half-decayed leaves which are swallowed in such large quantities by worms would, therefore, after they have been moistened and triturated in the alimentary canal, be apt to produce such acids. And in the case of several worms, the contents of the alimentary canal were found to be plainly acid, as shown by litmus paper. This acidity cannot be attributed to the nature of the digestive fluid, for

[19] De Vries, *Landwirth. Jahrbücher* (1881), p. 77.

pancreatic fluid is alkaline; and we have seen that the secretion which is poured out of the mouths of worms for the sake of preparing the leaves for consumption, is likewise alkaline. The acidity can hardly be due to uric acid, as the contents of the upper part of the intestine were often acid. In one case the contents of the gizzard were slightly acid, those of the upper intestines being more plainly acid. In another case the contents of the pharynx were not acid, those of the gizzard doubtfully so, while those of the intestine were distinctly acid at a distance of 5 cm. below the gizzard. Even with the higher herbivorous and omnivorous animals, the contents of the large intestine are acid. "This, however, is not caused by any acid secretion from the mucous membrane; the reaction of the intestinal walls in the larger as in the small intestine is alkaline. It must therefore arise from acid fermentations going on in the contents themselves. . . . In Carnivora the contents of the coecum are said to be alkaline, and naturally the amount of fermentation will depend largely on the nature of the food." [20]

With worms not only the contents of the intestines, but their ejected matter or the castings, are generally acid. Thirty castings from different places were tested, and with three or four exceptions were found to be acid; and the exceptions may have been due to such castings not having been recently ejected; for some which were at first acid were on the following morning, after being dried and again moistened, no longer acid; and this probably resulted from the humus acids being, as is known to be the case, easily decomposed. Five fresh castings from worms which lived in mould close over the chalk were of a whitish colour and abounded with calcareous matter; and these were not in the least acid. This shows how effectually carbonate of lime neutralises the intestinal acids. When worms were kept in pots filled with fine ferruginous sand it was manifest that the oxide of iron, with which the grains of silex were coated, had been dissolved and removed from them in the castings.

The digestive fluid of worms resembles in its action, as already stated, the pancreatic secretion of the higher animals; and in these latter, "pancreatic digestion is essentially alkaline; the action will not take place unless some alkali be present; and

[20] M. Foster, *A Text-Book of Physiology* (1878), 2nd edit., p. 243.

the activity of an alkaline juice is arrested by acidification, and hindered by neutralisation." [21] Therefore it seems highly probable that the innumerable calciferous cells, which are poured from the four posterior glands into the alimentary canal of worms, serve to neutralise more or less completely the acids there generated by the half-decayed leaves. We have seen that these cells are instantly dissolved by a small quantity of acetic acid, and as they do not always suffice to neutralise the contents of even the upper part of the alimentary canal, the lime is perhaps aggregated into concretions in the anterior pair of glands, in order that some may be carried down to the posterior parts of the intestine, where these concretions would be rolled about amongst the acid contents. The concretions found in the intestines and in the castings often have a worn appearance, but whether this is due to some amount of attrition or of chemical corrosion could not be told. Claparède believes that they are formed for the sake of acting as mill-stones, and of thus aiding in the trituration of the food. They may give some aid in this way; but I fully agree with Perrier that this must be of quite subordinate importance, seeing that the object is already attained by stones being generally present in the gizzards and intestines of worms.

[21] M. Foster, ibid., p. 200.

Two Letters

To Joseph Hooker

To J. D. Hooker, at Dr. Falconer's, Botanic Garden, Calcutta

Down, May 10th, 1848

I was indeed delighted to see your handwriting; but I felt almost sorry when I beheld how long a letter you had written. I know that you are indomitable in work, but remember how precious your time is, and do not waste it on your friends, however much pleasure you may give them. Such a letter would have cost me half a day's work. How capitally you seem going on! I do envy you the sight of all the glorious vegetation. I am much pleased and surprised that you have been able to observe so much in the animal world. No doubt you keep a journal, and an excellent one it will be, I am sure, when published. All these animal facts will tell capitally in it. I can quite comprehend the difficulty you mention about not knowing what is known zoologically in India; but facts observed, as you will observe them, are none the worse for reiterating. Did you see Mr. Blyth in Calcutta? He would be a capital man to tell you what is known about Indian zoology, at least in the Vertebrata. He is a very clever, odd, wild fellow, who will never do what he could do, from not sticking to any one subject. By the way, if you should see him at any time, try not to forget to remember me very kindly to him; I liked all I saw of him. Your letter was the very one to charm me, with all its facts for my species-book, and truly obliged I am for so kind a remembrance of me. Do not forget to make enquiries about the origin, even if only traditionally known, of any varieties of domestic quadrupeds, birds, silkworms, etc. Are there domestic bees? If so hives ought to be brought home. Of all the facts you mention, that of the wild [illegible], when breeding with the domestic, producing offspring somewhat sterile, is the most surprising; surely they must be different species. Most zoologists would absolutely disbelieve such a statement, and consider the result as a proof that they were distinct species. I do not go so far as that,

431

but the case seems highly improbable. Blyth has studied the Indian Ruminantia. I have been much struck about what you say of lowland plants ascending mountains, but the alpine not descending. How I do hope you will get up some mountains in Borneo; how curious the result will be! By the way, I never heard from you what affinity the Maldive flora has, which is cruel, as you tempted me by making me guess. I sometimes groan over your Indian journey, when I think over all your locked up riches. When shall I see a memoir on Insular floras, and on the Pacific? What a grand subject alpine floras of the world would be, as far as known; and then you have never given a *coup d'oeil* on the similarity and dissimilarity of arctic and antarctic floras. Well, thank heavens, when you do come back you will be *nolens volens* a fixture. I am particularly glad you have been at the Coal; I have often since you went gone on maundering on the subject, and I shall never rest easy in Down churchyard without the problem be solved by some one before I die. Talking of dying makes me tell you that my confounded stomach is much the same; indeed, of late has been rather worse, but for the last year, I think, I have been able to do more work. I have done nothing besides the barnacles, except, indeed, a little theoretical paper on erratic boulders, and Scientific Geological Instructions for the Admiralty Volume, which cost me some trouble. This work, which is edited by Sir J. Herschel, is a very good job, inasmuch as the captains of men-of-war will now see that the Admiralty cares for science, and so will favour naturalists on board. As for a man who is not scientific by nature, I do not believe instructions will do him any good; and if he be scientific and good for anything the instructions will be superfluous. I do not know who does the botany; Owen does the zoology, and I have sent him an account of my new simple microscope, which I consider perfect, even better than yours by Chevalier. N.B. I have got a $\frac{1}{8}''$ object-glass, and it is grand. I have been getting on well with my beloved Cirripedia, and get more skilful in dissection. I have worked out the nervous system pretty well in several genera, and made out their ears and nostrils, which were quite unknown. I have lately got a bisexual cirripede, the male being microscopically small and parasitic within the sack of the female. I tell you this to boast of my species theory, for the nearest closely allied genus to it is, as usual, hermaphrodite, but I had observed some minute parasites adhering to it, and these parasites I now can show are supplemental males, the male organs in the hermaphrodite being unusually small, though perfect and containing zoosperms; so we have almost a polygamous animal, simple females alone being wanting. I never should have made this out had not my species theory convinced me that an hermaphrodite species must pass into a bisexual species by insensibly small stages; and here we have it, for the male organs in the hermaphrodite are beginning to fail, and independent males ready formed. But I can hardly explain what I mean, and you will perhaps wish my barnacles and species theory *al Diavolo* together. But I don't care what you say, my species theory is all gospel. We have had only one party here: viz., of the Lyells, Forbes, Owen, and Ramsay, and we both missed you and Falconer very much. . . . I know more of your history than you will suppose, for Miss Henslow most good-naturedly sent me a packet of your letters, and she wrote me so nice a little note that it made me quite proud. I have not heard of anything in the scientific line which would interest you. Sir H. De la Beche gave a very long and rather dull address; the most interesting part was from Sir J. Ross. Mr. Beete Jukes figured in it very prominently; it really is a very nice quality in Sir Henry, the manner in which he pushes forward his subordinates. Jukes has since read what was considered a very valuable paper. The man, not content with moustaches, now sports an entire beard, and I am sure thinks himself like Jupiter tonans. There was a short time since a not very creditable discussion at a meeting of the Royal Society, where Owen fell foul of Mantell with fury and contempt about belemnites. What wretched doings come from the order of fame; the love of truth alone would never make one man attack another bitterly. My paper is full, so I must wish you with all my heart farewell. Heaven grant that your health may keep good.

To Charles Lyell

Torquay, Aug. 21st [1861].

. . . I have really no criticism, except a trifling one in pencil near the end, which I have inserted on account of dominant and important species generally varying most. You speak of "their views" rather as if you were a thousand miles away from such wretches, but your concluding paragraph shows that you are one of the wretches.

I am pleased that you approve of Hutton's review.[1] It seemed to me to take a more philosophical view of the manner of judging the question than any other review. The sentence you quote from it seems very true, but I do not agree with the theological conclusion. I think he quotes from Asa Gray, certainly not from me; but I have neither A. Gray nor *Origin* with me. Indeed, I have over and over again said in the *Origin* that Natural Selection does nothing without variability; I have given a whole chapter on laws, and used the strongest language how ignorant we are on these laws. But I agree that I have somehow (Hooker says it is owing to my title) not made the great and manifest importance of previous variability plain enough. Breeders constantly speak of Selection as the one great means of improvement; but of course they imply individual differences, and this I should have thought would have been obvious to all in Natural Selection; but it has not been so.

I have just said that I cannot agree with "which variations are the effects of an unknown law, ordained and guided without doubt by an intelligent cause on a preconceived and definite plan." Will you honestly tell me (and I should be really much obliged) whether you believe that the shape of my nose (eheu!) was ordained and "guided by an intelligent cause"? By the selection of analogous and less differences fanciers make almost generic differences in their pigeons; and can you see any good reason why the Natural Selection of analogous individual differences should not make new species? If you say that God ordained that at some time and place a dozen slight variations should arise, and that one of them alone should be preserved

---

[1] "Some Remarks on Mr. Darwin's Theory," by F. W. Hutton. *Geologist* (1861), Vol. IV, p. 132.

in the struggle for life and the other eleven should perish in the first or few first generations, then the saying seems to me mere verbiage. It comes to merely saying that everything that is, is ordained.

Let me add another sentence. Why should you or I speak of variation as having been ordained and guided, more than does an astronomer, in discussing the fall of a meteoric stone? He would simply say that it was drawn to our earth by the attraction of gravity, having been displaced in its course by the action of some quite unknown laws. Would you have him say that its fall at some particular place and time was "ordained and guided without doubt by an intelligent cause on a preconceived and definite plan"? Would you not call this theological pedantry or display? I believe it is not pedantry in the case of species, simply because their formation has hitherto been viewed as beyond law; in fact, this branch of science is still with most people under its theological phase of development. The conclusion which I always come to after thinking of such questions is that they are beyond the human intellect; and the less one thinks on them the better. You may say, Then why trouble me? But I should very much like to know clearly what you think.